Success in

CHEMISTRY

SECOND EDITION

Success Studybooks

Advertising and Promotion
Book-keeping and Accounts
Business Calculations
Chemistry
Commerce
Commerce: West African Edition
Communication
Electronics
European History 1815-1941
Information Processing
Insurance
Investment
Law
Managing People
Marketing
Politics
Principles of Accounting
Principles of Accounting: Answer Book
Psychology
Sociology
Statistics
Twentieth Century World Affairs
World History since 1945

Success in
CHEMISTRY
SECOND EDITION

John Bandtock, Ph.D.
and Paul Hanson, Ph.D.
Senior Lecturers in Chemistry,
Leicester Polytechnic

JOHN MURRAY

© John Bandtock, Paul Hanson 1988

First published 1974
by John Murray (Publishers) Ltd
50 Albemarle Street, London W1X 4BD
Reprinted 1975, 1977, 1978, 1980, 1983, 1986
Second edition 1988
Reprinted 1990, 1992, 1993, 1994, 1995, 1996, 1997, 1999

Printed and bound in Great Britain by
Biddles Ltd, Guildford and King's Lynn

British Library Cataloguing in Publication Data
Bandtock, John
  Success in chemistry. – 2nd ed –
  (Success studybooks).
  1. Chemistry
  I. Title   II. Hanson, Paul
  540   QD31.2

ISBN 0–7195–4448–3

Foreword

This book is designed as an introduction to chemistry for anyone who wants to acquire a sound, basic knowledge of the subject. Whether you are tackling it for the first time, or are using the book for revision, you will find the detailed, self-teaching presentation of the material of the greatest help in following and understanding every stage of the course.

The book covers most of the topics included in GCSE and BTEC First courses in chemistry and is therefore of especial value to students taking these courses. The text is divided into Units of study, each defined in its own area, with self-testing questions at the end of the Unit. At the back of the book (page 351) there is a comprehensive selection of questions taken from recent papers of British examining boards.

For many years the main emphasis in the teaching of chemistry has increasingly been on the 'discovery' approach: students are encouraged to follow their natural spirit of inquiry and carry out experiments for themselves. This has culminated in the excellent work of the Nuffield Foundation, which is reflected in large areas of present-day teaching and in examination syllabuses. However, the aim of *Success in Chemistry* is to enable a student working alone to achieve a certain standard of knowledge and we realize it is unlikely that such a student will have a well-equipped laboratory at his disposal. The experiments in this book are, therefore, confined to those that are most practicable and can be safely carried out at home. The more complex experiments, of which the results are vital to the student's understanding of a topic, are described in full and illustrated so that all processes can be followed in detail.

We live in a world of changing scientific nomenclature, but the tendency is always towards international systematization. Throughout the text we have adopted the recommendations of the Association for Science Education, based on those agreed by the International Union of Pure and Applied Chemistry. These recommendations are followed after their introduction in Unit Three but references to the traditional style of nomenclature have also been given where appropriate.

J. B. and P. H.

Acknowledgments

In bringing together the ideas and material for this book we have received help from many people to whom we would like to offer thanks:

the examining boards who co-operated at various stages and permitted us to use questions from their papers; Unilever PLC and Imperial Chemical Industries PLC who supplied us with details of industrial processes; United Kingdom Atomic Energy Authority for Figs 15.3 to 15.6; BP Educational Services for Fig. 15.2; Michael Kipps and Basil Lilley who read and commented on the book in its early stages; Jean Breward who so patiently typed the manuscript.

Our special gratitude goes to the editors Leslie Basford and Dr Jean Macqueen. Their knowledge and professional handling of the material brought inspiration not only to the book but to its authors. Finally, our thanks go to John Murray (Publishers) Ltd for their unfailing help and encouragement.

J.B. and P.H.

Contents

Introduction

Chemistry is a study of the properties of the materials of the world around us. Chemists look at and try to interpret *changes* in materials, from the simple burning of methane (natural gas) to the complex reactions taking place in the human body. At the same time they are also interested in *energy changes* which occur in chemical processes (many reactions in chemistry are carried out purely for the energy they yield).

Another important aspect is the elucidation of the *structure* of materials and the identification of their constituents. Thus chemists are analysts, employing both simple and complex techniques to determine the composition of every kind of substance. Applying their knowledge of existing structures, chemists are concerned with the discovery and development of *new* materials: drugs, dyestuffs, detergents, synthetic fibres and plastics are familiar examples.

Look around and you will see that almost everything is either connected with the chemical industry or has been treated with the products of that industry, e.g. the paper and ink used for printing this book, the glass in windows, the cement and paintwork of buildings, and the crops in the fields.

Nomenclature Used in this Book
In general the recommendations set out in the report *Chemical Nomenclature, Symbols and Terminology for Use in School Science* (The Association for Science Education, 3rd edition, 1985) have been adopted. Where the recommended name for a substance is fundamentally different from the common name, the latter is given in parentheses after the recommended name. For example:

(*a*) dilead(II) lead(IV) oxide (red lead);

(*b*) ethanoic acid (acetic acid);

(*c*) sodium chlorate(I) (sodium hypochlorite).

The concept of oxidation number required for this nomenclature is explained in Section 3.4.

Units in Chemistry
For their GCSE examination syllabuses most examining boards now recommend the use of SI units (Système International d'Unités), subject to certain qualifications. The recommendations set out in the report *SI Units, Signs, Symbols and Abbreviations for use in School Science* (The Association for Science Education, 1981) have been substantially adopted as far as they apply to these syllabuses. These are summarized below.

Quantity	Unit	Abbreviation
Length	metre centimetre kilometre	m cm km
Mass	kilogram gram	kg g
Amount of substance	mole	mol
Time	second	s
Area	square metre square centimetre	m^2 cm^2
Volume	cubic metre cubic decimetre cubic centimetre	m^3 dm^3 cm^3
Density	kilogram per cubic metre gram per cubic centimetre	$kg\,m^{-3}$ $g\,cm^{-3}$
Electric current	ampere	A
Charge	coulomb	C
Electromotive force	volt	V
Resistance	ohm	Ω
Force	newton	N
Work Energy Potential energy Kinetic energy	joule	J
Temperature: common thermodynamic (absolute)	degree Celsius kelvin	°C K

Quantity	Unit	Abbreviation
Pressure	newton per square metre (pascal)	Nm^{-2} (Pa)
Avogadro constant	number per mole	mol^{-1}
Faraday constant	coulomb per mole	$C\,mol^{-1}$

Writing of Abbreviations

(a) when two or more unit symbols are combined to indicate a derived unit, a space is left between them, e.g. J s;

(b) when unit symbols are combined as a quotient (the unit resulting from the division of one unit by another), e.g. kilogram per cubic metre, they are written $kg\,m^{-3}$;

(c) a full stop is not written after symbols for units (except when they are at the end of a sentence).

Relationship between Units

The following relationships may be of use in helping to understand the metric system:

Multiple		Prefix	Symbol
10^9 or	1 000 000 000	giga-	G
10^6 or	1 000 000	mega-	M
10^3 or	1 000	kilo-	k
10^{-1} or	0·1	deci-	d
10^{-2} or	0·01	centi-	c
10^{-3} or	0·001	milli-	m
10^{-6} or	0·000 001	micro-	μ
10^{-9} or	0·000 000 001	nano-	n
10^{-12} or	0·000 000 000 001	pico-	p

N.B.
$$1\,m = 100\,cm$$
$$1\,m^2 = 100 \times 100\,cm^2 = 10\,000\,cm^2$$
$$1\,m^3 = 100 \times 100 \times 100\,cm^3 = 1\,000\,000\,cm^3$$

Although the basic SI unit of volume is the cubic metre (m^3), gas and liquid volumes are more conveniently measured by the cubic centimetre (cm^3) and the cubic decimetre (dm^3). Note that, since 1 dm = 10 cm, $1\,dm^3 = (10\,cm)^3 = 1000\,cm^3$.

Unit One
The Nature of Matter

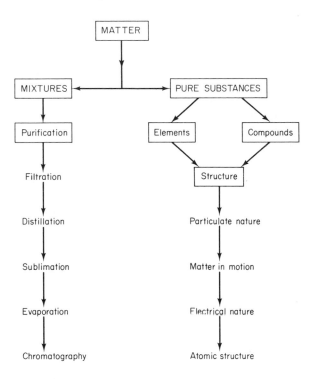

1.1 What is Matter?

Matter is anything that occupies space and possesses *weight*. The *amount* of matter contained in any object is known as its **mass**.

Matter is 'concrete' in that it can be seen, tasted, smelled, or felt. Even colourless, odourless gases have weight.

The matter we encounter every day is almost always a complex **mixture** of various substances. The air we breathe is a mixture of more than five gases; petrol is a mixture of several substances; all living things are extremely complicated mixtures of almost countless numbers of different substances.

A **substance** can be defined as a material all parts of which are chemically identical and all samples of which have the same composition.

Common examples of pure substances used in everyday life are distilled water for car batteries, copper used in electrical wiring, and washing soda crystals. All samples of any one of these—however they may have been prepared—are identical.

1.2 Changes of State

Energy in its various forms, e.g. mechanical (kinetic and potential), heat (thermal), atomic, electrical, magnetic and chemical, contrasts with matter in that it is abstract instead of concrete. Energy interacts with matter in many different and interesting ways, and a major portion of the chemist's time is spent in studying these interactions.

All substances exist in three different **physical states**: *solid*, *liquid* and *gas*.

A **solid** has a definite shape and distinct boundaries.

A **liquid** has a definite volume and takes up the shape of its container.

A **gas** takes up the shape and the volume of its container.

At normal temperatures copper is generally regarded as being a solid, water a liquid and oxygen a gas, but if the temperature is altered they can exist in different states. Which state a substance happens to occupy at any particular moment is governed by the *energy* it contains. If we alter this energy, by heating or by cooling, we can change the state of the substance.

A familiar change of state is that which occurs when a solid (e.g. the wax in a candle) melts and becomes a liquid. Other examples are *vaporization*, the change from liquid to gas; *freezing*, the change from liquid to solid; and *condensation*, the change from gas to liquid.

Melting Point

If a pure solid is heated it changes into a liquid at a particular temperature (provided no decomposition occurs). The temperature at which this change of state from a solid to a liquid occurs is called the **melting point**. The liquid will solidify again on cooling, and the temperature at which the liquid changes to a solid is called the **freezing point**: it is identical with the melting point.

Experiment 1.1   Determination of a melting point
A sample of 1,4-dichlorobenzene is placed in an ignition tube or small test tube. (1,4-dichlorobenzene, also known as *para*-dichlorobenzene, is widely used as a moth repellent; in fact a fragment of a moth-ball will serve as an acceptable substitute in this experiment.) The ignition tube is attached to a thermometer (110 °C) by means of a rubber band and placed in a large boiling tube of water (Fig. 1.1). The water is heated over a bunsen burner, using a tripod and gauze, stirring carefully with the thermometer and making sure that no water enters the ignition tube. The solid is observed as the temperature of the water slowly rises. Record the temperature at which the sample melts.

Devise a method for the determination of the *freezing point* of 1,4-dichlorobenzene. The melting point and freezing point should be identical: if they differ, the liquid may have cooled below its freezing point without any crystals forming. This is called *supercooling* and can often be avoided by effective stirring.

Boiling Point

If a pure liquid is heated, its temperature rises. Eventually a point is reached where the addition of heat produces no further rise in temperature, because all the energy

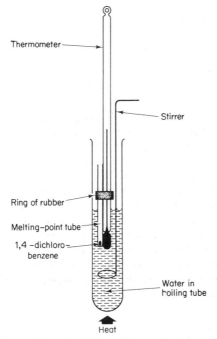

Thermometer

Stirrer

Ring of rubber

Melting-point tube

1,4 –dichloro-benzene

Water in boiling tube

Heat

Fig. 1.1   Apparatus used for the determination of melting points

supplied in heating is used to change the liquid to vapour. This temperature is called the **boiling point** of the liquid.

Evaporation is a process similar to boiling, but it occurs at *any* temperature (below the boiling point) and takes place only from the *surface* of the liquid. After a shower of rain, pools of water on the road disappear owing to evaporation. The process is accelerated by wind and higher temperatures.

The three states of matter and the various processes by which they are related

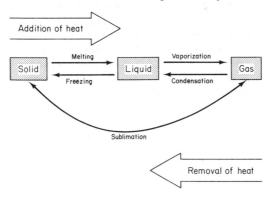

Addition of heat

| Solid | Melting → / ← Freezing | Liquid | Vaporization → / ← Condensation | Gas |

Sublimation

Removal of heat

Fig. 1.2   The three physical states of matter

are shown in Fig. 1.2. **Sublimation** is the *direct* change of state from a solid to a gas, or from a gas to a solid, without first becoming a liquid.

1.3   Mixtures and Pure Substances

The terms 'pure substance' and 'mixture' need to be explained in a little more detail.

A *pure substance* can be a chemical **element** or a chemical **compound**. *Elements* are the basic units of all matter. They cannot be formed from simpler substances or broken down into simpler forms of matter. Chemical *compounds* are formed when two or more elements combine together. For example, sodium (a soft, shiny, very reactive metal) and chlorine (a greenish, choking, poisonous gas) are both elements; they *combine* together to form white crystals of the chemical compound called sodium chloride, which we use in everyday life as common salt.

Sometimes a chemist's work involves the separation of **mixtures** containing two or more substances not chemically joined together. Methods of separation depend on the individual physical properties (e.g. boiling point, hardness, solubility) of the components of the mixture. The following examples illustrate the use of some of these physical properties in the separation of mixtures.

1.4   Filtration

We expect our car to perform satisfactorily under all conditions, and we take it for granted that the water we drink is pure. Solid material has been removed from engine oil, dirt from the air used in the car engine and solid particles from drinking water by a process called **filtration**. This is a method of separating a solid from a liquid (or a solid from a gas) by passing the mixture through a *filter*. The filter consists of a porous material (such as paper or glass wool or fine gravel) which prevents solid particles from passing through but does not retain the liquid.

Rock salt obtained commercially contains common salt (sodium chloride) contaminated with sand and other earthy impurities. On stirring rock salt with water the sodium chloride seems to disappear while the earthy impurities remain unaffected. The sodium chloride has *dissolved* in the water forming a **solution**.

In any solution the dissolved substance is called the **solute**, and the substance in which it is dissolved is called the **solvent**. The solute and solvent *cannot* be separated by filtration.

Experiment 1.2   Separation of salt and sand
A teaspoonful of impure rock salt is stirred into a beaker of warm water. (If rock salt is not available, a mixture of sand and common salt may be used.) The suspension is filtered into a beaker or test tube, as shown in Fig. 1.3. The clear colourless solution (called the **filtrate**) obtained in the test tube contains the salt; the sand and other impurities remain behind on the filter paper. Salt crystals can be obtained by warming the filtrate or by allowing it to stand for several days.

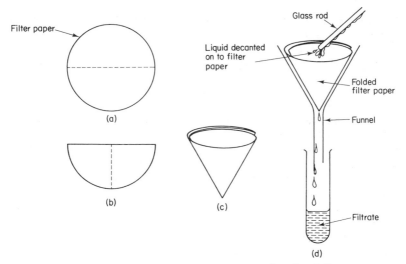

Fig. 1.3 Filtration in the laboratory, using a funnel and filter paper

Industrial applications of filtration processes are wide-ranging and important. They include not only the separation of liquid/solid mixtures, but also the removal of solids from gas samples.

(*a*) Domestic water supplies are filtered through a layer of sand and gravel in a filter bed (Fig. 1.4) before final purification.

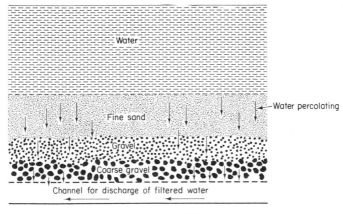

Fig. 1.4 Filtration of drinking water

(*b*) In the Solvay process (described in Section 14.5) for the manufacture of sodium carbonate, the precipitated sodium hydrogencarbonate is filtered by suction through flannel stretched on a slowly rotating cylindrical frame.

(c) A vacuum cleaner filters dust from the air. The dust particles are trapped in a fabric or paper bag, while the purified air passes through.

(d) In a car engine, the air is filtered through an oil-impregnated pad. Solid particles in the engine oil, which would cause excessive engine wear, are removed by a separate oil filter.

(e) In many industrial processes which require extremely pure, clean gases, dust is removed using electrostatic precipitators (Fig. 1.5). The dust particles become electrically charged and are attracted on to wires of opposite charge. Periodically the current is switched off to permit the dust, now in lumps, to fall to the bottom of the chamber.

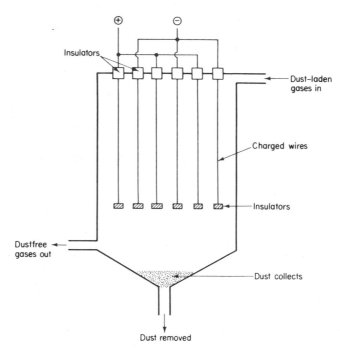

Fig. 1.5   Electrostatic precipitation of dust

1.5 Distillation

Distillation can be used to separate a *volatile* substance (i.e. one which evaporates fairly readily) from a non-volatile substance. It consists of the two basic processes of boiling and condensation. For example, impure water can be purified by distillation because the ordinary impurities in water are dissolved solids which are non-volatile, The impure water is heated to make it boil, its vapour is collected and cooled. As it cools, the vapour *condenses* into pure *distilled* water. The non-volatile solid impurities remain behind.

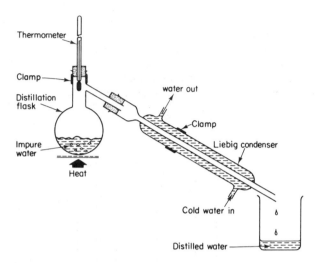

Fig. 1.6   Laboratory apparatus for distillation

Fractional Distillation

Fractional distillation is similar to the simple distillation described above, except that a *fractionating column* is fitted between the distilling flask and the condenser. This process is used to separate two or more volatile liquids which have different boiling points. The fractionating column is a tube with an irregular interior, e.g. a glass column packed with glass beads, which exposes a large cooling surface to ascending vapours. Some condensation occurs as the vapours cool.

On heating a mixture of, for example, benzene (boiling point 80 °C) and methylbenzene (toluene, boiling point 110 °C), the liquid mixture boils and its vapour rises into the fractionating column. This vapour mixture contains a higher proportion of the more-volatile benzene than methylbenzene. Both vapours begin to condense in the fractionating column, the methylbenzene more easily than the benzene. Thus the higher the vapour mixture rises in the column, the richer it becomes in benzene. Finally, at the top of the column, pure benzene can be collected.

Industrial Applications of Distillation

(*a*) Crude oil as obtained from the ground is not suitable for use. It is a mixture of many liquids which must be separated by several processes, the more important of which is fractional distillation (see Unit 11).

(*b*) On being cooled to a very low temperature, air can be liquefied. This liquid is a mixture containing mainly nitrogen and oxygen. Liquid nitrogen has a lower boiling point (and is therefore more volatile) than liquid oxygen. On

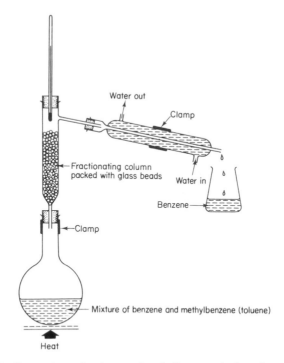

Fig. 1.7   Separation of a benzene/methylbenzene (toluene) mixture by fractional distillation

fractional distillation, the nitrogen is collected from the top of the column while the oxygen concentrates at the bottom.

(*c*) Distillation is a major process in the manufacture of spirits such as whisky, gin, rum and brandy. It serves both to separate the spirit from the raw materials and also to 'strengthen' it, the distillate (i.e. condensed vapour) having a higher percentage of alcohol than the original alcohol/water mixture.

1.6 Sublimation

When shiny black iodine crystals are heated gently they do not melt: instead they pass from the solid directly to the vapour state. On cooling, the purple vapour condenses to the solid without any liquid appearing. This process is called **sublimation**.

Experiment 1.3   Sublimation of iodine
A few crystals of impure iodine are placed in a conical flask (see Fig. 1.8), and the flask is heated very gently over a burner. First, a purple vapour is observed, and then small black lustrous crystals are seen forming on the water-cooled inner tube. The iodine passing directly from solid to gas has *sublimed*.

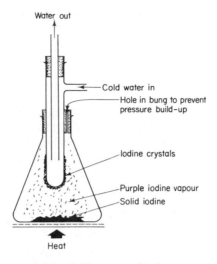

Fig. 1.8   Sublimation of iodine

Experiment 1.4   *Sublimation of ammonium chloride*
A little white solid ammonium chloride is placed in an evaporating basin, and a funnel is inverted over it (see Fig. 1.9). When the basin is heated gently a white vapour is observed and white crystals of ammonium chloride form on the inside of the funnel. The ammonium chloride has *sublimed*.

1.7 Chromatography

Chromatography is the technique of separation and identification of mixtures by a moving liquid (the *solvent*) on an absorbent material such as filter paper or

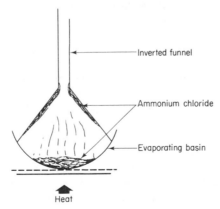

Fig. 1.9   Sublimation of ammonium chloride

blotting paper. A drop of a solution of the mixture is placed near one end of a strip of filter paper, and allowed to dry. Solvent is then allowed to flow along the paper, dissolving the mixture as it goes. Separation of the components takes place because each pure substance moves at a different rate.

Experiment 1.5   *Simple demonstration of chromatography*
Two parallel cuts are made from the edge of a circle of filter paper to near the centre so as to produce a 'wick' about 1 cm wide (see Fig. 1.10). A single drop of black ink is placed at the centre of the paper and allowed to dry. The filter paper is then arranged as shown, with its wick dipping into water in an evaporating basin. The 'chromatogram' should be left undisturbed to develop.

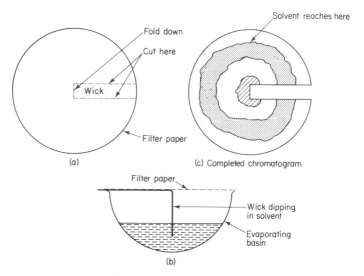

Fig. 1.10   Simple demonstration of chromatography

The effectiveness of the separation varies according to the solvent used in the evaporating basin. Determine the most suitable solvent mixture by developing chromatograms using methylated-spirit/water mixtures of varying proportions.

Experiment 1.6   *Separation of pigments from grass*
Small pieces of grass (or spinach) are ground with a little methylated spirit, and the resulting green liquid is filtered. The methylated spirit dissolves the pigments (chlorophylls, xanthophyll and carotene) from the grass. A chromatogram, using methylated spirit as the solvent, is now made and left to develop. The outer orange band is xanthophyll and the green band contains the two chlorophylls. If methylbenzene (toluene) is used as the solvent, an inner orange ring of carotene can be seen.

1.8 The Particulate Nature of Matter

So far we have seen that matter is composed of mixtures of pure substances, and we have investigated ways in which mixtures can be separated into pure substances. Let us now look more closely at pure substances.

We are faced with two entirely different ideas concerning the nature of matter. One suggests a continuous structure in which matter is infinitely divisible. The other proposes a 'particulate' structure in which the fundamental particle cannot be divided without altering the nature of the material. There is a considerable weight of scientific evidence in support of the latter idea, that all matter is composed of very small particles called **atoms**. All atoms of one element are alike, but they differ from those of any other element. Two or more atoms may combine to form a **molecule**.

Probably the most important piece of evidence to support the particulate nature of matter comes from the study of crystals.

1.9 Crystals

Most solid substances exist in a definite geometrical form—although a magnifying glass or microscope is often required to see this. Such substances are said to be *crystalline*, and the individual units of which they are composed are called **crystals**. Among the many examples of crystals which can be seen in the laboratory and in the home are the cube-like crystals of common salt (sodium chloride), blue crystals of copper sulphate, and the fern-like crystals of zinc which can be seen on the surface of galvanized iron. Different crystals of the same substance may not at first sight appear to be the same shape, but the *angles* between the faces remain constant. This can be illustrated by studying large crystals of substances such as quartz or calcite.

These regularities in crystal form are indications of an orderly arrangement based on particles from which matter is composed. The symmetry of crystals, the way crystals grow and the way they cleave, give confidence in a particulate idea. This orderliness of crystals can be illustrated by two simple experiments.

Experiment 1.7 Growth of crystals
A glass 35 mm slide is smeared with a drop of ammonium chloride solution and placed in a projector so that a magnified image of the smear can be viewed on a screen. The solution becomes more and more concentrated as the heat from the projector lamp warms the slide. Needle-like crystals of ammonium chloride eventually appear and their growth can be observed.

Experiment 1.8 Cleavage of crystals
Using a single-edged razor blade, attempts are made to split crystals of calcite, copper sulphate or an alum. This is done by striking the reinforced upper edge of the blade with a small hammer. It will be noticed that a crystal cleaves easily in certain directions, producing smaller identically shaped pieces, while cleavage in other directions is very difficult.

This regularity in crystal structure can be best appreciated by considering that matter is particulate in nature. Is it possible to estimate the size of these particles?

1.10 Size of Particles

A drop of oil poured on to water spreads out as far as possible and makes an extremely thin film. Chemists suggest on evidence that this film is *one molecule* thick. In the following experiment a very small measured drop of oil is placed on a clean water surface and the diameter of the oil patch is measured. The thickness of the film, which is an estimate of the size of an oil molecule, can then be calculated.

Experiment 1.9   *Estimation of molecular size*
This experiment only aims to give the order of magnitude of the quantity measured.

A loop of very thin steel wire is dipped into a sample of olive oil to collect a small drop (Fig. 1.11a). Excess oil is removed from the wire using filter paper. The diameter of the drop on the loop is measured using a 2 cm scale fitted with a magnifying lens (Fig. 1.11b). With practice a drop $\frac{1}{2}$ mm in diameter can be obtained.

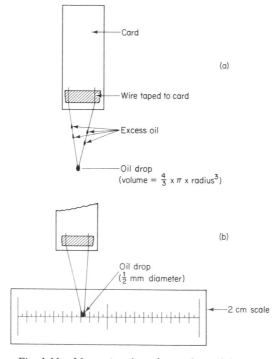

Fig. 1.11   Measuring the volume of an oil drop

A large waxed tray is filled with water to the brim and the surface is dusted with lycopodium or talcum powder. The drop of oil is transferred from the loop to the water by touching the lower end of the loop to the water surface (Fig. 1.12). Measure the diameter of the circular oil film produced.

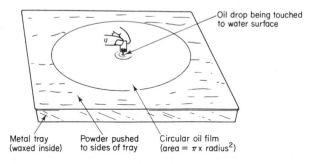

Oil drop being touched to water surface

Metal tray (waxed inside)

Powder pushed to sides of tray

Circular oil film (area = $\pi \times$ radius^2)

Fig. 1.12   The oil-film experiment

Results

Diameter of small drop on loop $= 0.05$ cm

radius of small drop on loop $= 0.025$ cm

volume of small drop on loop $= \frac{4}{3} \times \pi \times 0.025^3$ cm^3

Diameter of oil film on water $= 20$ cm

radius of oil film on water $= 10$ cm

volume of oil film on water $= \pi \times 10^2 \times$ thickness

Volume of film on water $=$ volume of oil drop on loop

$$\pi \times 10^2 \times \text{thickness} = \frac{4}{3} \times \pi \times 0.025^3$$

$$\therefore \text{ thickness} = \frac{4 \times 0.025^3}{3 \times 10^2}$$

$$= 0.000\,000\,2 \text{ cm}$$

Thus our *estimate* of the length of one molecule of olive oil is $0.000\,000\,2$ cm.

The evidence from experiments such as this suggests that matter is particulate, and that the particles are extremely small. The particles could either be at rest (a 'static' theory), or they could be in continuous movement (a 'kinetic' theory).

1.11 Matter in Motion

In a still room on a sunny day, dust or smoke particles are clearly visible in the air. They appear to be moving continuously in a completely haphazard way. A similar effect can be seen when cigarette smoke trapped in an illuminated glass container is viewed under a microscope. These and similar observations are consistent with the theory that invisible gas molecules in the air are moving about at high speed and jostling the larger dust particles.

Experiment 1.10   *Demonstration of Brownian motion*
A small drop of black printing ink is placed in water and observed under the high power of a microscope. The black particles are seen to be in irregular motion. The smaller black particles move in a more rapid and haphazard way. This constant, random, jerky motion can be explained by the large carbon particles of the ink being continually buffeted by enormous numbers of water molecules. Each single collision would have so small an effect as to be undetectable, but the imbalance produced by large numbers of collisions could produce the irregular motion observed.

This important type of observation in the elucidation of the structure of matter was first described by Robert Brown (1827) looking at pollen grains in water, and is thus referred to as **Brownian motion**. This movement is caused by the continuous irregular bombardment of the large particles by the smaller molecules of the surrounding medium.

Further evidence in support of this particulate motion comes from the phenomenon of **diffusion**.

(*a*) **Diffusion in solids.** With the development of extremely sensitive methods of analysis it has been found that when two metals are kept in very close contact over a long period of time, particles of each metal have migrated into the other. This can be explained if we assume that the metal particles are in motion, however slight, so that some drift from one metal to the other can occur.

(*b*) **Diffusion in liquids**. Movement of particles in a liquid can be illustrated by the following experiment.

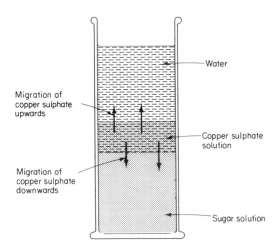

Fig. 1.13   Diffusion of copper sulphate in solution

Experiment 1.11   *Diffusion of copper sulphate in solution*
A very concentrated solution of sugar is poured into a gas jar until the jar is one-third full. A fairly concentrated solution of copper sulphate is carefully poured on to the sugar without allowing the two solutions to mix. A further layer of water is then poured on to the copper sulphate solution (see Fig. 1.13). Initially the boundaries between layers are sharp, but after a time it is seen that the blue copper sulphate is migrating both upwards and downwards. This observation is only explained by a *kinetic* theory.

(*c*) **Diffusion in gases.** If the cover is removed from a jar containing hydrogen sulphide, the unmistakable 'rotten-egg' smell of this gas can quickly be detected throughout the laboratory: surely strong evidence in favour of a kinetic theory.

Experiment 1.12   *Diffusion of bromine*
Liquid bromine is dark red in colour and evaporates spontaneously to give a heavy red-brown vapour. Bromine is *extremely corrosive and must be handled with great care.* Using rubber gloves and working in a fume cupboard, a little liquid bromine is placed in a crucible standing on a glass plate. A gas jar is inverted over the crucible. Notice the absence of colour in the jar. After a few hours the reddish colour of bromine vapour fills the jar—despite the fact that bromine is approximately five times as heavy as air. The bromine molecules have diffused throughout the air in the jar, clearly indicating that they are not static. In fact diffusion occurs whenever two or more gases come into contact.

Thus we have seen that the particles of a gas, a solid and a liquid are not stationary but are evidently in a state of continuous random motion. The **energy** in a solid, liquid or gas is partly due to this motion of particles (*kinetic* energy) and partly due to the forces between the particles (*potential* energy).

1.12 The Kinetic Theory of Matter

Kinetic means 'of movement' and the kinetic theory of matter explains the properties of matter in terms of the movement of particles. It postulates that as the temperature rises, so does the **kinetic energy**. Thus the motion of the particles increases as the temperature rises.

In a **solid**, the particles are extremely close together and this results in high density and incompressibility, i.e. the fact that high pressure has little effect on the volume. The highly ordered arrangement of the particles is called a **lattice**, and this leads to a regular geometrical shape in crystals of the solid.

Consider what happens when energy (heat) is added to a crystal. Initially the solid *expands*, because the kinetic energy of the particles increases and the molecules, vibrating more strongly, take up more space. Eventually the particles vibrate so much that they leave their fixed positions and the solid *melts*, becoming a liquid. This liquid can be poured into a container and it will take up the shape of the container. The particles of the liquid have more kinetic energy than those of the solid and do not have fixed positions in an ordered lattice (see Fig. 1.14). They are free to wander from one position to another.

When a liquid is converted into a gas, its energy increases and the distance between the molecules increases approximately one thousand fold. There are only very weak interactions between molecules in the gas, and each molecule is free to move in continuous chaotic motion at great speed throughout its containing vessel. Molecules collide with other molecules and with the walls of the vessel. The constant bombardment of molecules on the walls exerts a steady force, and this is called the **pressure** of the gas.

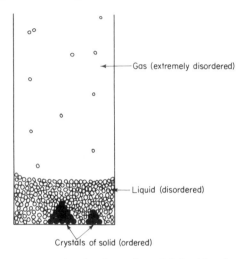

Fig. 1.14   Arrangement of molecules in the solid, liquid and gaseous states

1.13 The Structure of Atoms

The particulate concept of matter which we have discussed in Sections 1.8–1.12 may be traced back to the ideas of John Dalton. In 1808 Dalton published his theory that all matter is made up from individual solid particles called atoms. This theory was a very great step forward, but it failed to explain the following *electrical* phenomena:

(*a*) different materials when rubbed together generate electrical charge;

(*b*) there are only two types of charge, which scientists call positive and negative.

In the latter half of the 19th century and the beginning of the 20th century, the nature of the atom itself, its electrical properties and the idea that atoms might not be the ultimate particles of matter, became better appreciated.

Let us trace the development of scientists' ideas concerning the structure of the atom.

Period	Development

1808   John Dalton developed the idea of a solid atom (*a*) as the fundamental building block of all matter.

(a) Solid atom

1897   Sir J. J. Thomson carried out research at Cambridge on the discharge of electricity through gases. This led to the discovery of negatively charged particles called **electrons** (with charge −1) as constituents of all matter. He measured their charge-to-mass ratio, and described the atom as a solid sphere with electrons embedded in it (*b*).

(b) 'Plum–duff' or 'Currant–bun' atom

1909   R. Millikan determined the charge of a single electron, thus enabling the mass of the electron to be calculated.

1913   Lord Rutherford, one of the greatest scientific experimenters, confirmed his idea that an atom consisted of a positively charged central **nucleus** in which was concentrated most of the mass. He concluded that there must be a large area of empty space around the nucleus in which electrons revolved, their negative charge balancing the positive charge on the nucleus (*c*). The name **proton** was given to the basic unit of mass having a positive charge +1 and mass 1. Niels Bohr refined Rutherford's model. He worked out fixed orbits for each electron around the nucleus based on the calculated energies of the electrons.

(c) 'Solar–system' atom

1925   Louis Victor de Broglie, a Frenchman, suggested that electrons could be regarded as waves. This leads to a charge-cloud picture of the atom (*d*) where the density of the cloud varies with distance from the nucleus.

(d) Charge–cloud atom

1932   Sir James Chadwick discovered another nuclear particle with a mass similar to that of a proton, but which had no electrical charge. This was the **neutron**. Thus the model of the atom had to be extended to include neutrons as well as protons in the nucleus (*e*).

Present day   Recent work has led to the discovery of many new particles in the nucleus of an atom, with the electrons pictured either by the 'charge-cloud' model or the 'orbital' representation.

(e) Nuclear atom containing protons and neutrons

For our purposes, however, the atom can be considered as follows:
(*a*) a small central **nucleus**, containing
      (i)   **protons** (charge $+1$, mass 1)
      (ii) **neutrons** (charge 0, mass 1)
(*b*) **electrons** (charge -1, mass $\frac{1}{1836}$ that of a proton) which may be pictured as
  either (i)   particles orbiting the nucleus in definite paths
     or (ii) clouds of charge enveloping the nucleus.
Particles within the nucleus of an atom are collectively called 'nucleons'.

1.14   How Heavy are Atoms?

Mass and weight. A piece of metal weighing one kilogram on the earth would weigh much less on the moon, because the moon is smaller than the earth and has less 'pull' for the object. **Weight** is the result of the attraction of the earth's (or the moon's) gravity on an object and this will vary from place to place and indeed from planet to planet, whereas **mass** is dependent upon the *quantity* of matter in a substance and will not vary. Chemists are interested in reacting *masses* and these are obtained on a beam balance by comparison with a *standard mass*. Because the mass of an object remains constant it is more meaningful to speak of its mass than its weight.

Atomic Number and Mass Number

In the world of the atom the standard mass is the **mass of the proton** (1 unit) and the masses of the neutron and the electron can be compared with this standard. Obviously the mass of an atomic nucleus depends on the number of protons and neutrons it contains. The number of protons in the atomic nucleus is called the **atomic number** and is denoted by the symbol Z; the sum of the protons plus the neutrons in the nucleus is called the **mass number**, and is denoted by the symbol A. In the neutral atom the number of protons in the nucleus is equal to the number of electrons outside the nucleus (in fact *all* the electrons are outside the nucleus). The chemical behaviour of an atom depends to a very large extent on its electrons. Thus all atoms with the *same atomic number* are alike in chemical behaviour because they have the same number of electrons.

An **element** can now be defined as *a substance consisting entirely of atoms of the same atomic number*. There are at present more than 100 different elements (including some that are man-made) from which all matter is composed.

Isotopes

The number of *neutrons* present in the nucleus of an atom makes no difference to the atomic number or to the chemical behaviour. But the number of neutrons will affect the *mass* of individual atoms. For example, all carbon atoms (atomic number 6) contain six protons, and the majority (98·93%) contain six neutrons: their *mass number* is therefore $6+6 = 12$. A few carbon atoms (1·07%) have six protons and *seven* neutrons, yet they still behave chemically as carbon atoms: their mass number is $6+7 = 13$. Atoms of the same element having different numbers of neutrons are called **isotopes** of that element. In other words, isotopes have the same atomic number Z but differ in mass number A. The isotopes of hydrogen are shown in Fig. 1.15.

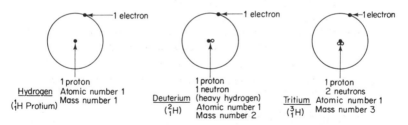

Fig. 1.15   *The isotopes of hydrogen*

Relative Atomic Mass

Measured in grams, the mass of an atom is an extremely small and inconvenient fraction. It is therefore preferable to express atomic masses by comparing them with the mass of a 'standard' atom. The atom chosen nowadays as the standard for comparison is that of carbon-12, i.e. the most common isotope of carbon, which has a mass of 12 units. The atomic masses of all other elements are determined relative to one-twelfth of the mass of this carbon-12 isotope. The term **relative atomic mass** replaces the term 'atomic weight' found in many older textbooks. Modern techniques using a *mass spectrometer* enable relative atomic masses to be measured with great accuracy.

From the way relative atomic mass is defined it might be thought that all elements would have whole-number masses. In fact atomic masses are rarely found to be exactly whole numbers. This is because most elements are composed of two or more naturally occurring isotopes, and the relative atomic mass takes into account the abundance of each isotope. For example, chlorine contains approximately 75% of an isotope with mass number 35, and approximately 25% of an isotope with mass number 37; consequently the relative atomic mass of an average sample of chlorine is 35·46.

Chemical Symbols

Each element has been given a chemical symbol consisting of either a single letter (e.g. C for carbon) or two letters (e.g. Cl for chlorine). The symbol represents *one atom* of an element. A table showing the symbols, atomic numbers and atomic masses is given in the Appendix near the end of this book.

If the mass number A or atomic number Z of each element is required, it is conventional to place these two numbers in front of the chemical symbol:

$$\text{mass number} \atop \text{atomic number} \; \text{CHEMICAL SYMBOL}$$

For example, the two isotopes of chlorine described above would be written $^{37}_{17}\text{Cl}$ and $^{35}_{17}\text{Cl}$. Similarly, $^{12}_{6}\text{C}$ represents the carbon-12 isotope.

Summary of Unit 1

1. **Matter** is anything which occupies space and possesses weight.
2. The *amount* of matter contained in any object is known as its **mass**.
3. Matter can exist in three *states*: **solid, liquid** and **gas**.
4. A pure substance has a distinct *melting point* and *boiling point*.
5. Most solids *melt* on heating but some *sublime*, i.e. change directly from solid to vapour without melting.
6. Matter is composed of (*a*) elements, (*b*) compounds and (*c*) mixtures of elements and compounds.

    (*a*) **Elements** are the basic units of matter. They cannot be formed from simpler substances or broken down into simpler forms of matter.

    (*b*) **Compounds** are formed when two or more elements combine together chemically.
7. **Mixtures** can be separated by physical techniques including: *filtration, distillation, sublimation* and *chromatography*.
8. A **solution** consists of one substance dissolved in another. The dissolved substance is called the *solute*, and the substance in which it dissolves is called the *solvent*.
9. All matter is composed of very small particles called **atoms**. Atoms can combine to form **molecules**.
10. Evidence supporting the particulate nature of matter comes from a study of *crystals*.
11. An estimation of the very small size of a molecule can be obtained by measuring the area covered by a drop of oil when it spreads on water.
12. *Brownian movement* and *diffusion* experiments support the idea that particles of matter are in a constant state of motion.
13. The *kinetic theory of matter* explains the movement of particles.
14. An atom has a small **nucleus** containing **protons** (mass 1, charge $+1$) and **neutrons** (mass 1, charge 0).
15. **Electrons** (mass $\frac{1}{1836}$, charge -1) surround the nucleus. They may be represented either as a charged cloud or as particles orbiting the nucleus in definite paths.
16. **Atomic number** (Z) is the number of protons in the nucleus of an atom.
17. The **mass number** (A) of an atom is the sum of the protons and neutrons in the nucleus.
18. An **element** is defined as a substance consisting entirely of atoms of the same atomic number.
19. **Isotopes** of an element have the same number of protons, but differ in the number of neutrons.
20. **Relative atomic mass** of an element is the mass of an atom of that element compared with $\frac{1}{12}$ the mass of a carbon-12 atom.

Test Yourself on Unit 1

1. The following diagram refers to solid/liquid/gas phase changes.

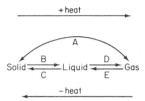

Name the changes represented by A, B, C, D and E.

2. The following techniques are all used for the separation of mixtures: *evaporation, chromatography, filtration, fractional distillation* and *sublimation*. Which of these is the *most suitable* technique for obtaining:

(*a*) Sodium chloride from a solution of sodium chloride.

(*b*) Benzene from a liquid mixture of benzene and methylbenzene (toluene).

(*c*) Ammonium chloride from a white powder composed of ammonium chloride and sodium chloride.

(*d*) Small pieces of metal from the engine oil of a car.

(*e*) The different pigments from an extract of flower petals.

3. Mark each of the following statements true or false.

(*a*) It is assumed that the layer of oil in the oil-film experiment is one molecule thick.

(*b*) The oil-film experiment gives the exact size of a molecule.

(*c*) Diffusion experiments support the theory that particles of matter are in continuous random motion.

(*d*) A study of crystals gives strong evidence that matter is particulate and not continuous.

4. Link the following scientists with the discoveries which helped to make them famous:

(*a*) Sir J. J. Thomson (i) Discovery of the neutron

(*b*) Sir James Chadwick (ii) Wave nature of electrons

(*c*) H. Becquerel (iii) Charge on a single electron

(*d*) Lord Rutherford (iv) Discovery of the electron

(*e*) R. Millikan (v) Discovery of radioactivity

(*f*) Louis Victor de Broglie (vi) Discovery of the nucleus and proton

5. Complete the following table:

Particle	Mass	Charge
Proton		
		0
	$\frac{1}{1836}$	-1

6. What do the numbers in the symbol $^{39}_{19}K$ indicate about the nucleus of the potassium atom?

7. The following symbols refer to atoms of nitrogen:

$$^{14}_{7}N, \ ^{15}_{7}N$$

Using these symbols, explain briefly the meaning of the term 'isotopes'.

Mark this test out of 30 with the answers provided on page 369.

Unit Two

The Organization of Matter

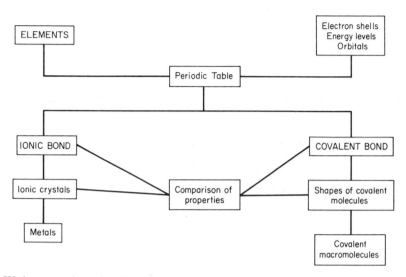

We have seen in Unit 1 that an atom is a particle of matter with clouds of electrons revolving about a small dense mass of protons and neutrons. Furthermore it has been mentioned that these planetary electrons are in some way responsible for the chemical properties of the atom. In this Unit we will look further into the role of the electrons, their arrangement around the nucleus, the organization of elements according to their electronic structure, and the ways in which atoms combine.

2.1 Electron Shells, Energy Levels and Orbitals

Planetary electrons, regarded either as a cloud or as particles, can be considered as revolving in concentric rings or **shells** about the nucleus. The first shell, i.e. the shell closest to the nucleus, can hold 2 electrons. The second shell, at a greater distance from the nucleus, can hold 8 electrons; the third can hold 18 electrons, and so on.

In general, the number of electrons in each *filled* shell is given by the formula $2n^2$, where n is the number of the shell.

Number of shell, n	1	2	3	4
$2n^2$	2×1^2	2×2^2	2×3^2	2×4^2
Number of electrons	2	8	18	32

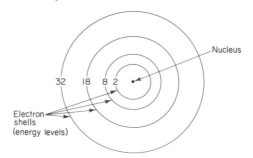

Fig. 2.1   Electron shells of an atom

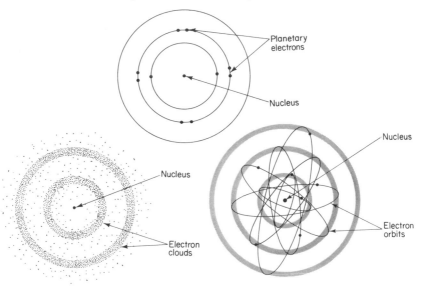

*Fig. 2.2   Representations of an atom: (top) flat two-dimensional model;
(left) charge-cloud model; (right) three-dimensional model*

No one diagram presents an exact picture of the atom with its electrons, and three of the different representations that have been used are shown in Fig. 2.2. For many purposes, however, it is sufficient to illustrate the electron shells of the atom schematically, as shown in Fig. 2.3.

It has been observed in the atoms of all elements that the electrons fill the shells in a systematic manner. The electrons in each shell are associated with different *energies*, and for this reason the shells or orbits are sometimes referred to as **energy levels**. In filled energy levels the electrons are associated in pairs. The shell nearest to the nucleus is the lowest energy level and electrons fill this shell first. Since these electrons are closer to the nucleus than those in other higher energy levels, they are held more tightly and are therefore more difficult to remove.

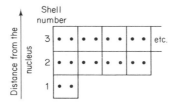

Fig. 2.3. Schematic representation of the electron shells of an atom

When the nearest shell contains two electrons it is filled and further electrons begin to enter the next energy level, the second shell.

In the charge-cloud picture of the atom, the electrons are thought of as clouds of electricity which have different shapes. Each type of cloud has a specific energy value. These clouds of electricity around the nucleus are called **orbitals**.

Regardless of whether we think of them as occupying shells or orbitals, the electrons in atoms have different energies depending on their distance from the nucleus. These energies determine which energy level or shell the electrons may occupy.

2.2 The Elements

It is now possible to give the electron structure of the elements showing their electron shells and atomic number. A list of the first 20 elements is given in Table 2.1 and a full list is given in the Appendix.

Table 2.1 Electronic structure of the elements hydrogen to calcium

Atomic number	Name	Symbol	Shell number			
			1	2	3	4
1	Hydrogen	H	1			
2	Helium	He	2			
3	Lithium	Li	2	1		
4	Beryllium	Be	2	2		
5	Boron	B	2	3		
6	Carbon	C	2	4		
7	Nitrogen	N	2	5		
8	Oxygen	O	2	6		
9	Fluorine	F	2	7		
10	Neon	Ne	2	8		
11	Sodium	Na	2	8	1	
12	Magnesium	Mg	2	8	2	
13	Aluminium	Al	2	8	3	
14	Silicon	Si	2	8	4	
15	Phosphorus	P	2	8	5	
16	Sulphur	S	2	8	6	
17	Chlorine	Cl	2	8	7	
18	Argon	Ar	2	8	8	
19	Potassium	K	2	8	8	1
20	Calcium	Ca	2	8	8	2

THE PERIODIC TABLE

(An alphabetical list of the elements

KEY

Relative atomic mass ———— ► 6.939

Atomic number ———— ► ₃Li ◄———— Symbol

Lithium ———— Name

1.008
₁H
Hydrogen

1	2	Transition metals						
6.939	9.012							
₃Li	₄Be							
Lithium	Beryllium							
22.99	24.31							
₁₁Na	₁₂Mg							
Sodium	Magnesium							
39.10	40.08	44.96	47.90	50.94	52.00	54.94	55.85	58.93
₁₉K	₂₀Ca	₂₁Sc	₂₂Ti	₂₃V	₂₄Cr	₂₅Mn	₂₆Fe	₂₇Co
Potassium	Calcium	Scandium	Titanium	Vanadium	Chromium	Manganese	Iron	Cobalt
85.47	87.62	88.91	91.22	92.91	95.94	(99)	101.1	102.9
₃₇Rb	₃₈Sr	₃₉Y	₄₀Zr	₄₁Nb	₄₂Mo	₄₃Tc	₄₄Ru	₄₅Rh
Rubidium	Strontium	Yttrium	Zirconium	Niobium	Molybdenum	Technetium	Ruthenium	Rhodium
132.9	137.3	(see below) 57–71	178.5	180.9	183.9	186.2	190.2	192.2
₅₅Cs	₅₆Ba		₇₂Hf	₇₃Ta	₇₄W	₇₅Re	₇₆Os	₇₇Ir
Caesium	Barium		Hafnium	Tantalum	Tungsten	Rhenium	Osmium	Iridium
(223)	(226)	(see below) 89–103						
₈₇Fr	₈₈Ra							
Francium	Radium							

	138.9	140.1	140.9	144.2	(145)	150.4	152.0
Lanthanides ►	₅₇La	₅₈Ce	₅₉Pr	₆₀Nd	₆₁Pm	₆₂Sm	₆₃Eu
	Lanthanum	Cerium	Praseodymium	Neodymium	Promethium	Samarium	Europium
	(227)	232.0	(231)	238.0	(237)	(244)	(243)
Actinides ►	₈₉Ac	₉₀Th	₉₁Pa	₉₂U	₉₃Np	₉₄Pu	₉₅Am
	Actinium	Thorium	Protactinium	Uranium	Neptunium	Plutonium	Americium

Fig. 2.4   The periodic table. (For a list of the elements in alphabetical order, see the Appendix near the end of this book.) Some relative atomic masses are cited in parentheses to indicate that they refer to the most stable or best-known isotope of the element concerned.

OF THE ELEMENTS

will be found in the appendix)

			3	4	5	6	7	O
								4.003
								$_2$He
								Helium
			10.811	12.01	14.00	16.00	19.00	20.18
			$_5$B	$_6$C	$_7$N	$_8$O	$_9$F	$_{10}$Ne
			Boron	Carbon	Nitrogen	Oxygen	Fluorine	Neon
			26.98	28.09	30.99	32.06	35.45	39.95
		1	$_{13}$Al	$_{14}$Si	$_{15}$P	$_{16}$S	$_{17}$Cl	$_{18}$Ar
			Aluminium	Silicon	Phosphorus	Sulphur	Chlorine	Argon
58.71	63.54	65.37	69.72	72.59	74.92	78.96	79.91	83.80
$_{28}$Ni	$_{29}$Cu	$_{30}$Zn	$_{31}$Ga	$_{32}$Ge	$_{33}$As	$_{34}$Se	$_{35}$Br	$_{36}$Kr
Nickel	Copper	Zinc	Gallium	Germanium	Arsenic	Selenium	Bromine	Krypton
106.4	107.9	112.4	114.8	118.7	121.8	127.6	126.9	131.3
$_{46}$Pd	$_{47}$Ag	$_{48}$Cd	$_{49}$In	$_{50}$Sn	$_{51}$Sb	$_{52}$Te	$_{53}$I	$_{54}$Xe
Palladium	Silver	Cadmium	Indium	Tin	Antimony	Tellurium	Iodine	Xenon
195.1	197.0	200.6	204.4	207.2	209.0	(209)	(210)	(222)
$_{78}$Pt	$_{79}$Au	$_{80}$Hg	$_{81}$Tl	$_{82}$Pb	$_{83}$Bi	$_{84}$Po	$_{85}$At	$_{86}$Rn
Platinum	Gold	Mercury	Thallium	Lead	Bismuth	Polonium	Astatine	Radon

157.3	158.9	162.5	164.9	167.3	168.9	173.0	175.0
$_{64}$Gd	$_{65}$Tb	$_{66}$Dy	$_{67}$Ho	$_{68}$Er	$_{69}$Tm	$_{70}$Yb	$_{71}$Lu
Gadolinium	Terbium	Dysprosium	Holmium	Erbium	Thulium	Ytterbium	Lutetium
(247)	(247)	(251)	(254)	(253)	(256)	(253)	(257)
$_{96}$Cm	$_{97}$Bk	$_{98}$Cf	$_{99}$Es	$_{100}$Fm	$_{101}$Md	$_{102}$No	$_{103}$Lw
Curium	Berkelium	Californium	Einsteinium	Fermium	Mendelevium	Nobelium	Lawrencium

The main points to notice in Table 2.1 are as follows:

(i) For the first 18 elements the electron is added to the outermost shell until the shell is complete.

(ii) The innermost shell is filled when it contains 2 electrons. This first occurs in the element helium, with atomic number 2.

(iii) Lithium (atomic number 3) is the first element to contain an electron in the second shell. This shell is completed when it contains 8 electrons, represented by the element neon (atomic number 10).

(iv) The third shell can contain 18 electrons. However, *the fourth shell begins to fill* before the third shell is complete. Thus the element potassium (atomic number 19) has one electron in the fourth shell and only eight electrons in the third shell.

(v) The maximum number of electrons in the outermost shell is never more than eight.

If, as suggested in Section 1.13, the chemical properties of an element are related to its electronic structure, it would seem probable that any regularities, or *periodicity*, in electronic structure should lead to regularities, or periodicity, in chemical properties. This is in fact true, and it can be demonstrated by arranging the elements systematically.

2.3   The Periodic Table

A Russian chemist, Dmitri Ivanovich Mendeleev, in 1869 first produced a table showing this regularity or periodicity of elements. He stated that 'the elements, if arranged according to their *atomic weight*, show a distinct periodicity of their properties'. Using this table Mendeleev was able to forecast the existence and properties of several undiscovered elements; a truly remarkable step.

Today we arrange elements in order of ascending **atomic number** with elements having similar properties and electronic structures at regular intervals. This is the **periodic table** of the elements and is shown in Fig. 2.4.

Horizontal rows of elements are called *periods*.

Vertical columns of elements are called *groups* or *families*.

Periods

Period 1 contains 2 elements only.

Periods 2 and 3 each contain 8 elements and are called *short periods*. They have one unfilled shell.

Periods 4, 5, 6 and 7 are called *long periods*. They contain additional sets of elements called 'transition' elements. For example, the first long period (period 4) contains 18 elements. It begins in the same way as periods 2 and 3 but contains 10 distinctly different elements produced when the third shell, already containing 8 electrons, increases from 8 to 18. These are the transition elements scandium (atomic number 21) to zinc (atomic number 30). They have some similarities in properties because of their similar outer electron structure.

Groups or Families

The elements within a *group* of the periodic table show marked chemical similarities with a gradation of properties.

The group on the extreme right of the periodic table (see Fig. 2.4) is interesting because of the reluctance of the elements in this family to undergo chemical reactions. Their electronic structures are shown in Table 2.2.

Table 2.2 Electronic structure of the noble gases

Atomic number	Name	Symbol	Shell number					
			1	2	3	4	5	6
2	Helium	He	2					
10	Neon	Ne	2	8				
18	Argon	Ar	2	8	8			
36	Krypton	Kr	2	8	18	8		
54	Xenon	Xe	2	8	18	18	8	
86	Radon	Rn	2	8	18	32	18	8

This chemically unreactive family of elements is called the **noble gases**. The outer shell of each of these elements is either filled, as in the case of helium and neon, or contains an 'octet' of eight electrons. It seems, therefore, that such an electronic structure is associated with chemical stability.

The family of elements on the extreme left of the periodic table is called the **alkali metals**. They all have one electron in the outer shell (see Table 2.3), and show remarkable chemical similarity.

Table 2.3 Electronic structure of the alkali metals

Atomic number	Name	Symbol	Shell number					
			1	2	3	4	5	6
3	Lithium	Li	2	1				
11	Sodium	Na	2	8	1			
19	Potassium	K	2	8	8	1		
37	Rubidium	Rb	2	8	18	8	1	
55	Caesium	Cs	2	8	18	18	8	1

Another interesting and important family of elements is the **halogens**. They each contain seven electrons in their outer shell, as shown in Table 2.4.

2.4 The Nature of the Chemical Bond

Elements are held together by forces which are called **bonds**. Some bonds are strong and require much energy to break them, while others are comparatively weak. *Chemical reactivity*, i.e. the ease of making and breaking bonds, is often the result of atoms trying to attain a more stable configuration of electrons.

The most stable electronic configuration of all is the 'noble-gas' structure with its outer shell of eight electrons. There are essentially two ways in which

Table 2.4 Electronic structure of the halogens

Atomic number	Name	Symbol	Shell number					
			1	2	3	4	5	6
9	Fluorine	F	2	7				
17	Chlorine	Cl	2	8	7			
35	Bromine	Br	2	8	18	7		
53	Iodine	I	2	8	18	18	7	
85	Astatine	At	2	8	18	32	18	7

other elements can achieve this stable electronic structure: (*a*) by *electron transfer* and (*b*) by *electron sharing*.

2.5 The Ionic Bond

An ionic bond (sometimes called an 'electrovalent' bond) is formed by the *complete transfer* of electrons from one atom to another, resulting in the formation of charged particles called **ions**. Atoms which have *lost* electrons will have a *positive* charge because the number of positive protons in the nucleus will outnumber the remaining negative electrons. Such *positively charged ions* are called **cations**. *Negatively charged ions* are called **anions**, and are the result of atoms gaining electrons (see Unit 5).

The alkali metals (Group 1 of the periodic table) have one electron in excess of a noble-gas structure, whereas the halogens (Group 7) all have a noble-gas structure less one electron. Consider the electronic structure of the sodium atom and the chlorine atom (Fig. 2.5).

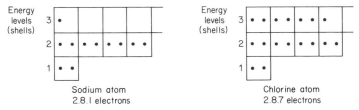

Sodium atom
2.8.1 electrons

Chlorine atom
2.8.7 electrons

Fig. 2.5   Schematic representations of the electrons in a sodium atom and a chlorine atom

When sodium metal is warmed gently with chlorine gas in a tube there is a vigorous chemical reaction, during which a white powder (sodium chloride) is formed. In this reaction one electron is transferred completely from each sodium atom to each chlorine atom, forming sodium *ions* and chloride *ions* (see Fig. 2.6). These ions now have the electron configuration of two noble gases, neon and argon. Both have a *stable octet* of eight electrons in their outer shell and show no further tendency to undergo chemical reaction. Furthermore these ions show no resemblance to the original elements, sodium and chlorine. Whereas we started with two reactive elements—a metal and a poisonous gas—we now have an unreactive white powder—common salt—which is neither metallic nor poisonous.

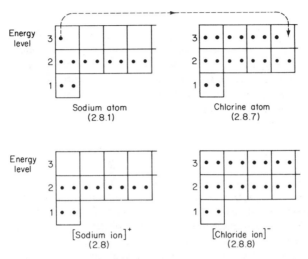

Fig. 2.6   *Schematic representation of the reaction between sodium metal and chlorine gas*

This process of electron transfer can be conveniently represented as follows:

$$Na \rightarrow Na^+ + e$$

and
$$Cl + e \rightarrow Cl^-$$

where Na represents an atom of sodium, Na^+ represents a sodium ion, Cl an atom of chlorine, Cl^- a chloride ion, and e is the symbol for an electron.

Between any pair of electrical charges of opposite sign there is an 'electrostatic' force of attraction (just as there is a force of attraction between unlike magnetic poles). Thus positively charged sodium ions and negatively charged chloride ions are held together by electrostatic attraction. We conclude, therefore, that an ionic bond is caused by the attraction between oppositely charged ions which result from the complete transfer of electrons from the outer electron shell of one atom to another.

The alkali metals in Group 1 of the periodic table achieve a stable octet in the outer shell by losing one electron to form a *unipositive* ion, X^+ (where X represents any Group 1 metal).

Group 2 metals (the alkaline-earth family) have two electrons in their outer shell. Thus to achieve a stable octet of outer electrons they must lose two electrons to produce a *dipositive* ion, X^{2+} (where X represents any Group 2 metal). For example, in the reaction between magnesium metal and fluorine gas (shown in

Fig. 2.7) each magnesium atom transfers two electrons, one to each of two fluorine atoms. The electronic configuration of all the resulting ions is that of the noble gas neon, but this is the only similarity. They have, of course, different charges and their own characteristic atomic number.

Ionic Crystals

In an ionic crystal, oppositely charged ions will attract one another, whereas ions of like charge will repel one another. Each positive cation therefore becomes surrounded by as many negative anions as can be grouped around it, and each negative anion tends to be surrounded by as many positive cations as possible. This produces a giant assembly of ions held in a rigid **crystal lattice**.

No individual molecules exist in the crystal. Each ion is associated with all its nearest neighbours of opposite charge. Although the type of lattice (i.e. the actual arrangement of ions) will depend upon the relative size of the anion and cation, the result is always a closely interlocked structure of ions arranged to reach a state of minimum energy. Let us consider by way of example the crystal lattices of sodium chloride (Na^+Cl^-) and caesium chloride (Cs^+Cl^-).

Because of its relatively small size each sodium cation can accommodate only six chloride ions around itself. To preserve *electrical neutrality*, each chloride ion must be similarly surrounded by six sodium ions. In contrast, the larger caesium ion can accommodate eight chloride ions around itself; so each chloride ion must be surrounded by eight caesium ions. This results in a *face-centred cubic lattice* for sodium chloride and a *body-centred cubic lattice* for caesium chloride (see Fig. 2.8).

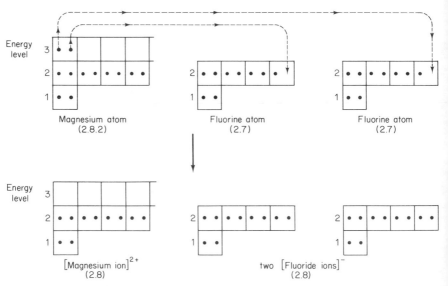

Fig. 2.7   Schematic representation of the reaction between magnesium metal and fluorine gas

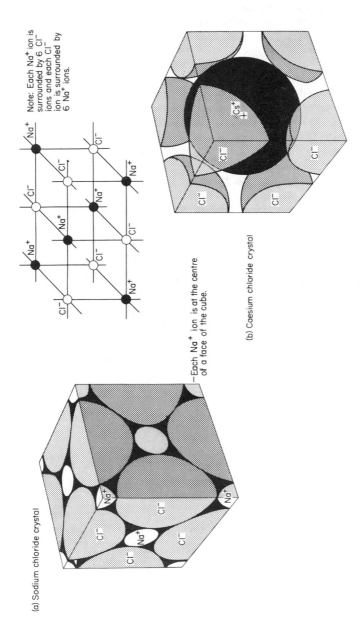

Note: Each Na$^+$ ion is surrounded by 6 Cl$^-$ ions and each Cl$^-$ ion is surrounded by 6 Na$^+$ ions.

—Each Na$^+$ ion is at the centre of a face of the cube.

(a) Sodium chloride crystal

(b) Caesium chloride crystal

Fig. 2.8   Lattice structures for (a) sodium chloride and (b) caesium chloride

The formulas Na^+Cl^- and Cs^+Cl^- which we give to these giant structures only imply that the ratio of anion to cation is $1:1$.

2.6 The Covalent Bond

To remove one electron from an atom requires energy. This process creates a positive ion, and the removal of any further electron or electrons from this ion requires even more energy (because of the increased attraction by the positive nucleus for the remaining negative electrons).

Carbon, in Group 4 of the periodic table, has an electronic structure comprising two electrons in the first shell and four electrons in the second shell. In the compound tetrachloromethane (carbon tetrachloride), each carbon atom is associated with four chlorine atoms. If the bonding in this compound was formed by electron transfer, the four outer electrons of carbon would be transferred to each of four chlorine atoms. Using symbols, this would be represented as:

$$C \rightarrow C^{4+} + 4e$$

and

$$4Cl + 4e \rightarrow 4Cl^-$$

The energy required to remove four electrons from a carbon atom is so large that this process is extremely unlikely. Instead, the carbon atom attains the electronic structure of the noble gas neon by *sharing* pairs of electrons: this requires much less energy than electron transfer. Each *shared pair* of electrons is a **covalent bond**, formed between two atoms. Thus a molecule of tetrachloromethane (carbon tetrachloride) contains four covalent bonds and can be represented:

$$\cdot \overset{\cdot}{\underset{\cdot}{C}} \cdot + 4 \cdot \overset{\cdot\cdot}{\underset{\cdot\cdot}{Cl}} : \rightarrow \quad :\overset{\cdot\cdot}{\underset{\cdot\cdot}{Cl}} : \overset{:\overset{\cdot\cdot}{Cl}:}{\underset{:\overset{\cdot\cdot}{Cl}:}{C}} : \overset{\cdot\cdot}{\underset{\cdot\cdot}{Cl}} :$$

This equation shows only the outermost electrons, which are represented by small dots. Note that, because of this sharing of electrons, the outer shells of the carbon atom and of the four chlorine atoms each achieve a stable *octet* of electrons.

All carbon–hydrogen compounds (hydrocarbons) contain covalent bonds. Methane, the major constituent of natural gas, is the simplest of these hydrocarbons, and each molecule of methane contains a carbon atom joined by four covalent bonds to four hydrogen atoms:

$$\cdot \overset{\cdot}{\underset{\cdot}{C}} \cdot + 4 \cdot H \rightarrow H : \overset{H}{\underset{H}{\overset{\cdot\cdot}{C}}} : H$$

By sharing four pairs of electrons, carbon has achieved a stable octet of electrons, giving it the noble-gas structure of neon; at the same time, each hydrogen atom has achieved a stable *duet* of electrons, giving it the noble-gas structure of helium.

For greater convenience the covalent bond can be represented by a straight line; thus the methane molecule is often shown as

$$
\begin{array}{c}
\text{H} \\
| \\
\text{H} - \text{C} - \text{H} \\
| \\
\text{H}
\end{array}
$$

or, in abbreviated form, CH_4.

In order to achieve a stable octet of electrons, atoms sometimes share more than one pair of electrons to form a bond. Carbon dioxide (CO_2), the gas which bubbles out of fizzy drinks, has in its molecule two pairs of electrons shared by two atoms

$$ \overset{..}{O} : C : \overset{..}{O} $$

This type of covalent bond, containing *two* shared pairs of electrons, is called the *double bond*. It can be represented as a double straight line; thus the carbon dioxide molecule is often shown as O=C=O where each line represents a pair of shared electrons.

2.7 Shapes of Simple Covalent Molecules

The shared pair of electrons, or covalent bond, holding two atoms together is extremely strong compared with the weak attractive forces between individual molecules. For this reason covalent compounds are usually composed of small individual molecules—in contrast to the giant lattice structure of ionic crystals.

(a) Methane

The two-dimensional representations for methane and tetrachloromethane (carbon tetrachloride) molecules given in Section 2.6 cannot show how these molecules are arranged in space. A more complete picture is given if we go back to the charge-cloud model of the atom (Section 1.12). The four outer electrons of carbon can be considered as pear-shaped clouds of negative charge. These clouds repel one another and become evenly distributed in space around the nucleus. The most stable shape that four mutually repelling pear-shaped clouds could adopt would be that in which each orbital points towards the corner of a a regular tetrahedron (Fig. 2.9).

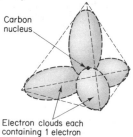

Carbon nucleus

Electron clouds each containing 1 electron

Fig. 2.9   Orbital clouds for the four outer electrons of carbon

The orbital cloud for the hydrogen electron is spherical (Fig. 2.10).

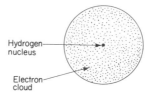

Hydrogen
nucleus

Electron
cloud

Fig. 2.10   Orbital cloud for the hydrogen electron

When the spherical orbitals of four hydrogen atoms overlap the pear-shaped orbitals of carbon, four covalent bonds are formed. These four bonds point towards the corners of a regular tetrahedron (Fig. 2.11).

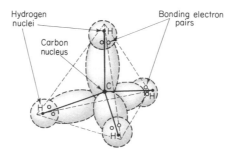

Hydrogen
nuclei

Carbon
nucleus

Bonding electron
pairs

Fig. 2.11   Charge-cloud model for methane

In a similar manner to the carbon atom in methane, the outer electrons of the nitrogen atoms in ammonia are distributed in four pear-shaped orbitals pointing towards the corners of a tetrahedron (Fig. 2.12).

(b) Ammonia

On opening a bottle of smelling salts the pungent smell of ammonia gas can quickly be detected. This gas contains the element nitrogen covalently bonded to three hydrogen atoms. Nitrogen atoms contain five electrons in their outer shell and can achieve the noble-gas structure of neon by sharing three electrons, one from each of three hydrogen atoms:

$$H : \overset{..}{\underset{\displaystyle H}{N}} : H \qquad \text{or} \qquad H - \overset{\displaystyle ..}{\underset{\displaystyle |}{N}} - H$$
$$\phantom{H : N : H \qquad \text{or} \qquad H - } H$$

Three of the orbitals contain only one electron and are thus available for bonding with three spherical, singly occupied, hydrogen orbitals. The fourth orbital of nitrogen is filled because it contains two electrons. Such a pair of electrons is called a *lone pair* (Fig. 2.13).

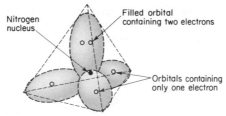

Fig. 2.12   *Orbital clouds for the five outer electrons of nitrogen*

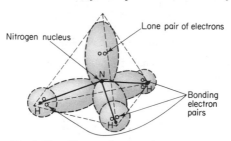

Fig. 2.13   *Charge-cloud model for ammonia*

Thus the tetrahedral model can be used for ammonia with a lone pair of electrons occupying one of the four positions, but the molecule is, of course, pyramidal.

(c) Water

Most people know the formula for a water molecule to be H_2O. Is this to be expected from its electron configuration? What is the *shape* of the water molecule?

An oxygen atom contains six electrons in its outer shell and can thus share two electrons from each of two hydrogen atoms to gain its stable octet:

$$: \overset{..}{\underset{..}{O}} : H \qquad \text{or} \qquad : \overset{..}{O} - H$$
$$\quad H \qquad\qquad\qquad\quad |$$
$$\qquad\qquad\qquad\qquad\qquad H$$

Once again the tetrahedral model can be used, with *two* lone pairs of electrons occupying two of the positions (Fig. 2.14).

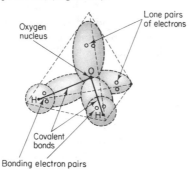

Fig. 2.14   *Charge-cloud model for water*

That part of the molecule which contains lone pairs of electrons becomes slightly negatively charged in comparison with the rest of the molecule. In consequence some other part of the molecule must possess a slight positive charge to make the whole molecule neutral.

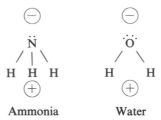

Ammonia   Water

Molecules which possess this charge separation are said to be **polar**, and such *polarity* has very important influences on the properties of these molecules.

2.8 The Co-ordinate Bond (Dative Covalent Bond)

We have seen that molecules can have orbitals containing lone pairs of electrons. Ammonia has one lone pair:

$$\overset{\displaystyle ..}{\underset{\displaystyle \overset{|}{H}}{\overset{H\diagdown \ \diagdown H}{N}}}$$

and water has two lone pairs:

$$\underset{H \qquad H}{\overset{..}{O}}$$

Ammonia will react with hydrogen cations, using its lone pair of electrons to form a covalent bond. The bond differs from normal covalent bonds only in that one atom supplies *both* electrons to form the shared-pair bond.

$$\begin{matrix} H \\ H \ \ddot{:}\ddot{N}\colon + H^+ \\ H \end{matrix} \rightarrow \begin{bmatrix} \quad H \quad \\ H \ :\ddot{N}: \ H \\ \quad \ddot{H} \quad \end{bmatrix}^{+}$$

Once formed, this co-ordinate bond between the nitrogen and hydrogen is identical and indistinguishable from the other nitrogen–hydrogen covalent bonds. The ion NH_4^+ produced in this reaction is called the ammonium ion.

2.9 Characteristics of Compounds containing Ionic and Covalent Bonds

Ionic compounds such as sodium chloride, with their huge rigid crystal lattices, might be expected to have markedly different properties from covalent compounds such as iodine and carbon dioxide which exist as separate small molecules (Fig. 2.15).

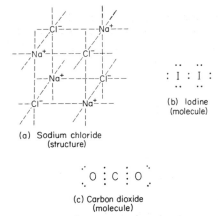

(a) Sodium chloride
(structure)

(b) Iodine
(molecule)

(c) Carbon dioxide
(molecule)

Fig. 2.15   Contrast between an ionic lattice and individual molecules: (a) lattice structure for sodium chloride; (b) an iodine molecule; (c) a carbon dioxide molecule

(a) Melting Point

Ionic compounds are solids with high melting point (e.g. sodium chloride); covalent compounds are often gases (e.g. carbon dioxide), liquids (e.g. tetrachloromethane) or low-melting solids (e.g. iodine). This is explained because the electrical forces between ions in an ionic compound are very strong and the thermal agitation of the ions must be great in order to break down the lattice. High temperatures are necessary to melt the crystal. In contrast, the forces *between* covalent molecules are weak (the actual covalent bond between atoms *in* each molecule is strong) and the thermal agitation necessary to separate the aggregate of molecules is small. In consequence the melting point is low, as in iodine (a crystalline covalent solid melting at 114 °C) and carbon dioxide (a gas at room temperature).

(b) Solubility

Ionic compounds are soluble in only a few solvents, notably water. These solvents must be capable of breaking down the crystal lattice into ions. Covalent compounds tend to be insoluble in the solvents which dissolve ionic compounds, but they do dissolve in covalently bonded solvents such as tetrachloromethane (carbon tetrachloride), petrol and paraffin.

(c) Electrical Properties

Solid ionic compounds do not conduct an electric current. However, when molten or dissolved in water the lattice is destroyed, the ions are free to move around, and they are then able to carry an electric current (see Section 5.1). Covalent compounds contain *no ions* even in the liquid state; therefore they are non-conducting. Pure water is a poor conductor of electricity, but a solution of an ionic solid dissolved in water is a good conductor. The water is able to break down the lattice so that the ions are free to move. A solution of a covalent compound in water is non-conducting, unless a chemical reaction between the compound and water takes place and produces ions.

2.10 Giant Molecules

(a) Diamond and Graphite

Unlike the simple covalent compounds we have considered so far, there are a few which do not appear to dissolve in any solvent and in addition have very high melting points. They are **giant molecules** in which the *whole crystal is one molecule*. Diamond and graphite, the two crystalline forms of carbon, are good examples of these giant covalent structures.

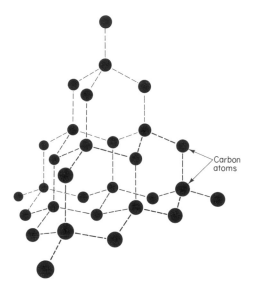

Fig. 2.16   The structure of diamond

Each carbon atom in **diamond** is tetrahedrally linked to four others by *single covalent* bonds. This arrangement is repeated throughout the whole molecule (Fig. 2.16) so that the structure is extremely strong and rigid. It is therefore very hard and is used for cutting-tools, drilling and in jewellery.

In contrast to diamond, **graphite** is soft, feels slippery and is used as a lubricant. It has a layer structure in which carbon atoms are joined by strong covalent bonds in a pattern of interlocking hexagons (Fig. 2.17). Each carbon atom is joined to three others in the layer. The bonding *between* the layers is much weaker than the bonding *in* the layer, and so the graphite is easily split into sheets.

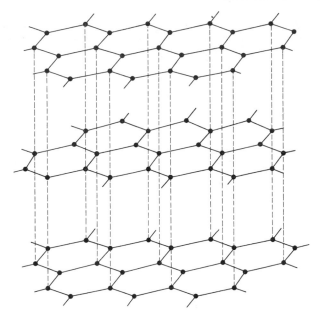

Fig. 2.17   The structure of graphite: widely separated layers of carbon atoms

(b) Polymers

We have seen in diamond and graphite that carbon is capable of forming multiple bonds. In everyday life there are numerous materials which contain similar multiple bonds: familiar examples include nylon, polyvinyl chloride (PVC), polythene, and polyurethane. These are called **polymers** (*poly* = many). Thus the polythene molecule is a very long chain of many carbon atoms, each covalently bonded to hydrogen:

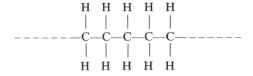

Polymeric molecules are discussed in detail in Unit 12.

(c) Ice

The polar nature of the water molecule produces attraction between separate molecules. This weak but important attractive force between the hydrogen of one molecule and the oxygen of an adjacent molecule is called a **hydrogen bond** (Fig. 2.18).

As water is cooled on a microscope slide ice begins to form on the surface. On examination under a microscope hexagonally shaped ice crystals can be seen.

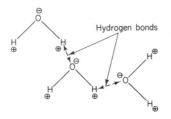

Fig. 2.18   The hydrogen bond in water

These crystals result from an open lattice arrangement produced by hydrogen bonding between the water molecules (Fig. 2.19).

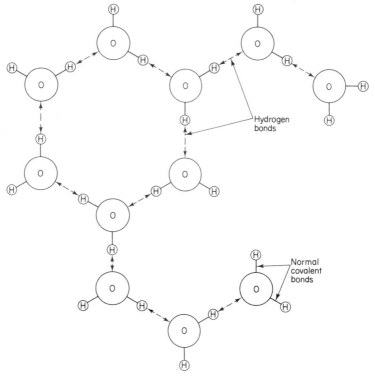

Fig. 2.19   The structure of ice

2.11 Metals

A metal can be considered as an arrangement of positive metal ions arranged in a regular three-dimensional crystal lattice, similar to those of sodium chloride and caesium chloride shown in Fig. 2.8. The electrons which once were charge-clouds attached to the individual metal atoms may be considered as having coalesced into a single cloud of charge which now surrounds the metallic ions in the crystal lattice. Each ion has an attraction for several of the outer electron charge-clouds of individual nuclei, and it is thought that these charge clouds are not bound to any single nucleus but *spread out*. These electrons are said to be *delocalized*. The ions are held together in a 'sea of mobile electrons' producing a giant structure (Fig. 2.20). It is these mobile 'free' electrons which account for the electrical properties of metals (see Unit 5).

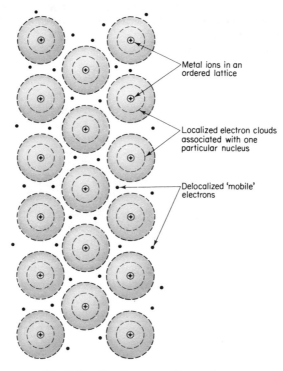

Metal ions in an ordered lattice

Localized electron clouds associated with one particular nucleus

Delocalized 'mobile' electrons

Fig. 2.20   The structure of a metal

Summary of Unit 2

1. Electrons are found in **shells** around the nucleus of an atom.
2. Each shell is associated with a particular *energy*. Shells are thus sometimes called **energy levels**. The lowest energy level is the one nearest the nucleus.
3. The maximum number of electrons in each energy level is given by $2n^2$, where n is the energy level number.

4. Electrons are found in **orbitals** within each energy level. These orbitals are variously shaped clouds of electronic charge.
5. Each element has a particular electron structure with its electrons arranged in specific energy levels.
6. The **periodic table** arranges elements in order of *ascending atomic number*. Horizontal rows of elements are called **periods**; vertical columns of elements are called **groups** or **families**.
7. The elements within a group or family (e.g. Group 1, the alkali metals) have similar chemical properties.
8. The *noble gases* are a family of elements each containing eight electrons (an octet) in their outer energy level. Such an outer octet of electrons is associated with chemical stability.
9. Elements react to achieve this stable electron structure by *electron transfer* or by *electron sharing*.
10. Electron transfer produces an **ionic bond** between positively charged cations and negatively charged anions.
11. Ionic compounds form *ionic crystals* in which the positively and negatively charged ions are arranged in a rigid **crystal lattice**.
12. Electron sharing produces a **covalent bond**. Each *shared pair* of electrons produces a *single* covalent bond; *two* shared pairs of electrons produce a *double* bond.
13. Covalent compounds exist as *molecules*. Each molecule has its own shape or spatial arrangement, e.g. a molecule of methane is tetrahedral.
14. Molecules in which there is some charge separation are said to be **polar**.
15. **Ionic compounds** tend to be solids with high melting points. They are soluble in polar solvents. When molten or in aqueous solution they are able to conduct an electric current.
16. **Covalent compounds** tend to be gases, volatile liquids or solids with low melting points. They are soluble in non-polar solvents and do not conduct electricity.
17. *Diamond* and *graphite* are giant molecules containing many carbon–carbon covalent bonds.
18. *Ice* is composed of small water molecules held together in a crystal lattice by **hydrogen bonds**.
19. **Metals** have a crystal lattice containing positive metal ions held together in a 'sea' of *mobile* electrons.

Test Yourself on Unit 2

1. Which of the following formulas gives the number of electrons in each complete shell of an atom, where n is the shell number?
   (a) n^2, (b) $2n^2$, (c) $(n+2)$, (d) $(n+8)$.

2. Mark each of the following statements true or false:
   (a) Electron shells are sometimes called energy levels.
   (b) The shell nearest to the nucleus has the highest energy.

(*c*) The shell nearest the nucleus can contain a maximum of two electrons.
(*d*) An orbital is a cloud of electricity formed by electrons around the nucleus of an atom.

3. Elements in the periodic table are arranged in order of their:
   (*a*) relative atomic mass,
   (*b*) atomic number,
   (*c*) mass number,
   (*d*) metallic character.

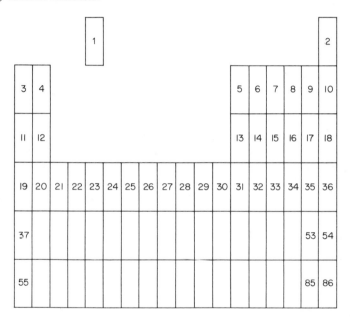

4. Consider the following elements from the outline of the periodic table above:
   (*a*) 3, 11, 19, 37, 55.
   (*b*) 9, 17, 35, 53, 85.
   (*c*) 2, 10, 18, 36, 54, 86.
   (*d*) 21, 22, 23, 24, 25, 26.
   (*e*) 3, 4, 5, 6, 7, 8, 9, 10.
   (i)   Which set of elements is called the noble-gas family?
   (ii) Which set of elements is called the halogen family?
   (iii) Which set of elements is called the alkali metal family?
   (iv) Which set contains only transition metals?
   (v)   Which set is a short period?

5. The following symbols refer to atoms of magnesium and chlorine:

$$^{24}_{12}Mg, \qquad ^{37}_{17}Cl$$

(*a*) State the number of electrons in successive electron shells of these atoms.

(b) Explain why chlorine forms an ion Cl^-.

(c) Write the symbol for the magnesium ion.

(d) Write the formula for the ionic solid formed when these two elements combine.

(e) Would you expect this solid to have a high or low melting point? Give a reason.

6. Why is it incorrect to refer to a molecule of sodium chloride while it is correct to refer to a molecule of ammonia?

7. Mark the following statements true or false:
   (a) A methane molecule is tetrahedral.
   (b) A water molecule is linear.
   (c) An ammonia molecule contains a lone pair of electrons.
   (d) An ammonium ion contains four equivalent covalent bonds.

8. Both methane and diamond are covalently bonded. Methane is a gas, diamond is a solid with a very high melting point. Why?

Mark this test out of 30 with the answers provided on page 370.

Matter and the Mole

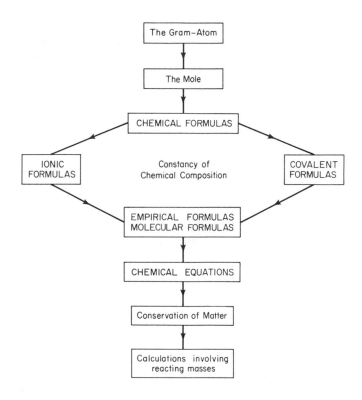

So far we have looked at the ways in which atoms combine: how electron transfer leads to the giant ionic-lattice structure of sodium chloride, for example, while electron sharing produces simple individual covalent molecules such as methane. In this Unit we move from the study of individual atoms, ions and molecules to the chemical laboratory where we use materials in measurable quantities. We see how chemical formulas and chemical equations are developed as a result of experiment.

3.1 The Gram-Atom and the Mole

(a) The Gram-Atom

We have discussed in Section 1.14 the idea of atomic masses relative to the mass of the carbon $^{12}_{6}C$ isotope taken as 12 units. *Relative atomic mass* is defined as the mass of one atom of the element compared with one-twelfth the mass of one atom of carbon $^{12}_6C$.

If we use grams as units to express these relative atomic masses we have the 'gram-atom'. One **gram-atom** of any element is the relative atomic mass of that element expressed in grams. For example, the relative atomic mass of the element copper (Cu) is 63·5 and that of sulphur (S) is 32·1. Therefore one *gram-atom* of copper would be 63·5 *grams* and one gram-atom of sulphur would be 32·1 *grams*.

The concept of the gram-atom is useful because, while it is impossible to see or weigh individual atoms very easily, we can actually see and weigh one gram-atom of an element.

One gram-atom of any element contains the same number of atoms. This fact is illustrated in Fig. 3.1 by analogy with children's marbles of masses 10 g, 20 g and 30 g. If we take three samples, A, B and C, each of 300 g, there will be 30 marbles in A, 15 marbles in B and 10 marbles in C.

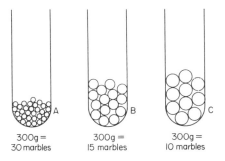

Fig. 3.1   Atoms represented as marbles: each sample has the same mass

However, if we take an amount of each type of marble proportional to its mass, say 100 g of A, 200 g of B and 300 g of C, then there will be the same number of marbles in each sample—in this case 10, as shown in Fig. 3.2.

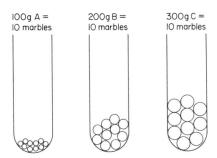

Fig. 3.2   Atoms represented as marbles: each sample contains the same number

Similarly, if we take masses of different *elements* proportional to their relative atomic masses, then each sample of the elements will contain the same number of atoms (see Fig. 3.3).

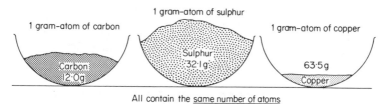

All contain the same number of atoms

Fig. 3.3   *One gram-atom of each element contains the same number of atoms*

Thus one gram-atom of each element contains the same number of atoms.

(b) The Mole

The actual number of atoms in every gram-atom of an element is, as we might expect, very large. It has been found to be 6.02×10^{23}. Thus 63·5 grams of copper metal, for example, contains 6.02×10^{23} atoms of copper.

This important number, 6.02×10^{23}, is known as the **Avogadro constant** or Avogadro's number in honour of Amadeo Avogadro (1776–1856). Avogadro was born and educated in Italy, where he practised law for a number of years. His interests turned towards physics, however, and he is remembered today for his famous hypothesis concerning molecules in gases (see Section 4.6).

The amount of substance which contains Avogadro's number (6.02×10^{23}) of particles is called **one mole**. It does not matter whether the particles are atoms, molecules or ions. Thus one mole of copper metal contains 6.02×10^{23} atoms of copper; one mole of chlorine gas (Cl_2) contains 6.02×10^{23} chlorine molecules or, since each chlorine molecule consists of two atoms, $2 \times (6.02 \times 10^{23})$ atoms of chlorine. One mole of sodium chloride (Na^+Cl^-) contains 6.02×10^{23} sodium ions (Na^+) and 6.02×10^{23} chloride ions (Cl^-).

3.2 Chemical Formulas through Experiment

Using the mole concept we are now able to determine chemical formulas by experiment in the laboratory. A *chemical formula* uses symbols to show the numbers of the the atoms or ions of the elements contained in one molecule or smallest portion of a compound. If there is more than one atom or ion of each element present, the actual number is indicated by a figure written as a subscript immediately after the symbol. For example, one molecule of carbon dioxide contains one atom of carbon and two atoms of oxygen: its *chemical formula* is therefore CO_2.

The following experiments illustrate laboratory methods of finding the chemical formula of a compound.

Experiment 3.1   *Determination of the chemical formula of black copper oxide*
In this experiment a pure sample of black copper oxide is analysed to find out how
many moles of copper combine with one mole of oxygen atoms.

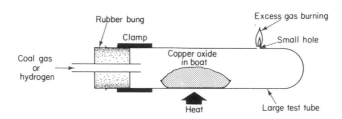

Fig. 3.4   Reduction of black copper oxide

A porcelain boat is weighed, first empty and then containing a little pure black
copper oxide. It is then placed in a hard-glass test tube having a small hole near
the closed end (see Fig. 3.4). A steady stream of coal gas, or hydrogen from a
cylinder, is passed through the apparatus until all the air has been displaced from
the test tube. *Great care must be taken with regard to safety whenever hydrogen is
being used, as mixtures of air and hydrogen are violently explosive.*

To check that all the air has been displaced, take samples of the gas in a small
test tube and ignite them at some distance from the apparatus. When a sample
burns quietly rather than explosively it is safe to ignite the gas issuing from the
small hole in the main test tube.

On heating the contents of the tube with a hot non-luminous bunsen flame, the
black colour of the oxide changes to the pink colour of copper. The hydrogen
removes the oxygen from the copper oxide, forming water vapour, and pure
copper is left.

The apparatus is then allowed to cool, but the stream of gas must be maintained
or oxygen from the air may enter the tube and re-form copper oxide with the hot
copper powder.

When the apparatus is cold, the boat containing the copper is removed and
reweighed.

Results

Mass of boat	=	8·32 g
Mass of boat and copper oxide	=	10·86 g
Mass of boat and copper	=	10·35 g
∴ mass of copper	=	2·03 g
∴ mass of oxygen	=	0·51 g

One mole of copper atoms have a mass of 63·54 g

$$\therefore \text{ number of moles of copper in the sample} = \frac{2 \cdot 03}{63 \cdot 54}$$

$$= 0 \cdot 032$$

One mole of oxygen atoms have a mass of 16 g

$\therefore$ number of moles of oxygen atoms in the sample $= \dfrac{0\cdot51}{16}$

$$= 0\cdot032$$

$\therefore$ 0·032 mole of oxygen atoms are combined with 0·032 mole of copper
$\therefore$ 1 mole of oxygen atom are combined with 1 mole of copper
$\therefore$ 6·02 × 10²³ oxygen atoms are combined with 6·02 × 10²³ copper atoms
$\therefore$ 1 oxygen atom is combined with 1 copper atom

Thus the chemical formula of copper oxide is Cu_1O_1, but as it is usual to ignore subscripts where only one atom of each element is present, the formula is generally written as CuO.

Experiment 3.2   *Determination of the chemical formula of water*
In Experiment 3.1 water was one of the products of the reaction. We now modify the procedure so that the water vapour produced can be collected and weighed. *Great care must be taken with regard to safety whenever hydrogen is being used, as mixtures of air and hydrogen are violently explosive.*

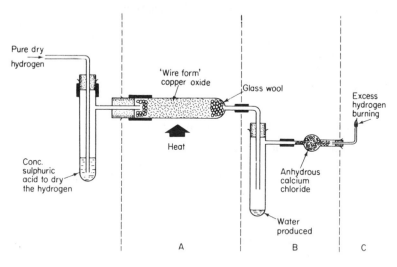

Fig. 3.5   Determination of the chemical formula of water

The apparatus is shown in Fig. 3.5. Tube A is filled with dry, pure, wire-form copper oxide and then weighed. It is connected to the weighed condensation and absorption tubes (B). The absorption tube contains anhydrous calcium chloride which absorbs water vapour. Pure dry hydrogen is passed through the apparatus until all the air is displaced. *Observe the same precautions as those described in Experiment 3.1.* The gas issuing from the jet (C) is ignited, and the tube A is heated with a hot non-luminous bunsen flame.

After 5–10 minutes the heating is stopped, but the flow of hydrogen is maintained until the apparatus is quite cold. Tubes A and B are then weighed separately.

Results

Mass of tube A and copper oxide before heating	=	108·97 g
Mass of tube A and contents after heating	=	104·17 g
∴ mass of oxygen taken from copper oxide and converted into water	=	4·80 g
Mass of tube B before experiment	=	94·49 g
Mass of tube B after experiment	=	99·89 g
∴ mass of water produced	=	5·40 g
∴ mass of hydrogen in the water	=	5·40 g − 4·80 g = 0·60 g

Calculation

1 mole of hydrogen atoms have a mass of 1 g

∴ number of moles of hydrogen atoms in the water produced $= \dfrac{0·60}{1}$

1 mole of oxygen atoms have a mass of 16 g

∴ number of moles of oxygen atoms in the water produced $= \dfrac{4·8}{16}$

$$= 0·30$$

∴ 0·60 mole of hydrogen atoms combine with 0·30 mole of oxygen atoms
∴ 2 moles of hydrogen atoms combine with 1 mole of oxygen atoms
∴ $2 \times 6·02 \times 10^{23}$ hydrogen atoms combine with $6·02 \times 10^{23}$ oxygen atoms
∴ 2 hydrogen atoms combine with 1 oxygen atom.

Thus the chemical formula of water is H_2O_1, usually written as H_2O.

Chemical formulas such as CuO for copper oxide and H_2O for water indicate the combining ratios of the elements but do not necessarily indicate the size of the basic unit. A chemical formula showing only the combining ratios of the elements in the compound is called the **empirical formula** (see Section 3.5).

3.3 Law of Constant Composition

Any sample of pure copper oxide, no matter how it is analysed, provided the analytical technique is efficient, always shows the same ratio of copper to oxygen. In fact this constancy in chemical composition is found for *all* pure chemical compounds, and we can say that *all pure samples of the same chemical compound contain the same elements in the same proportion by mass.* This statement is known as the **Law of Constant Composition**, sometimes called the Law of Definite Proportions.

3.4 Writing Chemical Formulas

Formulas for most compounds have been determined experimentally and details can be found in various books of chemical data. However, the chemist requires a simple and quick method for deducing these formulas without recourse to experiment.

(a) Ionic Compounds

Many different electrical experiments can be carried out on the quantity of electricity required to liberate one mole of an element from a chemical compound and the results used to calculate the charges on the ion(s) of that element (see Section 5.5). Some of the more common ions, both positive and negative, are listed in Table 3.1.

Ionic compounds result from a combination of cations and anions such that the resulting giant structure has no overall charge. For example:

(i)   one Na^+ and one Cl^- give Na^+Cl^-
(ii) two Na^+ and one SO_4^{2-} give $Na^+Na^+SO_4^{2-}$ or $Na_2^+ SO_4^{2-}$
(iii) one Ca^{2+} and one SO_4^{2-} give $Ca^{2+} SO_4^{2-}$
(iv) one Ca^{2+} and two NO_3^- give $Ca^{2+}NO_3^-NO_3^-$ or $Ca^{2+} (NO_3^-)_2$
(v)   one Al^{3+} and three Cl^- give $Al^{3+}Cl^-Cl^-Cl^-$ or $Al^{3+}Cl_3^-$
(vi) two Al^{3+} and three SO_4^{2-} give $Al^{3+}Al^{3+}SO_4^{2-}SO_4^{2-}SO_4^{2-}$ or $Al_2^{3+}(SO_4^{2-})_3$

Note that, in order to achieve electrical neutrality in example (vi), *two* Al^{3+} must unite with *three* SO_4^{2-}.

Nomenclature based on oxidation numbers. It will be noticed that a roman numeral appears after the names of many of the ions in Table 3.1. This is the **oxidation number**. The following rules show how this oxidation number is calculated:

(i) the oxidation number of hydrogen is defined as $+1$ (except in metallic hydrides when it is -1, and in hydrogen gas H_2 when it is zero);

(ii) the oxidation number of oxygen is defined as -2 (except in peroxides when it is -1, and in oxygen gas O_2 when it is zero);

(iii) the oxidation number of any other element is selected to make the sum of the oxidation numbers equal to the charge on the ion.

Where an element has more than one oxidation number, the oxidation number is stated directly after the name of the ion (see Table 3.1).

For *simple cations* the oxidation number of the element is its ionic charge.

For *simple anions* the systematic ending *-ide* is used (except for non-metal hydrides such as water, ammonia and methane).

For anions derived from the oxoacids the oxidation number of the element combined with oxygen is given together with the ending *-ate*. For example, the SO_4^{2-} anion is called tetraoxosulphate(VI) because the oxidation number of the sulphur in this ion is $+6$. This is calculated as follows:

the total charge on the ion is -2;
the oxidation number of each oxygen atom is -2, and

Table 3.1 Charges carried by ions

Cations (ions with a positive charge)				Anions (ions with a negative charge)		
Charge: +4	+3	+2	+1	−1	−2	−3
Sn^{4+} tin(IV)	Fe^{3+} iron(III)	Ca^{2+} calcium	Na^+ sodium	F^- fluoride	SO_4^{2-} sulphate	PO_4^{3-} phosphate
Mn^{4+} manganese(IV)	Al^{3+} aluminium	Mg^{2+} magnesium	K^+ potassium	Cl^- chloride	SO_3^{2-} sulphite	
		Ba^{2+} barium	Li^+ lithium	Br^- bromide	S^{2-} sulphide	
		Sr^{2+} strontium	Cs^+ caesium	I^- iodide	CO_3^{2-} carbonate	
		Cu^{2+} copper(II)	Cu^+ copper(I)	NO_3^- nitrate	O^{2-} oxide	
		Zn^{2+} zinc	Ag^+ silver	HCO_3^- hydrogencarbonate (bicarbonate)		
		Sn^{2+} tin(II)	Hg^+ mercury(I)	HSO_3^- hydrogensulphite (bisulphite)		
		Fe^{2+} iron(II)	NH_4^+ ammonium	HSO_4^- hydrogensulphate (bisulphate)		
		Pb^{2+} lead(II)	H^+ hydrogen cation (proton)	HS^- hydrogensulphide		
		Hg^{2+} mercury(II)		H^- hydride		
		Mn^{2+} manganese(II)		OH^- hydroxide		

the sum of the oxidation numbers of four oxygen atoms is -8;
therefore, in order to give the ion a total charge of -2, the sulphur must have an oxidation number of $+6$. The reasoning becomes clear if we write the oxidation numbers on the formula of the ion:

$$\left[\begin{array}{cc} \overset{+6}{S} & \overset{(4 \times -2)}{O_4} \end{array} \right]^{2-}$$

Similarly, the SO_3^{2-} anion is called trioxosulphate(IV) because the oxidation number of sulphur is $+4$:

$$\left[\begin{array}{cc} \overset{+4}{S} & \overset{(3 \times -2)}{O_3} \end{array} \right]^{2-}$$

The first two columns of Table 3.2 show the systematic nomenclature for a number of ionic compounds. However, IUPAC, national authorities and examination boards still prefer the more traditional names sulphate, sulphite, nitric, nitrous, nitrate, nitrite and thiosulphate, and these names, as shown in the third column of the table, will be used throughout the rest of this book.

Table 3.2 Use of oxidation numbers in naming ionic compounds

Formula	Systematic name	Adopted name
Na^+Cl^-	sodium chloride	sodium chloride
$Cu^{2+}SO_4^{2-}$	copper(II) tetraoxosulphate(VI)	copper(II) sulphate
$Cu_2^+SO_4^{2-}$	copper(I) tetraoxosulphate(VI)	copper(I) sulphate
$Na_2^+SO_3^{2-}$	sodium trioxosulphate(IV)	sodium sulphite
$H^+NO_3^-$	trioxonitric(V) acid	nitric acid
$H_2^+SO_4^{2-}$	tetraoxosulphuric(VI) acid	sulphuric acid
H^+Cl^-	hydrochloric acid	hydrochloric acid
$(NH_4)_3^+PO_4^{3-}$	ammonium tetraoxophosphate(V)	ammonium phosphate
$Ca^{2+}CO_3^{2-}$	calcium carbonate	calcium carbonate
$Fe^{3+}(OH^-)_3$	iron(III) hydroxide	iron(III) hydroxide

(b) Covalent Compounds

The formulas for covalent compounds are determined by the electron structure of the elements in the compound. The number of covalent bonds capable of being produced by an element is given in Table 3.3.

Table 3.3 Possible numbers of covalent bonds

Element	Number of covalent bonds
Hydrogen (H)	1
Chlorine (Cl) and other halogens	1
Oxygen (O)	2
Sulphur (S)	2, 4 or 6
Nitrogen (N)	3, 4 or (5)
Carbon (C)	4 or (2)

(*Note*: brackets indicate a number that is possible but rarely occurs)

Atoms of these and similar elements can combine with each other to form covalent **molecules**. Examples of covalent molecules are shown in Table 3.4.

Table 3.4 Structure of covalent molecules

Compound	Formula	Structure	Number of covalent bonds
Methane	CH_4	H—C—H with H above and below	4
Water	H_2O	H—O with H	2
Ammonia	NH_3	H—N—H with H below	3
Sulphur dioxide	SO_2	O=S=O	4
Sulphur trioxide	SO_3	O=S=O with O above	6
Carbon dioxide	CO_2	O=C=O	4
Tetrachloromethane (carbon tetrachloride)	CCl_4	Cl—C—Cl with Cl above and below	4

3.5 Empirical and Molecular Formulas

The chemical formulas Na^+Cl^- and CCl_4 represent quite different types of structure, but both show the *ratio* of the particles of the combining elements in the structure. A formula showing the ratio of the number of the respective atoms or ions in a compound is called an **empirical formula**. The formulas for all ionic compounds are empirical formulas, because they show the simplest ratio of the ions in the giant lattice.

Tetrachloromethane (carbon tetrachloride) is composed of separate molecules and the formula CCl_4 is a **molecular formula**. It shows the *actual* number of atoms in one molecule of the compound. In this case the molecular formula and empirical formula are identical.

The molecular formula of the gas ethane is C_2H_6, which shows that each molecule contains two carbon atoms covalently bonded to each other and to six hydrogen atoms:

$$\begin{array}{ccc} H & H \\ | & | \\ H-C-C-H \\ | & | \\ H & H \end{array}$$

The ratio of carbon atoms to hydrogen atoms is $1:3$, and the empirical formula is therefore CH_3. The molecule CH_3 does not exist. This *empirical* formula shows the simplest combining ratio of the elements hydrogen and carbon in this compound which has the *molecular* formula C_2H_6.

Other examples illustrating the differences between empirical and molecular formulas are listed in Table 3.5. The *relative molecular mass* is shown for those compounds having molecular formulas. It is obtained by summing all the relative atomic masses of the individual atoms of which its molecules are composed.

For example, ethane has a molecular formula C_2H_6; its relative molecular mass can be obtained by summing the relative atomic masses of two carbon atoms and six hydrogen atoms:

$$\underbrace{\underset{|}{C_2} \quad \underset{|}{H_6}}_{30}$$
$$(2\times 12)+(6\times 1)$$

Thus the relative molecular mass of ethane is 30.

Calculation of Empirical and Molecular Formulas

The masses of each element present in a compound can be obtained through experiment. These values are often expressed as percentages, or as the number of grams of each element contained in 100 g of the compound. The calculation of the empirical formula is similar to the calculations carried out in Experiments 3.1

Table 3.5 Molecular and empirical formulas

Substance	Molecular formula	Relative molecular mass	Empirical formula
Potassium nitrate			$K^+NO_3^-$
Magnesium sulphate			$Mg^{2+}SO_4^{2-}$
Nitric acid			$H^+NO_3^-$
Calcium oxide			$Ca^{2+}O^{2-}$
Benzene	C_6H_6	$(6 \times 12) + (6 \times 1)$ $= 78$	CH
Carbon dioxide	CO_2	$12 + (2 \times 16)$ $= 44$	CO_2
Ethene	C_2H_4	$(2 \times 12) + (4 \times 1)$ $= 28$	CH_2
Ethyne	C_2H_2	$(2 \times 12) + (2 \times 1)$ $= 26$	CH

and 3.2. The following examples illustrate the calculation of both empirical and molecular formulas.

Example 100 g of an oxide of sulphur contains 50 g of sulphur and 50 g oxygen. If the relative molecular mass of the compound is 64 and the relative atomic masses of sulphur and oxygen are 32 and 16 respectively, calculate the empirical and molecular formulas of the oxide.

Calculation
Number of moles of sulphur atoms is $\frac{50}{32} = 1·56$
Number of moles of oxygen atoms is $\frac{50}{16} = 3·12$
$\therefore$ 1·56 moles of sulphur atoms combines with 3·12 moles of oxygen atoms
    or 1 mole of sulphur atoms combines with 2 moles of oxygen atoms
$\therefore$ $6·02 \times 10^{23}$ atoms of sulphur combine with $2 \times (6·02 \times 10^{23})$ oxygen atoms
$\therefore$ 1 atom of sulphur combines with 2 atoms of oxygen
The *empirical formula* is therefore SO_2.

The sum of the relative atomic masses of this empirical unit (SO_2) is $32 + (2 \times 16) = 64$. This is identical with the relative molecular mass of the compound, and hence the *molecular formula* is also SO_2.

Example 100 g of a compound contains 32 g of carbon, 4 g of hydrogen and 64 g of oxygen. If its relative molecular mass is 150 and the relative atomic masses of carbon, hydrogen and oxygen are 12, 1, and 16 respectively, calculate the empirical and molecular formulas.

Calculation
Number of moles of carbon atoms $= \frac{32}{12} = 2\cdot67$
Number of moles of hydrogen atoms $= \frac{4}{1} = 4\cdot00$
Number of moles of oxygen atoms $= \frac{64}{16} = 4\cdot00$
$\therefore$ 2·67 moles of carbon combines with 4·00 moles of hydrogen and 4·00 moles of oxygen
We divide all of these values by the smallest (2·67) to obtain the simplest ratio.
$\therefore \frac{2\cdot67}{2\cdot67}$ moles of carbon combines with $\frac{4\cdot00}{2\cdot67}$ moles of hydrogen and $\frac{4\cdot00}{2\cdot67}$ moles of oxygen
$\therefore$ 1 mole of carbon combines with 1·5 moles of hydrogen and 1·5 moles of oxygen
$\therefore$ $(1 \times 6\cdot02 \times 10^{23})$ carbon atoms combines with $1\cdot5 \times 6\cdot02 \times 10^{23}$ hydrogen atoms and $1\cdot5 \times 6\cdot02 \times 10^{23}$ oxygen atoms
In simplest terms this would indicate that the ratio of carbon to hydrogen to oxygen *atoms* in the compound was 1 carbon : 1·5 hydrogen : 1·5 oxygen. However, since it is impossible to have a fraction of an atom, the empirical formula must be a *whole-number* ratio. The simplest whole-number ratio is 2 carbon : 3 hydrogen : 3 oxygen, and thus the empirical formula is $C_2H_3O_3$.

If this was the molecular formula, the relative molecular mass of the compound would be $(2 \times 12)+(3 \times 1)+(3 \times 16) = 75$. In fact the relative molecular mass is 150, and therefore the molecular formula must be $2 \times (C_2H_3O_3)$, which is $C_4H_6O_6$.

3.6 Law of Conservation of Matter

When a chemical reaction takes place the reacting substances (known as the *reactants*) are converted either wholly or partly into new substances (known as the *products*). During the process energy changes usually take place, but *the total mass of the reactants and products remains the same*. This is the **Law of Conservation of Matter**. It can be verified experimentally in a number of ways, most of which involve weighing reactants in a closed vessel before chemical reaction takes place and then weighing the products afterwards: no detectable change is observed.

3.7 Chemical Equations through Experiment

A chemical equation summarizes the results of quantitative investigations into a chemical reaction. It indicates the number of moles of reactants used and the number of moles of products formed. For convenience chemical formulas are used instead of writing out chemical names.

The following experiment illustrates the 'mole-ratio' method of determining the relative numbers of particles involved in chemical reactions. In this method, the concentration (number of moles) of one of the reactants (A) is kept constant while the concentration (number of moles) of the other reactant (B) is varied. During the reaction some change in physical property is measured for each different concentration of B. This change may be the height of a precipitate, the volume of a gas liberated, the weight of solid produced, or any other measurable change, provided that uniform conditions are used for each set of measurements. The number of moles of B reacting with the fixed concentration of A can then be obtained from a graph, like that shown in Fig. 3.6.

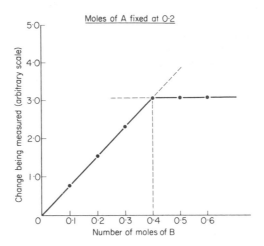

Fig. 3.6   Graph illustrating the 'mole-ratio' method

As B is added to a fixed amount of A, the quantity of A used up increases until, on the addition of 0·4 mole of B, all of A is reacted. This is indicated by the horizontal portion on the graph. Any further addition of B beyond 0·4 mole remains unreacted.

Thus 0·2 mole of A reacts exactly with 0·4 mole of B.
∴ 2 moles of A reacts exactly with 4 moles of B.
∴ 1 mole of A reacts exactly with 2 moles of B.

Hence the equation for this reaction is

$$1A + 2B \rightarrow \text{products}$$

Further analysis of the products formed when 0·2 mole of A reacts with 0·4 mole of B is necessary to complete the equation.

Experiment 3.3   Determination of the relative numbers of lead(II) ions reacting with iodide ions
In this experiment the concentration of lead(II) ions is kept constant while the concentration of iodide ions is varied. The change measured is the height of the yellow precipitate formed during the reaction. A solution of lead(II) nitrate $Pb^{2+}(NO_3^-)_2$ is prepared by dissolving 1 mole of it (331 g) in water and diluting

until the total volume is 1 dm³ (1000 cm³): this is *molar* solution of lead(II) nitrate. (A **molar solution** is simply one which contains one mole of the substance dissolved in one dm³ of solution.) A molar solution of potassium iodide K^+I^- is also required.

Two burettes are filled, one with the molar solution of lead(II) nitrate and one with the molar solution of potassium iodide. 2 cm³ of lead(II) nitrate solution are delivered into each of six test tubes of uniform internal diameter. Potassium iodide solution is then added to each of the tubes in turn: 1 cm³ to the first, 2 cm³ to the second, 3 cm³ to the third, and so on, 6 cm³ being added to the sixth tube. Each test tube is shaken and allowed to stand for an equal length of time, until the yellow solid (the *precipitate*) has settled to the bottom of the tube and a clear liquid appears above it. The height of each precipitate is measured (see Fig. 3.7).

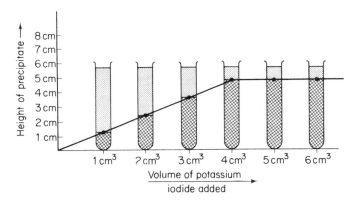

Fig. 3.7 The 'mole-ratio' method applied to the reaction between lead(II) ions and iodide ions. Pure water has been added to equalize the total volume in each tube

Results

∴ 2 cm³ of molar lead(II) nitrate reacts exactly with 4 cm³ of molar potassium iodide

∴ 2000 cm³ of molar lead(II) nitrate reacts exactly with 4000 cm³ of molar potassium iodide

∴ 2 moles of lead(II) nitrate reacts exactly with 4 moles of potassium iodide

∴ 1 mole of lead(II) nitrate reacts exactly with 2 moles of potassium iodide.

Thus the equation for the reaction is

$$Pb^{2+}(NO_3^-)_2 + 2K^+I^- \rightarrow \text{products}$$

Analysis of the products shows the yellow precipitate to be lead(II) iodide $(Pb^{2+}I_2^-)$ and the liquid above the precipitate to be potassium nitrate $(K^+NO_3^-)$. Thus the full equation for the reaction is

$$Pb^{2+}(NO_3^-)_{2(aq)}+2K^+I^-_{(aq)} \rightarrow Pb^{2+}I^-_{2(s)}+2K^+NO^-_{3(aq)}$$

(The symbol (aq) is used to indicate that a substance is in aqueous solution, while (s) indicates that it is a solid.)

3.8   Writing Chemical Equations

In theory it is possible to determine all chemical equations by experiment, but this would be very time-consuming. The following procedure provides a useful guide for balancing equations.

First, write down the formulas for all the products and the reactants in the equation. These formulas are constant and cannot be changed. For example, the formula for one molecule of carbon dioxide is CO_2, and in a chemical equation the formula of carbon dioxide must always be written as CO_2 or multiples of CO_2, e.g. $2CO_2$, $3CO_2$, etc.

Secondly, give a subscript to show whether the substance is a solid (s), a gas (g), a liquid (l), or dissolved in water (aq).

Thirdly, count the number of particles of each element on either side of the equation.

Finally, balance the equation by adjusting the number of particles of each substance on either side of the equation, without altering the formula for any substance. Use the smallest number for each substance necessary to balance the equation.

To illustrate this procedure, let us balance the equations for (a) magnesium burning in oxygen, and (b) methane burning in air.

(a) When ignited, magnesium ribbon burns in oxygen with a dazzling white light. The white powder produced in the reaction is magnesium oxide $(Mg^{2+}O^{2-})$.

Step (i)

$$Mg+O_2 \rightarrow Mg^{2+}O^{2-}$$

Step (ii)

$$Mg_{(s)}+O_{2(g)} \rightarrow Mg^{2+}O^{2-}_{(s)}$$

Step (iii)
Number of particles (atoms) on the left: one magnesium and two oxygen.
Number of particles (ions) on the right: one magnesium and one oxygen.

Step (iv)
The equation is not balanced. In order to balance the oxygen the formula for magnesium oxide must be multiplied by 2, giving $2Mg^{2+}O^{2-}$. Now the reactant magnesium must be multiplied by 2 in order to balance this. Therefore the balanced equation for this reaction is

$$2Mg_{(s)}+O_{2(g)} \rightarrow 2Mg^{2+}O^{2-}_{(s)}$$

(b) When the gas methane burns it reacts with the oxygen of the air, producing heat, carbon dioxide and water vapour.

Step (i)

$$CH_4 + O_2 \rightarrow CO_2 + H_2O$$

Step (ii)

$$CH_{4(g)} + O_{2(g)} \rightarrow CO_{2(g)} + H_2O_{(g)}$$

Step (iii)
Number of particles (atoms) on the left: one carbon, four hydrogen and two oxygen.
Number of particles (atoms) on the right: one carbon, two hydrogen and three oxygen (2 from CO_2 and 1 from H_2O).

Step (iv)
The equation is not balanced. The one carbon atom and the four hydrogen atoms in methane would produce, respectively, one carbon dioxide molecule and two water molecules:

$$CH_{4(g)} + O_{2(g)} \rightarrow CO_{2(g)} + 2H_2O_{(g)}.$$

Now balance the oxygen. There are four oxygen atoms in the products, therefore the reactants must also have four oxygen atoms, i.e. $2O_2$.
Therefore the balanced equation is

$$CH_{4(g)} + 2O_{2(g)} \rightarrow CO_{2(g)} + 2H_2O_{(g)}$$

3.9 Information given by an Equation

Consider the following balanced equation:

$$H_{2(g)} + Cl_{2(g)} \rightarrow 2HCl_{(g)}$$

This equation tells us that one mole of gaseous hydrogen reacts with one mole of gaseous chlorine to produce two moles of gaseous hydrogen chloride. In general, any balanced chemical equation contains the following information:
  (a) the reactants and the products;
  (b) the relative number of moles of each product and each reactant;
  (c) the physical state of the reactants and products.

However, a balanced equation does *not* tell us:
  (i) the conditions necessary for the reaction to take place;
  (ii) the concentration of the reactants;
  (iii) the reaction rate, i.e. how quickly the reaction proceeds;
  (iv) the extent or completeness of the reaction, i.e. whether or not all the reactants will be converted into products;
  (v) the mechanism by which the reaction takes place;
  (vi) the energy changes which occur during the reaction.

Consider the following balanced equation for the reaction between magnesium metal and hydrochloric acid:

$$Mg_{(s)} + 2H^+Cl^-_{(aq)} \rightarrow Mg^{2+}Cl^-_{2(aq)} + H_{2(g)}$$

This equation does not tell us that the reaction is spontaneous when the magnesium is added to the acid; that the reaction is rapid if the acid is concentrated, or give any indication of the rate or extent of the reaction. There is no suggestion in the equation as to the mechanism by which the reaction proceeds, nor any evidence that heat is produced as the magnesium dissolves.

3.10 Calculations involving Reacting Masses

The chemist in industry is constantly using chemical equations to calculate reacting quantities. A chemist needs to know, for example, the amount of raw materials that must be obtained for the manufacture of a specified amount of product. The method of calculation is the same whether the reaction is complex or simple, and is illustrated in the following discussion.

The soap industry requires large quantities of sodium hydroxide (caustic soda). This used to be manufactured by the Gossage process, in which calcium hydroxide suspended in water (milk of lime) was shaken with a solution of sodium carbonate (washing soda). The precipitate of calcium carbonate was filtered, leaving sodium hydroxide in solution. The equation for this reaction is

$$Ca^{2+}(OH^-)_{2(s)} + Na^+_2CO^{2-}_{3(aq)} \rightarrow Ca^{2+}CO^{2-}_{3(s)} + 2Na^+OH^-_{(aq)}$$

which tells us that one mole of calcium hydroxide reacts with one mole of sodium carbonate to produce one mole of calcium carbonate and two moles of sodium hydroxide. The relative atomic masses are Ca = 40, O = 16, H = 1, Na = 23, and C = 12. Hence

one mole of $Ca^{2+}(OH^-)_2$ has a mass of $40 + 2(16 + 1)$ =   74 grams
one mole of $Na^+_2CO^{2-}_3$ has a mass of $(2 \times 23) + 12 + (3 \times 16)$
                                                              = 106 grams
one mole of $Ca^{2+}CO^{2-}_3$ has a mass of $40 + 12 + (3 \times 16)$ = 100 grams
one mole of Na^+OH^- has a mass of $23 + 16 + 1$ =   40 grams

$$\underbrace{Ca^{2+}(OH^-)_{2(s)}}_{} \quad + \quad \underbrace{Na^+_2CO^{2-}_{3(aq)}}_{} \rightarrow \underbrace{Ca^{2+}CO^{2-}_{3(s)}}_{} + \underbrace{2Na^+OH^-_{(aq)}}_{}$$

74 g combines with 106 g   to give 100 g and 2×40 g

The chemist therefore knows that 74 g (or tonnes etc.) of calcium hydroxide will react with 106 g (tonnes etc.) of sodium carbonate, to produce 2×40 g (tonnes etc.) of sodium hydroxide. There will be 100 g (tonnes etc.) of calcium carbonate precipitated as a by-product of the reaction.

Example A student heats a mixture of iron filings and sulphur to produce iron(II) sulphide. What mass of iron will react with 10 g of sulphur, and what mass of iron(II) sulphide will be formed? Relative atomic masses of iron and sulphur are 56 and 32, respectively.

Solution The balanced equation for the reaction is

$$Fe_{(s)} + S_{(s)} \rightarrow Fe^{2+}S^{2-}_{(s)}$$

$$56 \quad 32 \quad \underbrace{56 + 32}_{88}$$

∴ 56 g iron reacts with 32 g sulphur to produce   88 g iron(II) sulphide

∴ $\frac{56}{32}$ g iron reacts with 1 g sulphur to produce $\frac{88}{32}$ g iron(II) sulphide

∴ $\frac{56}{32} \times 10$ g iron reacts with 10 g sulphur to produce $\frac{88}{32} \times 10$ g iron(II) sulphide

Thus the reacting mass of iron is $\frac{56}{32} \times 10 = 17\cdot5$ g, and the mass of iron(II) sulphide formed is $\frac{88}{32} \times 10 = 27\cdot5$ g.

Summary of Unit 3

1. **Relative atomic mass** is the mass of one atom of the element compared with one-twelfth the mass of one atom of carbon (^{12}C).
2. A **mole** is that amount of substance which contains Avogadro's number ($6\cdot02 \times 10^{23}$) of particles, whether they be atoms, ions or molecules.
3. A **chemical formula** uses symbols to show the numbers of the atoms of the elements contained in one molecule or smallest portion of a compound.
4. The **empirical formula** for a compound gives its elements written in their smallest whole-number ratio.
5. The **molecular formula** shows the actual number of atoms in one molecule of the compound.
6. The **law of constant composition** states that all pure samples of the same chemical compound contain the same elements in the same proportion by mass.
7. An ionic compound results from a combination of ions such that the resulting giant structure has no overall charge. The names of ionic compounds are based on the **oxidation numbers** of the elements in the compound.
8. The formula for a covalent compound is determined by the electron structure of the elements in the compound.
9. The **law of conservation of matter** states that during a chemical reaction the total mass of the reactants and products remains the same.
10. A **chemical equation** is a representation of a chemical reaction using symbols.
11. A **balanced** equation containing state symbols indicates the relative number of moles of each reactant and each product and the physical states of these reactants and products.
12. A balanced equation is used to calculate *reacting quantities*.

Test Yourself on Unit 3

1. 14·34 g of an oxide of lead was heated in a stream of dry hydrogen. 12·42 g of lead was obtained. Relative atomic masses of oxygen and lead are 16 and 207, respectively.
   (a) How many grams of oxygen were present in the oxide?
   (b) Calculate the number of moles of oxygen atoms.
   (c) Calculate the number of moles of lead.
   (d) How many moles of oxygen atoms were combined with 1 mole of lead?
   (e) Write down the formula of the oxide of lead used in this experiment.

2. Give the formula for the following compounds:
   (a) sodium carbonate
   (b) iron(II) sulphate
   (c) zinc nitrate
   (d) aluminium hydroxide
   (e) silver chloride
   (f) potassium sulphite

3. Name the following compounds:
   (a) $Na^+ HSO_4^-$
   (b) $Fe^{3+} (NO_3^-)_3$
   (c) $Mg^{2+} O^{2-}$
   (d) $(NH_4^+)_2 S^{2-}$
   (e) $Hg^{2+} SO_4^{2-}$

4. A hydrocarbon contains 14·3% by mass of hydrogen and has a relative molecular mass of 28. If the relative atomic masses of hydrogen and carbon are 1·0 and 12·0, respectively, calculate the molecular formula of the compound.

5. Molar solutions of silver nitrate $[Ag^+ NO_3^-]$ and of potassium chromate(VI) $[K_2^+ CrO_4^{2-}]$ were prepared. The following mixtures were shaken in test tubes:

	Volume of silver nitrate solution	Volume of potassium chromate(VI) solution	Volume of water
(a)	2 cm^3	3 cm^3	7 cm^3
(b)	3 cm^3	3 cm^3	6 cm^3
(c)	4 cm^3	3 cm^3	5 cm^3
(d)	5 cm^3	3 cm^3	4 cm^3
(e)	6 cm^3	3 cm^3	3 cm^3
(f)	7 cm^3	3 cm^3	2 cm^3
(g)	8 cm^3	3 cm^3	1 cm^3

After centrifuging the precipitates of silver chromate(VI) for the same length of time the height of the precipitate in each tube was measured and the results plotted as follows:

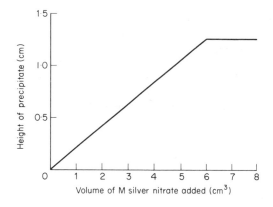

(i) How many cm^3 of M silver nitrate react exactly with 3 cm^3 of M potassium chromate(VI)?
(ii) How many moles of silver nitrate react exactly with 1 mole of potassium chromate(VI)?
(iii) Write a balanced equation for the reaction.
(iv) Why did the height of the precipitate stay constant in solutions (e), (f) and (g)?

6. Iron(III) sulphate solution reacts with sodium hydroxide solution to give a red-brown precipitate of iron(III) hydroxide. The following equation is not balanced:

$$Fe_2^{3+}(SO_4^{2-})_3 + Na^+OH^- \rightarrow Fe^{3+}(OH^-)_3 + Na_2^+ SO_4^{2-}$$

(a) Which set of state symbols is most suitable?
  (i) (aq), (aq), (aq), (aq)
  (ii) (aq), (aq), (s), (s)
  (iii) (s), (s), (s), (s)
  (iv) (aq), (aq), (s), (aq)
  (v) (aq), (s), (aq), (aq)

(b) Which set of numbers is most suitable to balance the equation?
  (i) 2, 6, 2, 3
  (ii) 1, 3, 1, 3
  (iii) 1, 6, 2, 3
  (iv) 1, 6, 1, 3
  (v) 1, 3, 2, 3

7. 1·3 g of zinc was added to an excess of copper(II) sulphate solution:

$$Zn_{(s)} + Cu^{2+}SO_{4(aq)}^{2-} \rightarrow Cu_{(s)} + Zn^{2+}SO_{4(aq)}^{2-}$$

If the relative atomic masses of copper and zinc are 63·5 and 65·0, respectively, what mass of copper would be produced?

(a) 1·3 g, (b) 6·5 g, (c) 6·35 g, (d) 1·27 g, (e) 1·5 g

Mark this test out of 30 with the answers provided on page 371.

Unit Four

Patterns in Gaseous Matter

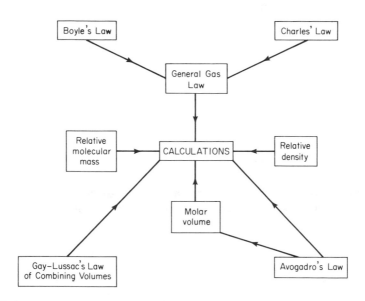

We have now seen how a chemical reaction may be represented by an equation, and how equations may be determined from the masses of the reactants and products. In this Unit we shall extend the discussion to reactions which involve gases.

With gases it is much easier to measure the reacting volume rather than the reacting mass. Unfortunately, the volume of a gas alone is insufficient to express the amount of matter present: it is necessary to know also the pressure and temperature of the gas.

4.1 Compressing Gases: Boyle's Law

Unlike solids and liquids, gases are easily compressible. This is apparent when a tyre is inflated: a large amount of air is compressed into a small space; if the valve is opened, or the tyre punctured, the air will expand into the atmosphere.

The effect of pressure on a fixed amount of gas at constant temperature is illustrated in Fig. 4.1.

The values of pressure and temperature in Fig. 4.1 are such that the product of pressure and volume gives the same result in each case:

$$1 \times 60 = 2 \times 30 = 3 \times 20$$

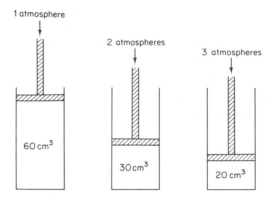

*Fig. 4.1   Variation of pressure on a fixed mass of gas at constant temperature
(Boyle's law)*

Repeated experiments suggest that, for any fixed mass of gas at a constant temperature, the product of pressure and volume is constant. So we can write

$$p_1 \times V_1 = p_2 \times V_2 = p_3 \times V_3 \quad \text{etc.}$$

where V_1 is the volume of the fixed mass of gas at a pressure p_1, and V_2 is its volume at a different pressure p_2.

The volume of a fixed amount of gas decreases as the pressure is increased, and increases as the pressure is decreased. Robert Boyle (1627–91) made quantitative measurements on the variation of gas volume with changes of pressure, and formulated a law now named after him. This is **Boyle's law** and can be stated as follows: *the volume of a given mass of gas is inversely proportional to the pressure upon it, provided the temperature remains constant.*

Using symbols, Boyle's law can be expressed as

$$V \propto \frac{1}{p}$$

where V is the volume of a fixed mass of gas at constant temperature, p is its pressure and $\propto$ means 'is proportional to'. In the more useful form of an equation we can write this as

$$pV = a\ constant$$

where the constant varies according to the units of V and p.

4.2   Heating and Cooling Gases: Charles' Law

There are three different temperature scales in common use. These are based on two constant temperatures: the melting point of pure ice under normal conditions, and the temperature of steam from water boiling at atmospheric pressure 101 325 N m^{-2}(Pa) = 1 atmosphere = 760 mm of mercury).

On the *Celsius* (or centigrade) scale the ice point is given the value 0 °C and the steam point the value 100 °C, while on the *Fahrenheit* scale these values are 32 °F and 212 °F, respectively.

The *Kelvin* scale is used when describing the variations of gas volume with change in temperature. When a gas is heated it expands, and when cooled it contracts. J. A. C. Charles (1746–1823), a French physicist, discovered that the volume of a fixed mass of gas increases (or decreases) by $\frac{1}{273}$ of its volume at 0 °C for every Celsius degree rise (or fall), the pressure remaining constant. As a gas is cooled, therefore, its volume will become less and less until at −273 °C

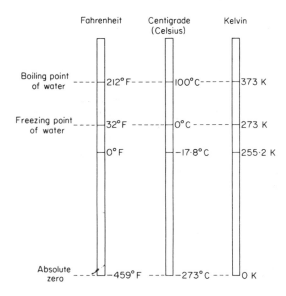

Fig. 4.2   Relationship between the Fahrenheit, Celsius and Kelvin temperature scales

it would theoretically be zero. (In practice all gases liquefy before this temperature is reached.) By this reasoning −273 °C must be the lowest temperature possible; it was therefore given the name *absolute zero*. On the Kelvin scale the value of absolute zero is 0 K, the ice point is 273 K and the steam point is 373 K. Thus to convert Celsius to Kelvin simply add 273 to the Celsius temperature.

The relationship between the Celsius, Fahrenheit and Kelvin temperature scales is illustrated in Fig. 4.2.

Fig. 4.3 illustrates the way in which the volume of a fixed mass of gas varies as the temperature is increased, pressure being constant.

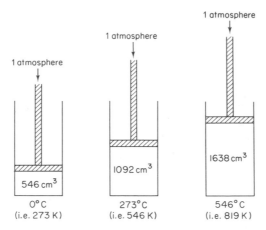

Fig. 4.3   Variation of temperature on a fixed mass of a gas at constant pressure (Charles' law)

These variations suggest that the volume divided by the temperature (measured in kelvins) always gives the same value:

$$\frac{546}{273} = \frac{1092}{546} = \frac{1638}{819}$$

Charles' work on gases can be summarized as follows: *the volume of a fixed mass of gas varies directly as the Kelvin temperature, pressure remaining constant.* This is known as **Charles' law**.

Using symbols, Charles' law can be expressed as

$$V \propto T$$

where V is the volume of a fixed mass of gas at constant pressure and T is its Kelvin temperature. In the more useful form of an equation we can write this as

$$V = \text{a constant} \times T$$

or
$$\frac{V}{T} = \text{a constant}$$

For a fixed mass of gas the ratio of volume to temperature, measured in kelvins, must be constant at constant pressure. Thus

$$\frac{V_1}{T_1} = \frac{V_2}{T_2} = \frac{V_3}{T_3} \quad \text{etc.}$$

where V_1 is the volume of the fixed mass of gas at a temperature T_1 K, and V_2 is its volume at temperature T_2 K.

4.3 The Combined Gas Law

Boyle's law and Charles' law can be combined in the expression

$$\frac{p_1 V_1}{T_1} = \frac{p_2 V_2}{T_2} = \frac{p_3 V_3}{T_3} \quad \text{etc.}$$

Thus, for a given mass of gas, the volume multiplied by the pressure and divided by the temperature measured in kelvins is a constant. This result is known as the 'combined gas law' or 'general gas law'.

4.4 Standard Conditions

The combined gas law is generally used by chemists to calculate gas volumes. It is convenient to be able to compare all gas volumes at one particular temperature and one particular pressure.

A *standard temperature* of 273 K (or 0°C) and a *standard pressure* of 101 325 N m^{-2}(Pa) (or 1 atmosphere or 760 mm of mercury) have been selected. These conditions are known as **Standard Temperature and Pressure**, usually abbreviated to STP.

The determination of gas volumes at STP is important in chemical reactions involving gases because it enables us to obtain information concerning the molar quantities reacting. Let us look at the following conversions.

Example A student finds that the volume of carbon dioxide liberated during a chemical reaction is 65 cm^3, measured dry at 20°C and 100 kPa pressure. What will the volume of carbon dioxide be at STP?

Calculation The temperatures must always be converted into kelvins:

$$20\,°C = (273+20)K$$

$$= 293 \text{ K}$$

Now we can apply the combined gas law,

$$\frac{p_1 \times V_1}{T_1} = \frac{p_2 \times V_2}{T_2}$$

where p_4, V_1 and T_1 apply to the conditions of the experiment (i.e. $p_1 = 100$ kPa, $V_1 = 65$ cm^3 and $T_1 = 293$ K) and p_2, V_2 and T_2 apply to the *standard* conditions (i.e. $p_2 = 101.3$ kPa and $T_2 = 273$ K).

In order to calculate V_2, the unknown volume at STP, we isolate V_2 in the above equation:

$$V_2 = V_1 \times \frac{p_1}{p_2} \times \frac{T_2}{T_1}$$

Substituting the given data in this new equation, we have

$$V_2 = 65 \text{ cm}^3 \times \frac{100}{101 \cdot 3} \times \frac{273}{293}$$

$$= 59 \cdot 78 \text{ cm}^3$$

Rounding off to the nearest whole number (since the initial volume was given as a whole number), the volume at STP will be 60 cm^3 (*Answer*).

4.5 Relative Molecular Masses

It has been shown that the relative molecular mass of a gas is calculated by summing the relative atomic masses of the individual atoms in each molecule of the gas. For example, the relative molecular mass of the gas sulphur dioxide (SO_2) is

$$32 + (2 \times 16) = 64$$

The relative molecular mass of a gas expressed in grams is the mass of *one mole* of the gas. Let us calculate the volume of one mole of gas under *standard conditions*. This can be achieved if the *density* of the gas is known. Density is mass per unit volume under standard conditions of temperature and pressure. If the unit of mass is the gram, then the volume is measured in cm^3; if the mass unit is the kilogram, then the volume is measured in m^3.

Example What is the volume of one mole of hydrogen?

Calculation The density of hydrogen gas (H_2) is 0·089 88 kilograms per cubic metre (kg m^{-3}), or 0·000 089 88 g cm^{-3}, and the mass of one mole is 2·016 g.

$$\text{density} = \frac{\text{mass}}{\text{volume}}$$

Rearranging this density formula, we have

$$\text{volume} = \frac{\text{mass}}{\text{density}}$$

Therefore the volume of 2·016 g of hydrogen at STP is

$$\frac{2 \cdot 016}{0 \cdot 000\,089\,88} \text{ cm}^3$$

$$= 22\,430 \text{ cm}^3$$

Thus one mole of hydrogen at STP occupies a volume of 22 430 cm^3.

The volumes occupied by one mole of other gases at STP can be similarly calculated and are shown in Table 4.1.

Table 4.1 Molar volumes of common gases

Gas	Formula	Density	Density	Volume of one mole at STP
		$g\ cm^{-3}$	$kg\ m^{-3}$	cm^3
Hydrogen	H_2	0·000 089 88	0·089 88	22 430
Helium	He	0·000 178 46	0·178 46	22 430
Nitrogen	N_2	0·001 250 51	1·250 51	22 400
Oxygen	O_2	0·001 428 98	1·428 98	22 390
Fluorine	F_2	0·001 696	1·696	22 400
Neon	Ne	0·000 899 91	0·899 91	22 430
Carbon monoxide	CO	0·001 250 03	1·250 03	22 410
Carbon dioxide	CO_2	0·001 976 96	1·976 96	22 260

Molar Volume

It will be seen from Table 4.1 that one mole of any gas occupies approximately the same volume at STP as one mole of any other gas. This volume 22 400 cm^3 is called the *molar volume* (V_m). It is normally expressed at STP, but the molar volume would have a constant value (not necessarily 22 400 cm^3) for all gases at any fixed temperature and pressure.

Calculation of Relative Molecular Mass

The concept of molar volume leads to a simple and effective method for finding relative molecular masses.

Example 0·56 g of a gas has a volume of 448 cm^3 at STP. Calculate the relative molecular mass of the gas.

Calculation

448 cm^3 of gas at STP has a mass of 0·56 g

$\therefore$ 1 cm^3 of gas at STP has a mass of $\frac{0·56}{448}$ g

$\therefore$ 22 400 cm^3 of gas at STP has a mass of $\frac{0·56}{448} \times 22\ 400$ g $= 28$ g

Thus the relative molecular mass is 28.

4.6 Avogadro's Law

We now know two important facts about *one mole of any gas*:
  (a) its volume at STP is 22 400 cm^3 (V_m),
  (b) it contains $6·02 \times 10^{23}$ molecules.
These two facts are summarized in **Avogadro's law**, which states that *equal volumes of all gases under the same conditions of temperature and pressure contain the same number of molecules.*

This law was first put forward in 1811 as a *hypothesis* by Amadeo Avogadro, an Italian physicist (see Section 3.1). A hypothesis differs from a law in that it refers to concepts which are incapable of direct experience. Since Avogadro could not count the number of molecules in a given volume of gas, his hypothesis only became a law when this was possible.

Avogadro presented his hypothesis partly as a result of his consideration of **Gay-Lussac's law of combining volumes** (1808), which states that *the volumes of gases taking part in a chemical change, either as reactants or products, bear a simple numerical relation to one another, provided that all measurements are made under the same conditions of temperature and pressure.* Gay-Lussac's law can be illustrated by the following experimental results:

(a) *One volume* of hydrogen reacts with *one volume* of chlorine to give *two volumes* of hydrogen chloride.

(b) *Two volumes* of hydrogen react with *one volume* of oxygen to give *two volumes* of steam.

(c) *Three volumes* of hydrogen react with *one volume* of nitrogen to give *two volumes* of ammonia.

(d) *One volume* of oxygen reacts with solid sulphur to give *one volume* of sulphur dioxide.

The law applies to *all* reactions in which gases take part and it will be observed that the volumes of the gaseous substances are in the ratios of small whole numbers.

Consider reaction (a) above: one volume of hydrogen reacts with one volume of chlorine to give two volumes of hydrogen chloride. Let one volume of gas contain n molecules, then, by Avogadro's law:

n molecules of hydrogen react with n molecules of chlorine to give $2n$ molecules of hydrogen chloride.

Therefore 1 molecule of hydrogen reacts with 1 molecule of chlorine to give 2 molecules of hydrogen chloride.

If we assume that molecules of hydrogen and chlorine each contain two atoms, then:

2 atoms of hydrogen react with 2 atoms of chlorine to give 2 molecules of hydrogen chloride.

Therefore 1 atom of hydrogen combines with 1 atom of chlorine to give 1 molecule of hydrogen chloride. From this we can conclude that one molecule of hydrogen chloride contains one atom of hydrogen and one atom of chlorine, i.e. its formula is HCl.

Consider reaction (c) above: three volumes of hydrogen react with one volume of nitrogen to give two volumes of ammonia. Let one volume of gas contain n molecules, then, by Avogadro's law:

$3n$ molecules of hydrogen react with n molecules of nitrogen to give $2n$ molecules of ammonia.

Therefore 3 molecules of hydrogen react with 1 molecule of nitrogen to give 2 molecules of ammonia

If we assume that molecules of hydrogen and nitrogen each contain two atoms, then:

6 atoms of hydrogen react with 2 atoms of nitrogen to give 2 molecules of ammonia.

∴ 3 atoms of hydrogen react with 1 atom of nitrogen to give 1 molecule of ammonia.

From this we can conclude that one molecule of ammonia contains three atoms of hydrogen and one atom of nitrogen, i.e. its formula is NH_3.

These two examples illustrate the use of the laws of Avogadro and Gay-Lussac in deducing the formulas for gaseous molecules.

4.7 Relative Density

The relative density (or vapour density) of a gas is defined as *the mass of a given volume of a gas compared with the mass of an equal volume of hydrogen measured under the same conditions of temperature and pressure.*

From a consideration of Gay-Lussac's law of volumes and Avogadro's law it can be shown that *the relative molecular mass of a gas (or vapour) is equal to twice its relative density.*

$$\text{Relative density} = \frac{\text{mass of } V \text{ cm}^3 \text{ of gas}}{\text{mass of } V \text{ cm}^3 \text{ of hydrogen}}$$

(provided that both masses are measured under the same conditions of temperature and pressure).

From Avogadro's law we know that V cm^3 of any gas contains the same number of molecules (say n molecules) as V cm^3 of hydrogen. Therefore

$$\text{Relative density} = \frac{\text{mass of } n \text{ molecules of gas}}{\text{mass of } n \text{ molecules of hydrogen}}$$

$$= \frac{\text{mass of 1 molecule of gas}}{\text{mass of 1 molecule of hydrogen}}$$

Since the relative molecular mass of the hydrogen molecule is 2 (approximately), we have

$$\text{Relative density} = \frac{\text{relative molecular mass of gas}}{2}$$

or   *Relative molecular mass of gas* $= 2 \times$ *relative density*

Thus the relative molecular mass is numerically equal to twice the relative density. Using this result we can determine the relative molecular mass of a gas, volatile liquid or a volatile solid simply by measuring its relative density.

Example The relative molecular mass of a gas was determined using the following information:

Mass of a glass sphere evacuated of air $= 102 \cdot 325$ g
Mass of a glass sphere filled with an unknown gas $= 104 \cdot 281$ g

Mass of a glass sphere filled with hydrogen at the same
temperature and pressure $= 102\cdot415$ g
What is the relative molecular mass of the gas?

Calculation

Mass of unknown gas	$= 1\cdot956$ g
Mass of hydrogen	$= 0\cdot090$ g
$\therefore$ relative density of unknown gas	$= \frac{1\cdot956}{0\cdot090} = 21\cdot7$
$\therefore$ relative molecular mass	$= 2 \times 21\cdot7$
	$= 43\cdot4$ (*Answer*)

4.8 Gas Calculations

The following worked examples are intended to show how the chemist makes use of the gas laws discussed in this Unit.

Example 4.1 Calculate the volume of carbon dioxide formed (*a*) at STP, (*b*) at 20°C and 99·3 kPa pressure, when 5 g of calcium carbonate is strongly heated until no further gas is evolved. The relative atomic masses of calcium (Ca), carbon (C) and oxygen (O) are 40, 12 and 16, respectively.

Calculation

(*a*) The equation for the reaction is:

$$\underbrace{Ca^{2+}CO_{3(s)}^{2-}}_{\underbrace{40+12+(3\times16)}_{100}} \xrightarrow{\text{heat}} \underbrace{Ca^{2+}O_{(s)}^{2-}+CO_{2(g)}}_{\underbrace{12+(2\times16)}_{44}}$$

$\therefore$ 100 g of calcium carbonate gives 44 g of carbon dioxide.
But 44 g is the mass of one mole of carbon dioxide and occupies 22 400 cm³ at STP.

$\therefore$ 100 g of calcium carbonate gives 22 400 cm³ of carbon dioxide at STP

$\therefore$ 1 g of calcium carbonate gives $\dfrac{22\ 400}{100}$ cm³ of carbon dioxide at STP

$\therefore$ 5 g of calcium carbonate gives $\dfrac{22\ 400}{100} \times 5$ cm³ of carbon dioxide at STP

$\therefore$ 5 g of calcium carbonate give 1120 cm³ of carbon dioxide at STP

Thus the volume of carbon dioxide at STP liberated from 5 g calcium carbonate $= 1120$ cm³ (*Answer*).

(*b*) In order to convert the volume at STP to the volume at 20°C and 99·3 kPa pressure, the general gas law

$$\frac{p_1 V_1}{T_1} = \frac{p_2 V_2}{T_2} \text{ is used.}$$

At STP: $p_1 = 101\cdot3$ kPa, $V_1 = 1120$ cm³, $T_1 = 273$ K

At 20°C: p_2 = 99·3 kPa, V_2 = unknown, T_2 = 20 + 273 = 293 K.

$$\therefore V_2 = V_1 \times \frac{p_1}{p_2} \times \frac{T_2}{T_1}$$

$$V_2 = 1120 \text{ cm}^3 \times \frac{101 \cdot 3 \text{ kPa}}{99.3 \text{ kPa}} \times \frac{293 \text{ K}}{273 \text{ K}}$$

$$= 1226 \text{ cm}^3.$$

Thus the volume of carbon dioxide at 20°C and 99·3 kPa pressure is 1226 cm^3 (*Answer*).

Example 4.2 Calculate the volume of gas remaining at 120 °C when 100 cm^3 of hydrogen is mixed with 100 cm^3 of oxygen and ignited.

Calculation Since all volumes are measured at the same temperature and pressure, there is no necessity to use the gas laws. Also water is present as a gas at 120 °C. The equation for the reaction is:

$$2H_{2(g)} \quad + \quad O_{2(g)} \quad \rightarrow \quad 2H_2O_{(g)}$$

2 molecules   1 molecule   2 molecules
$2n$ molecules $1n$ molecules $2n$ molecules

This can be converted to a volume relationship using Avogadro's law:
2 volumes of hydrogen react with 1 volume of oxygen to give 2 volumes of steam.
Relating this to the actual gas volumes: either
(i) 100 cm^3 of oxygen would react with 200 cm^3 of hydrogen to give 200 cm^3 steam,
or
(ii) 100 cm^3 of hydrogen would react with 50 cm^3 of oxygen to give 100 cm^3 steam.
It will be noted in (i) that 200 cm^3 of hydrogen is required but only 100 cm^3 is available. Thus alternative (ii) is relevant, and (100 − 50) cm^3 = 50 cm^3 of oxygen is unused.
At the end of the experiment the gas mixture would therefore consist of the steam formed (100 cm^3) and the excess oxygen (50 cm^3). Thus the total volume of gas at 120 °C is 150 cm^3 (*Answer*).

Example 4.3 The first stage in the manufacture of sulphuric acid involves burning sulphur with oxygen in the air to produce sulphur dioxide. A plant designer needs to know the volume of sulphur dioxide produced at 1 atmosphere pressure and 500 °C when large quantities of sulphur are burned. Calculate this volume for 50 kg of sulphur. The relative atomic masses of sulphur and oxygen are 32 and 16, respectively.

Calculation
The balanced equation for the reaction is:

$$S_{(s)} + O_{2(g)} \rightarrow SO_{2(g)}$$

$$\underset{32}{} \qquad \underbrace{32 + (2 \times 16)}_{64}$$

$\therefore$ 32 g sulphur produces 64 g sulphur dioxide.

Step 1: calculate the volume of sulphur dioxide at STP.
64 g (1 mole) of sulphur dioxide occupies 22 400 cm^3 at STP
$\therefore$ 32 g sulphur produces 22 400 cm^3 sulphur dioxide at STP
$\therefore$ 1 g sulphur produces $\dfrac{22\,400}{32}$ cm^3 sulphur dioxide at STP

$\therefore$ 50 000 g (50 kg) sulphur produces $\dfrac{22\,400}{32} \times 50\,000$ cm^3 sulphur dioxide at STP

$$= 35 \times 10^6 \text{ cm}^3$$
$$= 35 \text{ m}^3$$

Step 2: convert this volume at STP to a volume at 1 atmosphere and 500 °C using the general gas law

$$\frac{p_1 V_1}{T_1} = \frac{p_2 V_2}{T_2}$$

At 500 °C: $p_1 = 1$ atmosphere, $V_1 = ?$, $\qquad T_1 = (500 + 273) = 773$ K
At STP: $\quad p_2 = 1$ atmosphere, $V_2 = 35$ m^3, $T_2 = 273$ K

$$\therefore V_1 = \frac{p_2 V_2}{T_2} \times \frac{T_1}{p_1}$$

$$= \frac{1 \times 35 \text{ m}^3}{273} \times \frac{773}{1}$$

$$= 99 \cdot 1 \text{ m}^3 \text{ (\textit{Answer})}$$

Note: the volume of oxygen used would also be 99·1 m^3 at 1 atm pressure and 500 °C.

Summary of Unit 4

1. The **volume** of a gas varies with **temperature** and **pressure**.
2. **Boyle's law** states that the volume of a fixed mass of gas is inversely proportional to the pressure upon it, provided the temperature retains constant:
$$V \propto \frac{1}{p} \; .$$

3. The volume of a fixed mass of gas varies directly with temperature measured in degrees kelvin, pressure remaining constant: $V \propto T$.

4. *Absolute zero* is the lowest attainable temperature and is zero on the Kelvin scale (0 K).

5. *STP* (standard temperature and pressure) is 273 K and 101 325 Pa (1 atmosphere).

6. One mole of any gas occupies approximately the same volume (22 400 cm^3) at STP.

7. **Avogadro's law**: Equal volumes of all gases under the same conditions of temperature and pressure contain the same number of molecules.

8. **Gay-Lussac's law**: The volumes of gaseş taking part in a chemical change (either as reactants or products) bear a simple numerical relationship to one another, provided that all measurements are made under the same conditions of temperature and pressure.

9. **Relative density** of a gas is the mass of a given volume of a gas compared with the mass of an equal volume of hydrogen under the same conditions of temperature and pressure.

10. **Relative molecular mass** of a gas is equal to twice its relative density.

Test Yourself on Unit 4

1. If the pressure on 100 cm^3 of air is halved, will its volume (at the same temperature) be:
   (*a*) 50 cm^3, (*b*) 1 000 cm^3, (*c*) 200 cm^3, (*d*) 25 cm^3, (*e*) 100 cm^3?

2. A gas has a volume of 27·3 cm^3 at 0 °C. What would its volume be at 10 °C if the pressure remained unchanged?
   (*a*) 273 cm^3, (*b*) 2·73 cm^3, (*c*) 283 cm^3, (*d*) 28·3 cm^3, (*e*) 37·3 cm^3

3. 8·40 g of sodium hydrogencarbonate was heated until its mass remained constant. The white residue, sodium carbonate, had a mass of 5.30 g. In a separate experiment 0·005 mole of sodium hydrogencarbonate was treated with excess dilute hydrochloric acid. The volume of dry carbon dioxide corrected to STP was 112 cm^3. The relative atomic mass of hydrogen is 1, of carbon is 12, of oxygen is 16 and of sodium is 23. Use the above information to answer the following:
   (*a*) What fraction of a mole of sodium hydrogencarbonate has a mass of 8·40 g?
   (*b*) What fraction of a mole of sodium carbonate has a mass of 5·30 g?
   (*c*) How many moles of sodium hydrogencarbonate are required to produce 1 mole of sodium carbonate on heating?
   (*d*) Write a balanced equation for this reaction.
   (*e*) How many moles of gas were collected in the second part of the experiment?
   (*f*) Does the answer to (*e*) agree with the following equation:

$$Na^+HCO_{3(aq)}^- + H^+Cl_{(aq)}^- \rightarrow Na^+Cl_{(aq)}^- + CO_{2(g)} + H_2O_{(l)}$$

4. Calcium reacts with water as shown in the following equation:

$$Ca_{(s)}+2H_2O_{(l)} \rightarrow Ca^{2+}(OH^-)_{2(aq)}+H_{2(g)}$$

What volume of hydrogen measured at STP would be liberated when 8 g of calcium reacts with excess water? The relative atomic mass of calcium is 40.

(a) 4480 cm^3 (b) 2240 cm^3 (c) 1120 cm^3 (d) 0·4 cm^3 (e) 0·2 cm^3

5. A gas cooker burns pure gaseous methane to form carbon dioxide gas and water vapour according to the equation

$$CH_{4(g)}+2O_{2(g)} \rightarrow CO_{2(g)}+2H_2O_{(g)}$$

When 32 g of methane burns:
(a) How many moles of water are produced?
(b) How many moles of carbon dioxide are produced?
(c) What is the volume of this carbon dioxide at STP?
(d) What volume of oxygen at STP is required to burn the methane?
(Relative atomic masses: C = 12, H = 1)

6. The hydrocarbon butane C_4H_{10} burns in oxygen to give carbon dioxide and water vapour only.

(a) Write a balanced equation for this reaction.

(b) Calculate the minimum volume of oxygen needed for the complete combustion of 100 cm^3 of butane.

(c) Calculate the volume of gaseous products formed by the complete combustion of 100 cm^3 of butane.

Assume all gas volumes are measured at the same temperature and pressure.

7. In an experiment to determine the relative molecular mass of an unknown hydrocarbon, the following results were obtained:

Mass of evacuated flask   = 84·640 g
Mass of flask filled with nitrogen   = 84·920 g
Mass of flask filled with hydrocarbon = 85·420 g
Volume of flask   = 240 cm^3
(Relative atomic mass of nitrogen = 14)

All observations were made at room temperature and pressure.

(a) Calculate the number of moles of nitrogen.

(b) What is the volume occupied by 1 mole of nitrogen under these experimental conditions?

(c) What is the relative molecular mass of the unknown hydrocarbon?

Mark this test out of 20 with the answers provided on page 372.

Unit Five

Matter and Electricity

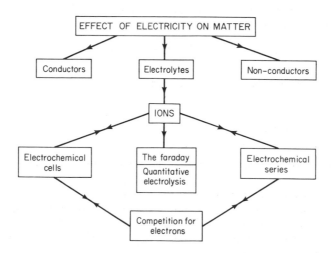

It has already been established in Unit 2 that matter is electrical in nature. Would we therefore expect matter to be affected by an electric current?

In answering this question, Unit 5 explores the effects of electricity on different types of matter and interprets the information in terms of ions and electron flow. Quantitative studies lead to the determination of the relative charges on ions.

Investigations into simple electrochemical cells develop the concept of an 'activity series' for the elements, and first ideas and applications of an electrochemical series follow from the cells investigated.

5.1 The Effect of Electricity on Matter

(a) **Solids**

The circuit shown in Fig. 5.1 can be used to test the conductivity of different solid materials.

The source of electricity is arranged to give 6 volts: it can be a laboratory low-voltage supply, a dry cell, or a battery pack.

An ammeter (or a torch bulb) is used to detect the flow of an electric current. Two test probes (*electrodes*) made of graphite are used to test materials (two steel nails are an acceptable substitute). When the probes are touched together the circuit is completed, and the presence of a current is indicated by a reading on the ammeter, or the torch bulb becoming illuminated. Various solid materials

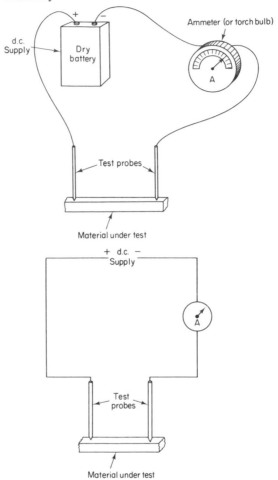

Fig. 5.1   Circuit used to test the electrical conductivity of solids

are then placed as shown in Fig. 5.1 in contact with both probes, to find out whether they are *conductors* or *non-conductors* (*insulators*) of electricity. A current will be observed only if the material is a conductor. Some typical results are shown in Table 5.1.

With the exception of graphite, good conductors could almost all be classified as metals. *The most distinctive property of metals is their high electrical conductivity.*

The fact that metals conduct an electric current is in agreement with their structure, which we discussed in Section 2.11. Metals are regarded as having a sea of mobile electrons surrounding positive nuclei, and it is these mobile electrons which give rise to electrical conductivity. An electric current, i.e. a

Table 5.1 Conductivity of solids

Solid conductors	*Solid insulators*

1. *All metals*
    e.g. iron
         copper
         lead
         aluminium
         sodium
         potassium
         etc.
         (graphite, a non-metal, is an
             exception)
	1. *Most non-metals*
	    e.g. sulphur
	         phosphorus
	         iodine
	         diamond
	         etc.
	
	2. *Most solid compounds*
	    e.g. sodium chloride
	         lead(II) bromide
	         copper(II) sulphate
	         potassium iodide
	         naphthalene
	         paraffin wax
	         plastics and rubbers
	         sugar
	         etc.

flow of electrons, in a conductor causes no change in the chemical nature of the conductor.

Non-conductors do not possess these mobile electrons and therefore do not conduct an electric current.

(b) Liquids

Liquids are tested directly; solid non-conductors are heated until they *first* melt before testing.

The apparatus required is shown in Fig. 5.2. It is similar to that used with solids, except that the two probes are mounted in an insulating support at a set distance apart and the materials are tested in a crucible.

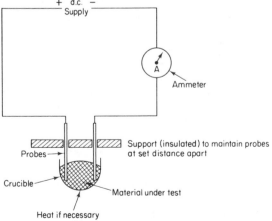

Fig. 5.2   Apparatus used in testing the effect of electricity on liquids

Table 5.2 Conductivity of liquids

Conductors	Electrolytes	Non-electrolytes
1. *Mercury*, the only common liquid metal at room temperature	1. *All ionic compounds in the fused (molten) state* e.g. molten sodium chloride molten lead(II) bromide molten potassium iodide molten sodium hydroxide	1. *Covalent liquids* e.g. tetrachloromethane (carbon tetrachloride) benzene petrol
2. *All molten metals* e.g. molten lead		2. *Covalent substances in the fused state* e.g. molten plastics molten sulphur

Table 5.2 classifies the conductivity of liquids in three groups:
  (i) those that pass an electric current and are not decomposed by it (*conductors*)
  (ii) those that pass an electric current and are decomposed by it (*electrolytes*)
  (iii) those that do not pass an electric current (*non-electrolytes*).

(c) Solutions in Water

Water does not appear in Table 5.2 although it is a covalent liquid. It does not readily fit into any of the three categories because it is a slight conductor.

Because of its polar nature (see Unit 2) water is a very good solvent, particularly for ionic compounds. The conductivity of various substances dissolved in water can be classified as shown in Table 5.3. It will be noted that electrolytes are either *weak* or *strong* according to the ease with which they pass an electric current, and

Table 5.3 Conductivity of aqueous solutions

Strong electrolytes	Weak electrolytes	Non-electrolytes
1. *Solutions of ionic compounds* e.g. sodium chloride solution potassium iodide solution copper(II) chloride solution etc.	e.g. ethanoic acid (acetic acid) aqueous ammonia	*Solutions of most covalent compounds* e.g. sugar solution starch solution
2. *Substances which react with water to produce ions* e.g. hydrogen chloride gives hydrochloric acid with water		

therefore it is necessary to use an ammeter in the circuit to differentiate them. A large deflection on the ammeter indicates a strong electrolyte, whereas a small deflection shows a weak electrolyte.

(d) **Gases**
Gases do not conduct an electric current under normal conditions.

5.2 Decomposition of Substances by an Electric Current

We have seen that metals and graphite conduct electricity without any change, whereas molten electrolytes and those dissolved in water are decomposed when a current flows through them. How do we account for this decomposition?

(a) **Molten Electrolytes**
Let us consider the action of electricity on lead(II) bromide. Lead(II) bromide is a white crystalline salt which has the advantage (for our present purpose) of a relatively low melting point, but in fact any other ionic salt would give similar results. Testing the salt in an apparatus like that of Fig. 5.2, we find that no current flows while the lead(II) bromide remains a solid. As soon as it melts current flows and bubbles of red-brown bromine gas can be seen at the *anode*, i.e. the electrode connected to the positive side of the battery. After heating is discontinued, current continues to flow until the liquid solidifies. A bead of metallic lead is found at the *cathode*, i.e. the electrode connected to the negative side of the battery.

Can these observations be explained in the light of our knowledge about ionic solids such as lead(II) bromide? Solid lead(II) bromide consists of *di*positive lead ions and negative bromide ions arranged in a rigid crystal-lattice framework. On heating lead(II) bromide, the ions vibrate more and more violently as the temperature rises. At a particular temperature, the *melting point*, the vibration becomes so strong that the crystal lattice is destroyed and the ions become free to move. The positively charged lead(II) ion is attracted towards the negative *cathode*, where it loses its positive charge by gaining two electrons to become a lead atom. The collection of many of these lead atoms results in the formation of a bead of lead. At the same time negative bromide ions are attracted towards the positive *anode*, where they each lose one electron to become bromine atoms. Two such reactive atoms combine immediately to produce molecules of red-brown bromine gas. For electrical neutrality it is necessary for *two* bromide ions to be discharged for every *one* lead ion.

Electrode Reactions
(a) CATHODE: $Pb^{2+} + 2e \rightarrow Pb_{(l)}$

(b) ANODE: $2Br^- - 2e \rightarrow 2Br$ ⎫
$\qquad\qquad 2Br \rightarrow Br_{2(g)}$ ⎬
⎭

Such a process, where an electrolyte is decomposed by an electric current, is called **electrolysis**.

The electrolysis of molten electrolytes liberates metals at the cathode and non-metals at the anode. This is summarized in Fig. 5.3.

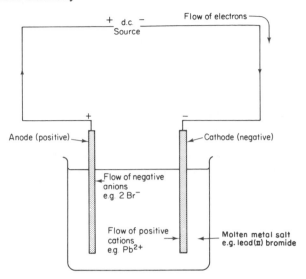

Fig. 5.3   Particle flow during electrolysis

(b) Solutions of Electrolytes in Water

When ionic solids dissolve, the process can be pictured as the removal of ions from the crystal lattice by molecules of water. Once the ions become surrounded by water molecules, i.e. become *hydrated*, they are free to move. A water molecule is polar, having an electrical imbalance (Fig. 2.18), and it is this imbalance which causes the formation of hydrated ions, as shown in Fig. 5.4.

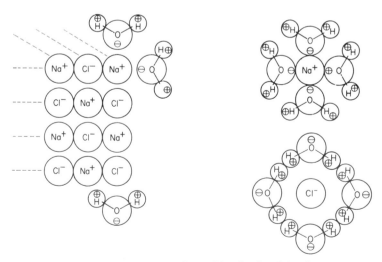

Fig. 5.4   An electrolyte, sodium chloride, dissolving in water

Thus aqueous solutions of electrolytes possess electrical conductivity because of these mobile hydrated ions. Electrolysis of aqueous solutions results in the formation of metals and hydrogen at the cathode, and non-metals at the anode. Water itself produces a few hydrogen and hydroxide ions

$$H_2O_{(l)} \rightarrow H^+_{(aq)} + OH^-_{(aq)}$$

These ions play an important part in the electrolysis of aqueous solutions and are often liberated at the electrodes. We will consider this effect in more detail in Section 5.10.

In metals electrons move independently, but in solutions and melts they move only as constituents of ions. An electron leaving a wire or electrode to enter a solution must become associated with a *cation* at the *cathode*. An electron leaving a solution or melt must come from an *anion* at the *anode*. These processes occur simultaneously whenever current flows.

5.3 Movement of Ions during Electrolysis

What evidence is there to show that ions actually move towards the electrodes during electrolysis? The following experiments provide visual confirmation for this movement of ions.

Experiment 5.1   Demonstration of the mobility of ions
A strip of filter paper about 3×8 cm is completely moistened with a little dilute potassium nitrate solution and placed on a clean dry watch glass. The paper is then connected to a source of direct current (d.c.) of about 15 volts, using a pair of crocodile clips as shown in Fig. 5.5.

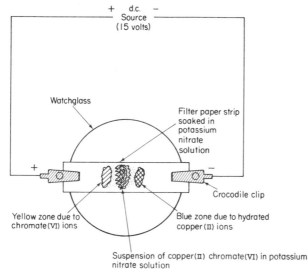

Fig. 5.5   Movement of copper(II) and chromate(VI) ions

A suspension of yellow copper(II) chromate(VI) in potassium nitrate solution is placed in the centre of the filter-paper strip. After a short while the blue colour of the hydrated copper(II) ion $Cu^{2+}_{(aq)}$ is seen to move in the direction of the cathode, while the yellow chromate(VI) ion $CrO^{2-}_{4(aq)}$ moves towards the anode.

The experiment can be repeated using other coloured water-soluble ionic crystals such as potassium manganate(VII) (potassium permanganate) $K^+MnO^-_4$, nickel(II) chloride $Ni^{2+}Cl^-_2$, cobalt(II) nitrate $Co^{2+}(NO^-_3)_2$ etc. In each case the coloured ion or ions move to the appropriate electrode.

Experiment 5.2   Demonstration of the movement of hydrogen ions and hydroxide ions

This experiment provides evidence for the movement of colourless ions by allowing them to react with a coloured dye (bromothymol blue) solution as they move towards the electrodes. Hydrogen $H^+_{(aq)}$ ions produce a yellow colour while hydroxide $OH^-_{(aq)}$ ions produces a blue colour when reacting with bromothymol blue.

A dilute solution of potassium nitrate containing $K^+_{(aq)}$ and $NO^-_{3(aq)}$ ions with a few $H^+_{(aq)}$ and $OH^-_{(aq)}$ ions from the water, is warmed and then stirred with a little gelatin. When the gelatin is completely dissolved a little bromothymol blue is added until the solution has a pronounced green colour. The warm liquid is then poured into a narrow U tube until it is 2–3 cm from the top of each limb,

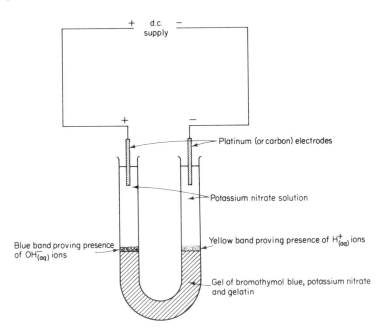

Fig. 5.6   Apparatus for showing the movement of $H^-_{(aq)}$ ions and $OH^-_{(aq)}$ ions

and allowed to set. Potassium nitrate solution is poured on top of the gel (jelly) to fill both sides of the tube. Two unreactive (e.g. platinum or carbon) electrodes are placed in the colourless potassium nitrate solution, one in each limb of the U tube, as shown in Fig. 5.6.

On connecting to a direct-current (d.c.) supply of about 50 volts, a yellow band appears above the gel at the cathode and a blue band appears above the gel at the anode side of the U tube. This shows that $H^+_{(aq)}$ ions are moving towards the cathode and $OH^-_{(aq)}$ ions are moving towards the anode.

5.4 The Faraday

An electric current, the flow of electrons round a circuit, is measured in units called amperes, often abbreviated to amps. The *quantity* of electricity passing when a current of 1 amp flows for 1 second is called a *coulomb*. Thus when I amps flow for t seconds the quantity of electricity is $I \times t$ coulombs.

(a) Quantitative Electrolysis of Lead(II) Bromide

When molten lead(II) bromide is electrolysed, lead is formed at the negative cathode and bromine at the positive anode. Is there any relationship between the mass of lead formed and the quantity of electricity passed through the melt? In order to answer this question we need to modify the circuit used in Fig. 5.2 by introducing a rheostat (a variable resistor). This enables a steady current to be maintained during the electrolysis.

The melt is cooled after the passage of a known quantity of electricity, enabling the bead of lead to be extracted, cleaned and weighed. Comparison of various masses of lead produced by known quantities of electricity shows that the mass of lead produced is proportional to the quantity of electricity flowing. Thus if twice the current is passed, twice the amount of lead is liberated. In general it is found that *the mass of a substance liberated during electrolysis is proportional to the quantity of electricity flowing*. This is often referred to as **Faraday's first law of electrolysis**.

(b) Quantitative Electrolysis of Aqueous Solutions

Electrolysis of aqueous solutions usually leads to the liberation of hydrogen or a metal at the cathode. If we electrolyse silver nitrate solution using silver electrodes, and copper(II) sulphate solution using copper electrodes, silver and copper are deposited at the respective cathodes. See Fig. 5.7.

Experiment 5.3   Determination of the mass of silver and copper liberated by the same quantity of electricity

In Fig. 5.7, R is a rheostat with the aid of which the current is controlled so that the ammeter A shows a steady reading.

The clean dry cathodes are weighed and then connected as shown. Beaker X contains dilute silver nitrate solution with silver electrodes, while beaker Y contains dilute copper(II) sulphate solution with copper electrodes. A steady

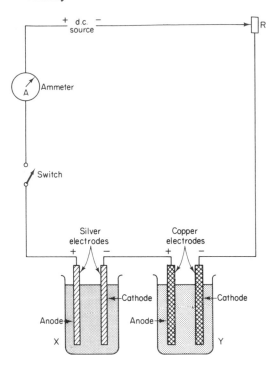

Fig. 5.7   Determination of the masses of silver and copper liberated by the same quantity of electricity

current (maintained at a constant value by adjustment of the rheostat) is passed through the solutions for a known length of time (0·2 amps for 30 minutes would be appropriate). The electrodes are removed from the solution, washed by dipping in a beaker containing distilled water, and air-dried. Great care must be taken to ensure that metal is not lost from the electrodes.

Results

Mass of silver cathode before experiment	= 2·054 g
Mass of silver cathode after experiment	= 2·457 g
mass of silver deposited	= 0·403 g
Mass of copper cathode before experiment	= 6·430 g
Mass of copper cathode after experiment	= 6·549 g
mass of copper deposited	= 0·119 g

A current of 0·2 amps was passed through both solutions for 30 minutes. Hence the *quantity* of electricity (in coulombs) which passed was (0·2 amps) × (30 × 60 seconds) = 360 coulombs.

Calculation

In order to calculate the number of coulombs of electricity required to liberate 107·9 g of silver, i.e. one mole of silver, the results of this experiment can be used in the following way.

By experiment:

0·403 g of silver is liberated by 360 coulombs

$$\therefore 107\text{·}9 \text{ g of silver is liberated by } \frac{360}{0\text{·}403} \times 107\text{·}9 \text{ coulombs}$$

$$= 96\,390 \text{ coulombs}$$

Repeated experiments suggest that an accurate value for the quantity of electricity required to liberate one mole of silver is 96 487 coulombs. This quantity is called the *faraday*. Thus one faraday of electricity liberates one mole of silver.

A similar calculation enables us to find the number of coulombs required to liberate 1 mole (63·54 g) of copper.

0·119 g of copper is liberated by 360 coulombs

$$\therefore 63\text{·}54 \text{ g of copper is liberated by } \frac{360}{0\text{·}119} \times 63\text{·}54 \text{ coulombs}$$

$$= 192\,200 \text{ coulombs}$$

This value, 192 200 coulombs, is approximately *two* faradays. Thus two faradays of electricity are required to liberate *one* mole of copper under the conditions of the experiment.

Within the limits of experimental error it is found that the quantity of electricity required to liberate *one mole* of any element is never less than one faraday: it is always either exactly one faraday or a whole-number multiple (2, 3 or 4) of one faraday.

5.5 Quantity of Charge on an Ion

One faraday is the least quantity of electricity that will liberate one mole of any metal from a solution or melt of its ions. Since an electron is the least possible quantity of electricity, we may assume that one faraday is one mole of electrons (thinking of electrons as charges rather than as particles). In the deposition of one mole of silver at a cathode during electrolysis one faraday is required. Therefore *one mole of silver ions* reacts with *one mole of electrons* to produce *one mole of silver atoms*. Thus *one* silver ion reacts with *one* electron to produce *one* silver atom:

$$Ag^{+}_{(aq)} + e \rightarrow Ag_{(s)}$$

Two faradays are required to liberate *one* mole of copper from a solution containing copper(II) ions. Thus by similar reasoning *one* copper(II) ion will react with *two* electrons to produce *one* copper atom:

$$Cu^{2+}_{(aq)} + 2e \rightarrow Cu_{(s)}$$

Thus the number of faradays required to liberate one mole of any element tells us the charge on the ion of that element.

5.6 Faraday's Laws of Electrolysis

We have found in Sections 5.4 and 5.5 that

(i) *The mass of substance liberated during electrolysis is proportional to the quantity of electricity flowing.*

(ii) *The liberation of one mole of any element during electrolysis requires an integral number of faradays, i.e. 1, 2, 3, or 4.*

These conclusions are known as Faraday's laws of electrolysis.

5.7 An Electrochemical Cell

The electrodes, electrolyte solution and container used in our electrolysis experiments constitute an 'electrochemical cell'. Is it possible to construct such a cell in which, instead of a current decomposing chemicals, chemicals react to produce an electric current? Towards the end of the eighteenth century, Luigi Galvani and Alessandro Volta acquired fame by their discovery that this is indeed possible. Volta found that electricity was produced by chemical reaction between two different metals, and made the first practical battery.

The two metals used by Volta in his battery were copper and zinc. It can be shown by experiment that if copper is added to zinc sulphate solution no reaction takes place, whereas if zinc is added to copper(II) sulphate solution the blue colour of the hydrated copper(II) ion gradually disappears and a reddish brown sludge of copper is deposited.

Equations

$$Cu_{(s)} + Zn^{2+}SO_{4(aq)}^{2-} \rightarrow \text{no reaction}$$

$$Zn_{(s)} + Cu^{2+}SO_{4(aq)}^{2-} \rightarrow Cu_{(s)} + Zn^{2+}SO_{4(aq)}^{2-}$$

The reaction above can be explained in terms of electron loss and gain. Each atom of zinc loses two electrons to each copper ion in the solution:

$$Zn \rightarrow Zn^{2+} + 2e$$

$$Cu^{2+} + 2e \rightarrow Cu$$

These are called **half-reactions**. The half-reaction in which electrons are lost is called *oxidation*, and the half-reaction in which electrons are gained is called *reduction*. Although there is a transfer of electrons, an electric current is not produced unless the electrons are able to flow from the zinc *through a circuit* (e.g. a wire) to the copper(II) ions in solution. A chemical device for creating a flow of electrons in a circuit is called a **primary cell**. See Fig. 5.8.

The two electrolytes must be in contact and yet not allowed to mix. The apparatus shown in Fig. 5.8 uses a cardboard partition which, on becoming soaked, provides electrical contact without mixing of the electrolytes. However, cardboard is not very strong and a more satisfactory partition is the porous pot used in the **Daniell cell** (Fig. 5.9). (This cell is described here because it helps in the understanding of the way in which chemical cells operate. It is not a practical cell in everyday use.)

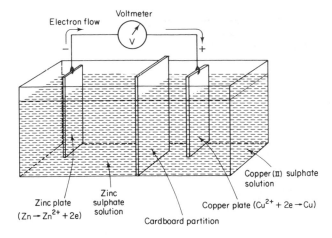

Fig. 5.8   *A simple primary cell*

The zinc rod in the Daniell cell gradually dissolves as atoms of zinc lose electrons to become zinc ions. These electrons pass through an external wire circuit to the copper container where copper(II) ions from the copper(II) sulphate solution are converted into metallic copper. Copper(II) sulphate

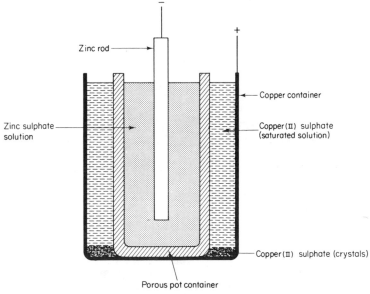

Fig. 5.9   *The Daniell cell*

crystals are present to maintain a saturated solution. A voltmeter connected across the terminals of the cell registers 1·1 volts. This voltage, recorded when there is no current flowing, is known as the *electromotive force* (e.m.f.) of the cell.

A shorthand notation has been developed for a cell: by convention, the half-cell in which oxidation occurs (zinc in this case) is written on the left-hand side and the half-cell in which reduction occurs (copper in this case) is written on the right-hand side (remember: reduction on the right). For the cell illustrated in Figs. 5.8 and 5.9 this notation, termed a **cell diagram**, would thus be:

$Zn_{(s)}$	$Zn^{2+}_{(aq)}$	$\|\|$	$Cu^{2+}_{(aq)}$	$Cu_{(s)}$
↑	↑	↑	↑	↑
metal terminal	aqueous solution in contact with metal terminal	salt bridge or porous pot	aqueous solution in contact with metal terminal	metal terminal

By using different combinations of metals in a suitable electrolyte (often a solution of the metal salt), it is possible to construct a variety of electrochemical cells delivering different voltages. How can we account for these different voltages?

5.8 Competition for Electrons

The half-reactions in the Daniell cell are

$$Zn_{(s)} \rightarrow Zn^{2+}_{(aq)} + 2e$$

and

$$Cu^{2+}_{(aq)} + 2e \rightarrow Cu_{(s)}$$

These half-reactions can be viewed as a competition between two kinds of metal atoms for electrons. In this case zinc atoms are the more reactive and their tendency to lose electrons is greater than that of copper.

Experiment 5.4   Investigation of electrochemical cell reactions
A 250 cm³ beaker is almost filled with 0·1 M copper(II) sulphate solution and a similar beaker is filled with 0·1 M silver nitrate solution. A silver rod is placed in the silver nitrate solution and a copper rod in the copper(II) sulphate solution. The circuit is connected as shown in Fig. 5.10a. A glass U tube containing potassium nitrate solution provides electrical contact between the two solutions. This is called a *salt bridge*. The U tube is plugged with cotton wool to prevent mixing of the solutions.

Observations
As soon as the circuit is complete a current is recorded on the ammeter, and the direction of this current shows that the flow of electrons through the circuit is from the copper to the silver. The copper rod dissolves and the silver rod grows by deposition of silver atoms.

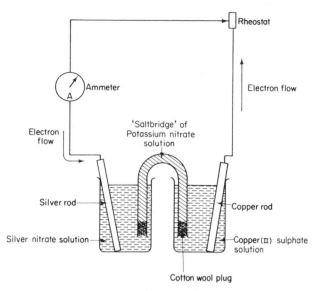

Fig. 5.10a   Investigating the reactions which occur in an electrochemical cell

In a similar cell to that in Fig. 5.10a, using 0·1 M zinc nitrate and 0·1 M copper(II) sulphate, the electron flow in the circuit is from zinc to copper. Zinc dissolves and copper is deposited on the copper plate.

The voltages of these two cells can be measured using a voltmeter, see Fig. 5.10b.

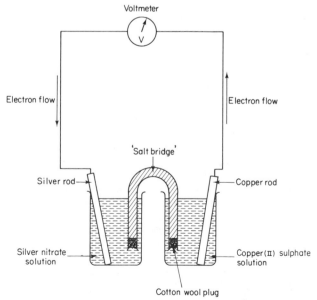

Fig. 5.10b   Measuring the voltage produced by an electrochemical cell

In the copper/silver cell the electron flow indicates that copper metal has a greater tendency to lose electrons than silver metal. Thus the reaction

$$Cu_{(s)} \rightarrow Cu^{2+}_{(aq)} + 2e$$

will occur more readily than the reaction

$$Ag_{(s)} \rightarrow Ag^{+}_{(aq)} + e$$

In the copper/zinc cell the electron flow indicates that the reaction

$$Zn_{(s)} \rightarrow Zn^{2+}_{(aq)} + 2e$$

will occur more readily than the reaction

$$Cu_{(s)} \rightarrow Cu^{2+}_{(aq)} + 2e$$

We have thus established a **reactivity series** for the elements zinc, copper and silver. It is

$$Zn_{(s)} \rightarrow Zn^{2+}_{(aq)} + 2e$$
$$Cu_{(s)} \rightarrow Cu^{2+}_{(aq)} + 2e$$
$$Ag_{(s)} \rightarrow Ag^{+}_{(aq)} + e$$

increasing reactivity →

From our list we would expect that a cell constructed from zinc and silver would react to produce zinc ions and silver metal. This is indeed so. Cell

Table 5.4 The electrochemical series

Name	Half-reaction	Standard reduction potential	
		volts	
Potassium	$K^{+}_{(aq)} + e \rightarrow K_{(s)}$	$-2\cdot92$	
Calcium	$Ca^{2+}_{(aq)} + 2e \rightarrow Ca_{(s)}$	$-2\cdot87$	
Sodium	$Na^{+}_{(aq)} + e \rightarrow Na_{(s)}$	$-2\cdot71$	
Magnesium	$Mg^{2+}_{(aq)} + 2e \rightarrow Mg_{(s)}$	$-2\cdot34$	
Aluminium	$Al^{3+}_{(aq)} + 3e \rightarrow Al_{(s)}$	$-1\cdot67$	
Zinc	$Zn^{2+}_{(aq)} + 2e \rightarrow Zn_{(s)}$	$-0\cdot76$	
Iron(II)	$Fe^{2+}_{(aq)} + 2e \rightarrow Fe_{(s)}$	$-0\cdot44$	
Tin(II)	$Sn^{2+}_{(aq)} + 2e \rightarrow Sn_{(s)}$	$-0\cdot14$	increasing
Lead(II)	$Pb^{2+}_{(aq)} + 2e \rightarrow Pb_{(s)}$	$-0\cdot13$	ease of
Hydrogen	$2H^{+}_{(aq)} + 2e \rightarrow H_{2(g)}$	$0\cdot00$	reaction
Copper(II)	$Cu^{2+}_{(aq)} + 2e \rightarrow Cu_{(s)}$	$+0\cdot34$	(more
Hydroxide ion	$2H_2O_{(l)} + O_{2(g)} + 4e \rightarrow 4OH^{-}_{(aq)}$	$+0\cdot40$	positive
Iodide ion	$I_{2(s)} + 2e \rightarrow 2I^{-}_{(aq)}$	$+0\cdot54$	potential)
Silver	$Ag^{+}_{(aq)} + e \rightarrow Ag_{(s)}$	$+0\cdot80$	
Mercury(II)	$Hg^{2+}_{(aq)} + 2e \rightarrow Hg_{(l)}$	$+0\cdot85$	
Bromide ion	$Br_{2(l)} + 2e \rightarrow 2Br^{-}_{(aq)}$	$+1\cdot06$	
Chloride ion	$Cl_{2(g)} + 2e \rightarrow 2Cl^{-}_{(aq)}$	$+1\cdot36$	
Gold	$Au^{+}_{(aq)} + e \rightarrow Au_{(s)}$	$+1\cdot68$	
Sulphate ion	$SO_4 + 2e \rightarrow SO^{2-}_{(aq)}$	$+1\cdot90$	
Fluoride ion	$F_{2(g)} + 2e \rightarrow 2F^{-}_{(aq)}$	$+2\cdot87$	

voltage indicates the tendency for a reaction to occur: the greater the voltage, the more likely the reaction. One would expect the cell voltage for the zinc/silver cell to be greater than either the zinc/copper or copper/silver cells. This is also true. The reactivity series for all the elements is called an **electrochemical series**.

5.9 The Electrochemical Series and Reduction Potentials

The reactions taking place in a cell can be expressed in terms of two *half-reactions* in two *half-cells*. In one half-cell electrons are lost: this is the *oxidation* half-reaction. In the other half-cell electrons are gained: this is the *reduction* half-reaction. When combined, these two half-reactions express the overall reaction in an **oxidation– reduction** or **redox** reaction. In all such reactions the number of electrons lost is equal to the number of electrons gained.

It is impossible to measure the actual voltage (e.m.f.) for any half-cell by itself because both half-cell reactions occur simultaneously. Chemists have adopted the quite arbitrary method of choosing the *hydrogen* half-reaction

$$2H^+_{(aq)} + 2e \rightarrow H_{2(g)}$$

and calling its e.m.f. *zero* when the effective concentration of hydrogen ions is one mole in 1000 g of water at 25 °C. This serves as a standard for comparing other half-cell e.m.f.s. These values can be positive or negative according to whether the half-reaction is more likely (in which case it is positive) or less likely (in which case it is negative) to take place than the hydrogen half-reaction. The e.m.f.s. of these half-reactions can be termed **reduction potentials** and a list of such potentials is called the **electrochemical series**, see Table 5.4.

5.10 Using the Electrochemical Series

Positive e.m.f. values in the electrochemical series (Table 5.4) indicate that the half-reaction will readily occur. The greater this positive value, the more readily the half-reaction will occur. Thus fluorine will form fluoride ions (reduction potential $+2\cdot87$ V) more readily than chlorine will form chloride ions (reduction potential $+1\cdot36$ V).

Similarly the reduction of potassium ions to potassium metal ($-2\cdot92$ V) will be much more difficult than the reduction of zinc ions to zinc metal ($-0\cdot76$ V). The converse of this is also true, in that potassium metal will form potassium ions much more readily than zinc metal will form zinc ions.

We say that fluorine is more reactive than chlorine, and that potassium is more reactive than zinc.

(a) Predicting the Voltage of a Cell
Consider the Daniell cell (see Section 5.7) in which there are two possible half-reactions:

Half-reaction	e.m.f.
$Zn^{2+}_{(aq)} + 2e \rightarrow Zn_{(s)}$	$-0\cdot76$V
$Cu^{2+}_{(aq)} + 2e \rightarrow Cu_{(s)}$	$+0\cdot34$V

The reduction reaction with the most positive e.m.f. is more likely, that is, copper ions will be reduced to copper metal. Thus zinc metal must also be oxidized to zinc ions for complete reaction. In any cell there must always be a reduction half-reaction and an oxidation half-reaction.

The cell diagram is constructed with the oxidation half-reaction on the left-hand side and the reduction half-reaction on the right:

$$Zn_{(s)} \mid Zn^{2+}_{(aq)} \parallel Cu^{2+}_{(aq)} \mid Cu_{(s)}$$

The e.m.f. of such a cell is always calculated:

$$
\begin{aligned}
\text{e.m.f. (cell)} &= \text{e.m.f. (right-hand-side)} - \text{e.m.f. (left-hand-side)} \\
&= [+0.34 - (-0.76)] \text{ V} \\
&= +1.10 \text{ V}
\end{aligned}
$$

Note:

(i) The *cell reaction* is obtained by adding the oxidation half-reaction to the reduction half-reaction:

$$
\begin{array}{ll}
Zn_{(s)} \rightarrow Zn^{2+}_{(aq)} + 2e & \text{Oxidation} \\
Cu^{2+}_{(aq)} + 2e \rightarrow Cu_{(s)} & \text{Reduction} \\
\hline
\end{array}
$$

Adding: $Zn_{(s)} + Cu^{2+}_{(aq)} \rightarrow Zn^{2+}_{(aq)} + Cu_{(s)}$ Cell reaction

(ii)   Electrons do not appear in the overall cell reaction.

(iii) A positive value for the e.m.f. indicates that the cell *will* operate in the direction indicated by the overall cell reaction.

(iv) This calculated value for the e.m.f. (1·10 V) agrees with the measured value.

Another cell we have considered was the copper/silver cell:

Half-reaction	e.m.f.
$Cu^{2+}_{(aq)} + 2e \rightarrow Cu_{(s)}$	$+0.34$ V
$Ag^{+}_{(aq)} + e \rightarrow Ag_{(s)}$	$+0.80$ V

The reduction half-reaction (the one with the most positive e.m.f.) must be:

$$Ag^{+}_{(aq)} + e \rightarrow Ag_{(s)}$$

Hence the oxidation reaction must be:

$$Cu_{(s)} \rightarrow Cu^{2+}_{(aq)} + 2e$$

The cell diagram would thus be:

$$Cu_{(s)} \mid Cu^{2+}_{(aq)} \parallel Ag^{+}_{(aq)} \mid Ag_{(s)}$$

and its e.m.f. is given by:

$$
\begin{aligned}
\text{e.m.f.} &= \text{e.m.f. (right-hand-side)} - \text{e.m.f. (left-hand-side)} \\
&= +0.80 - 0.34 \\
&= +0.46 \text{ V}
\end{aligned}
$$

Note:

(i)   The cell reaction is obtained by adding the oxidation half-reaction to the reduction half-reaction.

$$Cu_{(s)} \rightarrow Cu^{2+}_{(aq)} + 2e \qquad \text{Oxidation}$$
$$2 \times [Ag^{+}_{(aq)} + e \rightarrow Ag_{(s)}] \qquad \text{Reduction}$$

Adding: $Cu_{(s)} + 2Ag^{+}_{(aq)} \rightarrow Cu^{2+}_{(aq)} + 2Ag_{(s)}$ Cell reaction

(ii)   As no electrons must appear in the overall cell reaction, the silver half-reaction needs to be doubled.

(iii)   The e.m.f. of the silver half-reaction remains $+0.80$ V, because this does not depend upon how many moles we consider:

$$Ag^{+}_{(aq)} + e \rightarrow Ag_{(s)} \qquad \text{e.m.f.} = +0.80 \text{ V}$$
$$2Ag^{+}_{(aq)} + 2e \rightarrow 2Ag_{(s)} \qquad \text{e.m.f.} = +0.80 \text{ V}$$

(iv)   The e.m.f. of the cell, $+0.46$ V, shows that the cell reaction proceeds spontaneously in the direction indicated.

(b) Predicting Chemical Reactions

The electrochemical series can be used by a chemist to predict not only cell reactions, but also what chemical reactions can occur spontaneously.

Suppose we wish to know whether any reaction will take place if an iron pen-knife blade is dipped into a solution of copper(II) sulphate.

The half-reactions are:

$$Fe^{2+}_{(aq)} + 2e \rightarrow Fe_{(s)} \qquad \text{e.m.f.} = -0.44 \text{ V}$$
$$Cu^{2+}_{(aq)} + 2e \rightarrow Cu_{(s)} \qquad \text{e.m.f.} = +0.34 \text{ V}$$

This tells us that it is easier to reduce $Cu^{2+}_{(aq)}$ ions to copper metal (greater positive e.m.f.) than it is to reduce $Fe^{2+}_{(aq)}$ ions to iron metal. Therefore copper(II) ions are reduced to copper metal and iron metal must be oxidized to iron(II) ions.

$$Cu^{2+}_{(aq)} + 2e \rightarrow Cu_{(s)} \qquad \text{Reduction}$$
$$Fe_{(s)} \rightarrow Fe^{2+}_{(aq)} + 2e \qquad \text{Oxidation}$$

Adding: $Cu^{2+}_{(aq)} + Fe_{(s)} \rightarrow Cu_{(s)} + Fe^{2+}_{(aq)}$ Cell reaction

(c) Selective Discharge of Ions

How do we account for the fact that when an aqueous solution of sodium hydroxide is electrolysed using inert platinum electrodes, hydrogen is liberated at the cathode and oxygen at the anode? The electrochemical series helps us to predict which ion is capable of accepting or yielding an electron when several different ions may be present at an electrode.

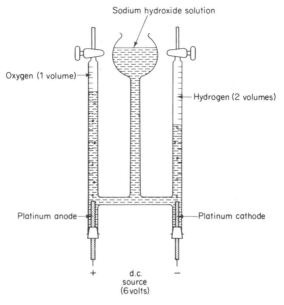

Fig. 5.11   The Hofmann voltameter

(i) Let us consider the electrolysis of sodium hydroxide solution in a Hofmann voltameter (Fig. 5.11) where the electrodes are made of platinum and take no part in the reaction.

Ions at the cathode
$Na^+_{(aq)}$ from the $Na^+OH^-_{(aq)}$
$H^+_{(aq)}$ from the water

Ions at the anode
$OH^-_{(aq)}$ from the $Na^+OH^-_{(aq)}$
$OH^-_{(aq)}$ from the water

Cathode reaction
The possible half-reactions are *either*
$Na^+_{(aq)}+e \rightarrow Na_{(s)}$
(e.m.f. $= -2\cdot71$ volts)
or
$2H^+_{(aq)}+2e \rightarrow H_{2(g)}$
(e.m.f. $= 0\cdot00$ volts)
The second alternative (evolution of hydrogen) has the more positive (i.e. less negative) e.m.f. and is therefore more probable than the deposition of sodium.

Anode reaction
Since only $OH^-_{(aq)}$ ions are present the only possible reaction is the evolution of oxygen:
$4OH^-_{(aq)} \rightarrow 2H_2O_{(1)} + O_{2(g)} + 4e$

Net result

$$4H^+_{(aq)} + 4e \rightarrow 2H_{2(g)} \quad \text{and} \quad 4OH^-_{(aq)} \rightarrow 2H_2O_{(1)} + O_{2(g)} + 4e$$

(The cathode half-reaction is doubled so that electron gain at the cathode balances electron loss at the anode.)

Thus two moles of hydrogen are produced for each mole of oxygen.

(ii) Let us now consider a case where the electrodes are not inert. During the electrolysis of copper(II) sulphate solution using copper electrodes, copper is deposited at the cathode and the anode dissolves.

Ions at the cathode
$Cu^{2+}_{(aq)}$
$H^+_{(aq)}$ from the water

Ions at the anode
$SO^{2-}_{4(aq)}$
$OH^-_{(aq)}$ from the water

Cathode reaction
The possible half-reactions are *either*
$Cu^{2+}_{(aq)} + 2e \rightarrow Cu_{(s)}$
(e.m.f. $= +0.34$ volts)
or
$2H^+_{(aq)} + 2e \rightarrow H_{2(g)}$
(e.m.f. $= 0.00$ volts)
The first alternative (deposition of copper) has the more positive e.m.f. and is therefore more probable than the evolution of hydrogen.

So copper is deposited.

Anode reaction
Three half-reactions are possible:
$SO^{2-}_{4(aq)} \rightarrow SO_4 + 2e$
(e.m.f. $= -1.90$ volts)
or
$4OH^-_{(aq)} \rightarrow 2H_2O_{(1)} + O_{2(g)} + 4e$
(e.m.f. $= -0.40$ volts).
or
$Cu_{(s)} \rightarrow Cu^{2+}_{(aq)} + 2e$
(e.m.f. $= -0.34$ volts)
The third alternative (dissolution of copper) has the most positive (i.e. least negative) e.m.f. and is therefore the most probable.

So copper dissolves.

Net result $Cu^{2+}_{(aq)} + 2e \rightarrow Cu_{(s)}$ and $Cu_{(s)} \rightarrow Cu^{2+}_{(aq)} + 2e$

Copper is deposited at the cathode and the same amount of copper dissolves from the anode; the concentration of the solution remains unchanged. This reaction is the basis for the purification of crude copper, described in Section 9.23.

(Note that if the electrodes are inert, e.g. platinum, hydroxide ions are discharged at the anode and the solution loses its blue colour.)

5.11 Cells in Everyday Use

The cells we have considered so far suffer from two major disadvantages: they are not easily recharged and, since they contain liquids, they are difficult to transport. The two most important cells in everyday use each overcome one of these difficulties.

(a) The Dry Cell
The dry cell (or torch battery) is a primary cell which is cheap to produce and contains no dangerous liquids. It is sealed and the reactive substances are present in the form of pastes (see Fig. 5.12).

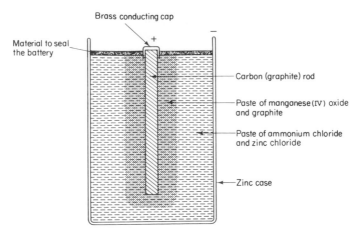

Fig. 5.12   The torch battery or dry cell

In the centre of the cell is a graphite rod which serves as the positive electrode. This is surrounded by a paste of manganese(IV) oxide and graphite, which is itself surrounded by a paste of ammonium chloride and zinc chloride. The rod and pastes are contained in a zinc case which acts as the negative electrode.

The chemical reactions taking place within the cell are complex and produce an e.m.f. of 1·5 volts. As these reactions are irreversible, the dry cell cannot be recharged. It is found that the life of the cell is increased if it is used intermittently.

(b) The Lead Storage Cell
Although this secondary cell contains a corrosive liquid (moderately concentrated sulphuric acid), its outstanding advantage is that it can be recharged because the two half-reactions in the cell are reversible. Six of these cells constitute a motor-car battery delivering 12 volts. Fig. 5.13 shows part of the lead storage cell.

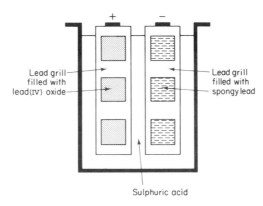

Fig. 5.13   Part of a lead storage battery

The positive plate is a lead grill filled with lead(IV) oxide, while the negative plate consists of a similar lead grill filled with spongy lead. The grills are immersed in sulphuric acid.

During **discharge** of electricity from the cell, the two half-reactions are:

$$Pb_{(s)} + SO_{4(aq)}^{2-} \rightarrow Pb^{2+}SO_{4(s)}^{2-} + 2e$$

$$Pb^{4+}O_{2(s)}^{2-} + SO_{4(aq)}^{2-} + 4H_{(aq)}^{+} + 2e \rightarrow Pb^{2+}SO_{4(s)}^{2-} + 2H_2O_{(l)}$$

Adding: $Pb_{(s)} + Pb^{4+}O_{2(s)}^{2-} + 4H_{(aq)}^{+} + 2SO_{4(aq)}^{2-} \rightarrow 2Pb^{2+}SO_{4(s)}^{2-} + 2H_2O_{(l)}$

White lead(II) sulphate is deposited on the metallic lead and the sulphuric acid becomes more dilute.

The two half-reactions described can be reversed by applying a suitable voltage to the terminals of the cell. When a car engine is running, the dynamo produces a greater voltage than that produced by the battery, and electrical energy is forced back into the cells. During this **recharge** white lead(II) sulphate dissolves and sulphuric acid is produced:

$$2Pb^{2+}SO_{4(s)}^{2-} + 2H_2O_{(l)} \rightarrow Pb_{(s)} + Pb^{4+}O_{2(s)}^{2-} + 4H_{(aq)}^{+} + 2SO_{4(aq)}^{2-}$$

The specific gravity of the sulphuric acid is a measure of the effectiveness of the recharge. Fig. 5.14 illustrates the process of charge and discharge in the lead storage cell.

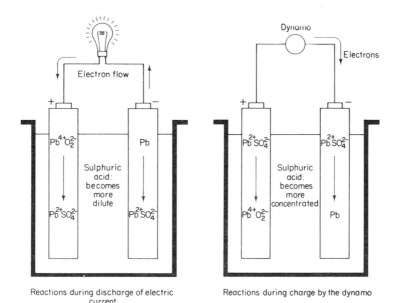

Reactions during discharge of electric current

Reactions during charge by the dynamo

Fig. 5.14   Charging and discharging processes in the lead storage battery

Summary of Unit 5

1. Substances are either *conductors* or *non-conductors* (insulators) of electricity.
2. The most distinctive property of *metals* is their *high electrical conductivity*.
3. Liquids may be classified as (*a*) *conductors*, (*b*) *electrolytes* and (*c*) *non-electrolytes*.
4. An **electrolyte** is a substance which when molten or in aqueous solution will conduct an electric current and be decomposed by it.
5. An *electric current* is a *flow of electrons*.
6. The *decomposition of an electrolyte* by an electric current is called **electrolysis**.
7. During electrolysis anions travel towards the positive *anode*, while cations travel towards the negative *cathode*.
8. Cations are discharged at the cathode by electron gain, whereas anions are discharged at the anode by electron loss.
9. Quantity of electricity is measured in *coulombs* (coulombs = current in amps × time in seconds); 96 487 coulombs is one faraday.
10. The mass of substance liberated during electrolysis is proportional to the quantity of electricity flowing.
11. The liberation of one mole of any element during electrolysis requires an integral (whole) number of *faradays*, i.e. 1, 2, 3 or 4.
12. A **faraday** is *one mole* of *electrons*.
13. An **electrochemical cell** uses chemical reaction to produce a flow of electrons in a circuit.
14. The reactions taking place in a cell can be expressed in terms of two **half-reactions** in two **half-cells**.
15. The half-reaction in which electrons are *lost* is called **oxidation**, while that in which electrons are *gained* is called **reduction**.
16. The *e.m.f.* of the cell indicates the *tendency for a reaction to occur*.
17. The e.m.f. of reduction half-reactions (or oxidation half-reactions) can be obtained by comparison with the hydrogen half-reaction, for which the e.m.f. is arbitrarily assigned a value of zero under stated conditions.
18. The **electrochemical series** is a list of *reduction potentials* obtained from the *reduction half-reaction*. It lists the elements *in order of reactivity*.
19. The electrochemical series can be used (*a*) to calculate the voltage of a cell, (*b*) to predict whether or not a chemical reaction will occur spontaneously, (*c*) to explain the products of electrolysis.
20. The *dry cell* and the *lead storage cell* are two important electrochemical cells in everyday use.

Test Yourself on Unit 5

1. Complete the following paragraph by inserting *one* of the following words in each space: heated, lead, ions, electrolysis, lattice, melts, electrodes, bromine.

Solid lead(II) bromide will not conduct an electric current. The .... are held in a rigid crystal .... and are not free to move to the .... When the solid is ...., it .... and allows the passage of an electric current. .... is liberated at the cathode and .... at the anode. The decomposition of an electrolyte by an electric current is called ....

2. Mark the following statements true or false.

(*a*) A metal will only allow the passage of an electric current when it is solid.

(*b*) Sugar is a non-electrolyte and when dissolved in water will allow the passage of an electric current.

(*c*) All anions are coloured.

(*d*) All coloured anions migrate to the positive electrode during electrolysis.

(*e*) Sulphur is an insulator in both the solid and liquid states.

3. A steady current of 1·6 amps was passed for 10 minutes through two voltameters arranged in series. The first had silver electrodes immersed in silver nitrate solution; the second had a metal X immersed in a solution of a salt of X. After the experiment the cathodes in each voltameter had increased in weight: that in the first voltameter by 1·08 g and that in the second by 0·09 g.

(1 faraday = 96 000 coulombs)

(*a*) How many coulombs of electricity were passed?

(*b*) How many faradays of electricity were passed?

(*c*) Assuming the silver ion to be Ag^+ calculate the relative atomic mass of silver.

(*d*) If the relative atomic mass of X is 27, how many moles of X were liberated?

(*e*) Calculate the charge on an ion of X.

(*f*) Write an equation for the reactions taking place at the cathodes in both voltameters.

(*g*) State whether these cathode reactions are oxidations or reductions.

4. In an experiment concerning the displacement of one metal from an aqueous solution of its salt by another metal the results were tabulated as follows:

	Metal A	Metal B	Metal C	Metal D
Solution of a salt of A	—	W	reaction	X
Solution of a salt of B	reaction	—	reaction	reaction
Solution of a salt of C	no reaction	no reaction	—	Y
Solution of a salt of D	reaction	no reaction	Z	—

The table shows whether or not reaction occurs between the metal and a solution of another metal salt.

   (a) Arrange the metals in order of reactivity, giving the most reactive one first.

   (b) State whether reaction will take place in the spaces labelled W, X, Y and Z.

5. This question relates to the following simple electrochemical cell.

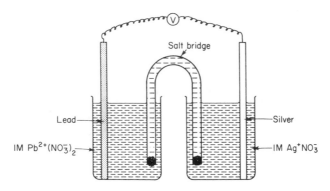

The half-cell reactions with their appropriate oxidation potentials are:

$$Pb^{2+} + 2e \rightarrow Pb \text{ (e.m.f. } = -0 \cdot 13 \text{ V)}$$
$$Ag^+ + e \rightarrow Ag \text{ (e.m.f. } = +0 \cdot 80 \text{ V)}$$

   (a) Which of the following reactions take place?

      (i) $Pb^{2+} + 2Ag \rightarrow 2Ag^+ + Pb$

      (ii) $Pb^{2+} + Ag \rightarrow Ag^+ + Pb$

      (iii) $Ag^+ + Pb \rightarrow Ag + Pb^{2+}$

      (iv) $2Ag^+ + Pb \rightarrow 2Ag + Pb^{2+}$

   (b) What voltmeter reading do you expect?

      (i) 1·73 volts

      (ii) 0·67 volts

      (iii) 0·93 volts

      (iv) 1·47 volts.

Mark this test out of 30 with the answers provided on page 373.

Unit Six

The Matter we Breathe

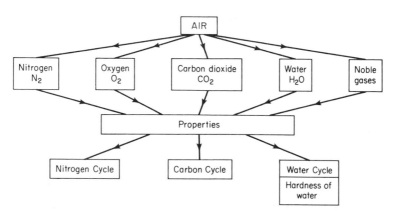

Now that we are becoming increasingly aware of the term *pollution*, the air we breathe is more than ever a matter of concern to us all. This Unit studies the nature and properties of the constituent gases of the air and compounds related to them. It examines some of the major natural cycles which are responsible for sustaining the earth's thin film of living matter.

6.1 Composition of the Air

From the time of the Greeks, for some 2000 years, the air was considered to be a single substance. It was not until the 17th century that this supposition was questioned. In fact dry air is a mixture of predominantly two gases, **oxygen** and **nitrogen**. It also contains small amounts of *carbon dioxide, argon* and other *noble gases.*

Table 6.1 Composition of the air

Gas	Composition by volume	Composition by mass
	%	%
Nitrogen	78·09	75·51
Oxygen	20·95	23·15
Carbon dioxide	0·03	0·04
Argon	0·93 ⎫	
Other noble gases	>0·003 ⎭	1·30

In addition to the gases indicated in Table 6.1, air contains a variable amount of *water vapour*, on average 2%. Wind and industry pollutes the air with *dust* and chemicals such as *sulphur dioxide, nitrogen oxides, ammonia* and many others. Man pumps into the air *smoke* from fires, *carbon monoxide* and *lead* from his car, and a variety of industrial pollutants.

The composition of the air is found to vary from place to place, confirming the fact *that air is a mixture*. Gases present in the air exhibit their own characteristic properties and can be separated by fractional distillation (see Section 1.5).

6.2 Evidence for the Composition of the Air

(a) Water Vapour in the Air

Anhydrous calcium chloride is a dry white powder which on standing for some time in the air turns into a liquid. If this liquid is distilled, a vapour can be condensed which boils at 100 °C and freezes at 0 °C, confirming that it is indeed water. A white, solid residue of the original anhydrous calcium chloride remains. The calcium chloride has absorbed water from the atmosphere and dissolved in it.

Any solid which absorbs water from the atmosphere and dissolves in it forming an aqueous solution is said to be **deliquescent**. Examples of such substances are:

calcium chloride	$Ca^{2+}Cl_2^-$
sodium hydroxide	Na^+OH^-
iron(III) chloride	$Fe^{3+}Cl_3^-$

Certain substances absorb water from the atmosphere but do not dissolve in it. They are said to be **hygroscopic**, and a common example is copper(II) oxide.

The amount of water vapour present in air can be determined by passing a known volume of air over a weighed quantity of anhydrous calcium chloride in a U tube (see Fig. 6.1) and measuring the increase in mass of the U tube and its contents.

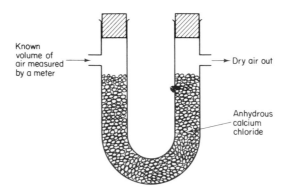

Fig. 6.1   Determination of the amount of water vapour in the air

(b) Carbon Dioxide in the Air

When air is drawn through a wash bottle containing lime-water (see Fig. 6.2), the colourless lime-water turns milky owing to the formation of insoluble calcium carbonate. This is a test for carbon dioxide.

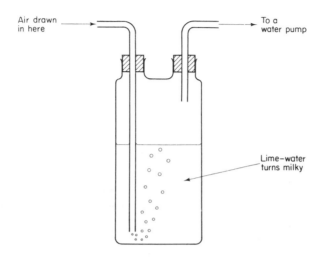

Air drawn in here

To a water pump

Lime–water turns milky

Fig. 6.2   Demonstration of the presence of carbon dioxide in the air

When a known volume of dry air is passed through a U tube containing pellets of sodium hydroxide, the carbon dioxide reacts with the sodium hydroxide forming sodium carbonate and is removed from the air:

$$2Na^+OH^-_{(s)} + CO_{2(g)} \rightarrow Na_2^+CO_{3(s)}^{2-} + H_2O_{(l)}$$

The U tube is weighed before and after the experiment, and the increase in mass indicates the amount of carbon dioxide present.

(c) Oxygen in the Air

A solution of pyrogallol (benzene-1,2,3-triol) containing sodium hydroxide rapidly absorbs oxygen. When a known volume of air is shaken with this solution in a graduated tube, oxygen is absorbed and the pyrogallol turns brown. After a few minutes the tube is inverted with its mouth under water. The levels of the water inside and outside the tube are made equal so that the gas in the tube is at atmospheric pressure. The volume of gas remaining in the tube is recorded. See Fig. 6.3.

The difference between this new volume and the original volume indicates the volume of oxygen in the sample. (Carbon dioxide is also absorbed by the sodium hydroxide solution but its volume is negligible in comparison with the volume of the oxygen: see Table 6.1.)

A more accurate determination of the percentage by volume of oxygen in the

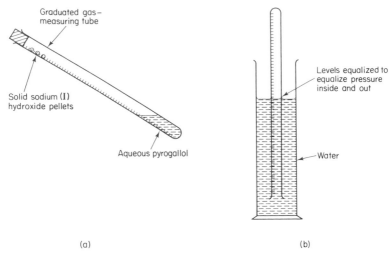

Fig. 6.3   Using alkaline pyrogallol to determine the percentage of oxygen in the air: (a) the start of the experiment; (b) bringing the unabsorbed gas to atmospheric pressure

air makes use of the reaction between heated copper and oxygen in which black copper(II) oxide is formed:

$$2Cu_{(s)} + O_{2(g)} = 2Cu^{2+}O_{(s)}^{2-}$$

Experiment 6.1   Determination of the percentage by volume of oxygen in the air
Some pure copper powder is placed in a silica tube which is connected horizontally between two 100 cm³ gas syringes. One of the syringes (A) contains 100 cm³ of air, the other (B) has its plunger level with the 0 cm³ mark. See Fig. 6.4.

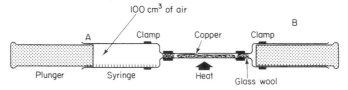

Fig. 6.4   Determination of the percentage by volume of oxygen in the air

The tube containing the copper is heated strongly. When the copper is hot the measured volume of air is pushed slowly backwards and forwards over it using the plungers. After a few minutes the heating is stopped and plunger B pushed to the 0 cm³ mark. When cool, the volume of air in syringe A is recorded. It is less than 100 cm³. The copper is reheated and the air passed over it once again. This procedure is continued until, on cooling, the volume reading remains constant, and it may be assumed that all the oxygen in the air has been removed.

Results

Volume of air in syringe A before experiment	$= 100 \text{ cm}^3$
Volume of air in syringe A after experiment	$= 79 \text{ cm}^3$
Volume of oxygen used in forming copper(II) oxide	$= 21 \text{ cm}^3$
Fraction of oxygen by volume in the air	$= \frac{21}{100}$
Percentage of oxygen by volume in the air	$= 21\%$

(d) Nitrogen and the Noble Gases in the Air

After the removal of water vapour, carbon dioxide and oxygen from the air, the gas remaining (about 80%) is mainly nitrogen with the exception of some 1% of argon and other noble gases. These can be separated by fractional distillation of liquid air.

6.3 Fractional Distillation of Liquid Air

The separation of liquid mixtures by fractional distillation is discussed in Section 1.5. It relies on the fact that different liquids have different boiling points, those having the lower boiling point distilling over first.

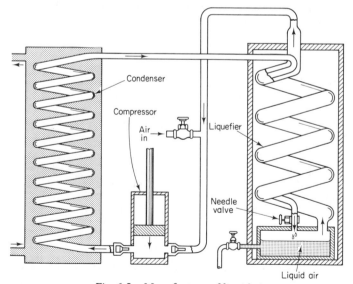

Fig. 6.5   Manufacture of liquid air

To prevent blockage caused by solid materials at these low temperatures, air is first freed from carbon dioxide, water vapour and dust impurities. It is then compressed to about 200 atmospheres, cooled and allowed to expand through a fine jet. See Fig. 6.5. This sudden expansion causes further cooling and the gas eventually liquefies. (You can observe this cooling effect by measuring the fall in temperature when a thermometer is placed near the opened valve of an inflated car tyre.) The liquid is tapped off through a valve, while gas which has escaped liquefaction returns to the compressor.

Liquid air is primarily a mixture of nitrogen and oxygen with small amounts of the noble gases. The boiling point of nitrogen is $-196\,°C$ (77 K) and that of oxygen is $-183\,°C$ (90 K). For this reason, when liquid air is allowed to warm up, the nitrogen boils first and the remaining liquid becomes richer in oxygen. By refractionating the oxygen fraction several times, the oxygen produced is of high purity. Most oxygen is not required in a state of high purity and is simply stored by pumping into steel cylinders under high pressure after only one fractionation.

6.4 Uses of Oxygen, Nitrogen and Argon

The gases obtained by fractional distillation of liquid air are produced in large quantities for a variety of uses. Oxygen is used principally in the steel industry and in the production of nitric acid, methanol and other chemicals. Nitrogen is required for the synthesis of ammonia and as a coolant in refrigeration. The other important gas obtained from liquid air is argon. Argon is used mainly to provide an inert atmosphere during arc welding of metals such as stainless steel and titanium which would otherwise react with the oxygen of the air at high temperatures.

6.5 Oxygen

Oxygen is the most abundant element in the earth's outer surface. It exists as a gas in the air; it is a major constituent of water (H_2O) and it occurs, combined with metals and other elements, in rocks and clays.

Preparation of Oxygen
By far the most important commercial source of pure oxygen is the fractional distillation of liquid air, see Section 6.3. There are, however, a number of chemical reactions by which the gas can be prepared in the laboratory.

(*a*) **From hydrogen peroxide.** When heated, hydrogen peroxide decomposes into water and oxygen:

$$2H_2O_{2(l)} \rightarrow 2H_2O_{(l)} + O_{2(g)}$$

This decomposition takes place at room temperature if a **catalyst** such as manganese(IV) oxide is present. A catalyst is a substance which alters the rate of a chemical reaction but itself *remains unchanged* at the end of the reaction.

The strength of a hydrogen peroxide solution is usually denoted by the volume of oxygen that it will evolve on complete decomposition. Thus '20 volume' hydrogen peroxide gives off 20 times its own volume of oxygen, e.g. 10 cm³ of a '20 volume' solution gives off 200 cm^3 of oxygen, measured at standard temperature and pressure.

Oxygen is liberated in a vigorous reaction when 20 volume hydrogen peroxide is dropped on to finely divided manganese(IV) oxide in a conical flask. The gas can be collected in jars by displacement of water, as shown in Fig. 6.6.

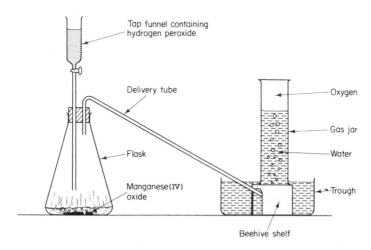

Fig. 6.6   Laboratory preparation of oxygen from the decomposition of hydrogen peroxide, using manganese(IV) oxide as a catalyst

(b) **From the oxides of metals low in the electrochemical series.** These oxides yield oxygen on heating. For example, dilead(II) lead(IV) oxide ('red lead'):

$$2Pb_3O_{4(s)} \rightarrow 6Pb^{2+}O^{2-}_{(s)} + O_{2(g)}$$

Another example is mercury(II) oxide:

$$2Hg^{2+}O^{2-}_{(s)} \rightarrow 2Hg_{(l)} + O_{2(g)}$$

(c) **From potassium chlorate(V).** Oxygen is evolved on heating a mixture of three parts potassium chlorate(V) with one part manganese(IV) oxide (here again the manganese(IV) oxide acts as a catalyst):

$$2K^+ClO^-_{3(s)} \rightarrow 2K^+Cl^-_{(s)} + 3O_{2(g)}$$

(d) **From the nitrates of certain metals high in the electrochemical series.** Oxygen is evolved when potassium nitrate is decomposed by strong heating:

$$2K^+NO^-_{3(s)} \rightarrow 2K^+NO^-_{2(s)} + O_{2(g)}$$
$$\text{potassium} \quad\quad \text{potassium}$$
$$\text{nitrate} \quad\quad\quad \text{nitrite}$$

A similar reaction occurs with sodium nitrate.

6.6 Properties of Oxygen

(a) **Physical**

Oxygen is a colourless, odourless, tasteless gas. Its relative molecular mass is 32. The solubility of oxygen in water is low but nevertheless important to underwater life.

(b) **Chemical**

With the exception of the noble gases and certain metals such as gold and platinum, all elements combine with oxygen to produce **oxides**. (Thus an oxide is a compound which contains oxygen combined with one other element. Unit 5 described oxidation as a reaction in which electrons are lost. Here *combination of an element with oxygen* provides a further example of oxidation.)

The reaction between oxygen and many of the elements, e.g. phosphorus, is vigorous and accompanied by the evolution of light and heat. Any such reaction is an example of ordinary **combustion** (or burning), i.e. the combination of oxygen with some *combustible* material. Oxygen is said to 'support the combustion' of the combustible material. In general, combustion (or burning) is *any* chemical reaction which occurs so rapidly that heat and light are produced: it need not necessarily involve oxygen. In the case of hydrogen burning in an atmosphere of chlorine we could speak of hydrogen as being combustible and the chlorine as *supporting combustion*.

Antoine Laurent Lavoisier (1743–94) is famous for having established the modern theory of combustion. He proved that air contained two components, one of which supports combustion and one of which does not. The *active* part of the air he named *oxygen*.

Reactions

(i) **Hydrogen.** A stream of hydrogen burns readily in oxygen with a clear blue flame to produce water vapour:

$$2H_{2(g)} + O_{2(g)} \rightarrow 2H_2O_{(g)}$$

Droplets of water are formed if the flame is directed on to a cold surface. *Caution*: hydrogen–oxygen and hydrogen–air mixtures are explosive (see Experiment 3.1).

(ii) **Sodium.** (This metal tarnishes rapidly in air and, since it also reacts violently with water, it is stored under liquid paraffin.) A small freshly cut piece of sodium is heated in a deflagrating spoon until it ignites and then lowered into a gas jar of oxygen (see Fig. 6.7). A bright yellow flame, characteristic of sodium, is observed and white solid sodium peroxide remains:

$$2Na_{(s)} + O_{2(g)} \rightarrow Na_2^+ O_{2(s)}^{2-}$$

Note that the peroxide ion has a formula O_2^{2-} and is not to be confused with the oxide ion O^{2-}

(iii) **Magnesium.** When ignited, magnesium burns in air with a white light. On

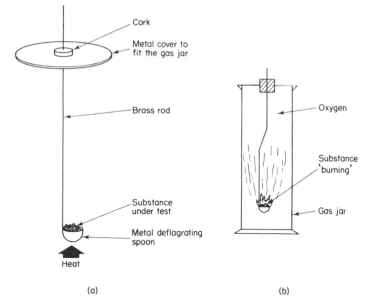

Fig. 6.7   Testing the action of oxygen on solids: (a) the solid is heated in a deflagrating spoon until it ignites, melts or becomes red hot; (b) the spoon and its contents are then plunged into a jar of oxygen

lowering into a gas jar of oxygen the light becomes intensely bright, and white solid magnesium oxide is formed:

$$2Mg_{(s)} + O_{2(g)} \rightarrow 2Mg^{2+}O^{2-}_{(s)}$$

(iv) **Iron**. Some iron wire is heated on a deflagrating spoon until it is red hot and then lowered into a gas jar containing oxygen. A shower of sparks is produced as iron burns to leave black iron(II) di-iron(III) oxide:

$$3Fe_{(s)} + 2O_{2(g)} \rightarrow Fe_3O_{4(s)}$$

Iron will react slowly with oxygen and water vapour in the air, producing *rust*. This and other forms of corrosion are discussed in Unit 9.

(v) **Carbon**. Red-hot charcoal glows brightly when lowered into a gas jar containing oxygen, and burns away to carbon dioxide:

$$C_{(s)} + O_{2(g)} \rightarrow CO_{2(g)}$$

(vi) **Phosphorus**. When lowered into oxygen, burning phosphorus produces a bright white flame and dense white clouds of phosphorus(v) oxide (phosphorus pentoxide):

$$4P_{(s)} + 5O_{2(g)} \rightarrow P_4O_{10(s)}$$

Not all elements undergo combustion with oxygen but many can be made to react under special conditions. Nitrogen, for example, combines with oxygen when a mixture of the gases is subjected to an electric discharge; the product is nitrogen monoxide (NO).

6.7 Nitrogen

Elementary nitrogen is very resistant to chemical attack, owing to a strong covalent bond between the two nitrogen atoms in the N_2 molecule. Apart from nitrate deposits there are few nitrogen compounds in the earth's crust. However, it is an essential element in **proteins** (vital constituents of animal and plant cells), and one of the major problems facing the world is that the natural conversion of atmospheric nitrogen to protein and other compounds may some day prove inadequate to meet the demand.

The Nitrogen Cycle

People and animals cannot use atmospheric nitrogen to manufacture their body-building protein because there is no body mechanism for this conver-

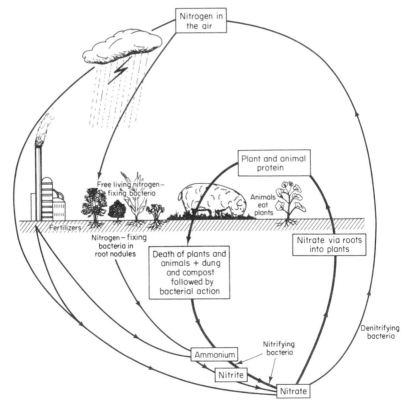

Fig. 6.8   The nitrogen cycle

sion. They must obtain their usable or *fixed* nitrogen by eating plants and animals. Plants, too, need nitrogen for growth. But plants in general cannot utilize atmospheric nitrogen; they absorb soluble nitrogen compounds such as nitrates through their roots.

The circulation of nitrogen in nature is called the **nitrogen cycle.**

It will be seen from Fig. 6.8 that the three main processes involved in the nitrogen cycle are: (*a*) conversion of atmospheric nitrogen to nitrates, (*b*) conversion of nitrogen compounds in the soil to nitrates, and (*c*) removal of nitrates from the soil and return of nitrogen to the atmosphere.

(*a*) Nitrogen in the atmosphere is converted into nitrates in three ways:

(i) *lightning flashes* (electrical discharge) cause nitrogen to combine with oxygen, forming oxides of nitrogen. These oxides dissolve in rain and are washed into the soil as nitric acid. It is estimated that 250 000 tons of nitric acid are produced every 24 hours in this way.

(ii) *free-living nitrogen-fixing bacteria* convert atmospheric nitrogen into nitrates.

(iii) *nitrogen-fixing bacteria* occur in the root nodules of certain leguminous plants such as clover. When crops of such plants are grown they replace the nitrogen in the soil.

(*b*) The death and decay of plants and animals, animal urine and faeces all return nitrogen to the soil. Eventually through bacterial action the protein nitrogen compounds in animal excreta are converted into ammonium salts. These salts are in turn converted first to nitrites and then to nitrates by *nitrifying bacteria.*

(*c*) Nitrogen is lost from the soil in two ways:

(i) plants absorb nitrates through their roots for use in building plant protein,

(ii) *denitrifying bacteria* convert ammonium salts, nitrites and nitrates into nitrogen gas which is released to the atmosphere.

Human intervention in the nitrogen cycle. The harvesting of crops for human use removes nitrogen from the natural cycle. When sewage is discharged into the sea, naturally produced *urea*—one of the best soluble nitrogenous fertilizers—is lost to the land.

It is essential that the nitrogen lost to the soil is replenished in some way, otherwise subsequent crops will give lower yields. This is done by adding both natural and artificial **fertilizers** to the land. These include farmyard manure, ammonium sulphate, ammonium nitrate, nitro-chalk (ammonium nitrate mixed with chalk), urea and ammonium phosphates.

Most of the artificial fertilizers are produced indirectly from atmospheric nitrogen via ammonia manufactured by the Haber process (see Section 14.8). Vast quantities of nitrogen are thus removed from the atmosphere, and for this reason the time is quickly approaching when we must look closely at the adequacy of atmospheric nitrogen as a source of protein nitrogen.

6.8 Preparation of Nitrogen

As with oxygen, the most convenient way of obtaining nitrogen in the laboratory is from a cylinder of the gas which has been produced commercially by fractionation of liquid air.

Nitrogen can be prepared by heating ammonium nitrite solution:

$$NH_4^+NO_{2(aq)}^- \rightarrow N_{2(g)} + 2H_2O_{(l)}$$

Ammonium nitrite is made *in situ* by mixing concentrated solutions of ammonium chloride and sodium nitrite which provide ammonium ions and nitrite ions.

6.9 Properties of Nitrogen

(*a*) **Physical**

Nitrogen is a colourless, odourless, tasteless gas. Its relative molecular mass is 28. The gas is only slightly soluble in water (less soluble than oxygen).

(*b*) **Chemical**

The *inert* (or unreactive) nature of nitrogen is due to the strong covalent bond between the two nitrogen atoms in the molecule N_2. In the air it neither burns nor supports combustion and acts mainly as a diluent for the oxygen.

Reactions

(i) **Magnesium**. Burning magnesium will continue to burn in nitrogen, producing magnesium nitride:

$$3Mg_{(s)} + N_{2(g)} \rightarrow Mg_3^{2+}N_{2(s)}^{3-}$$

Magnesium nitride is a whitish solid, and when treated with water or a solution of sodium hydroxide, the characteristic pungent odour of ammonia gas can be detected:

$$Mg_3^{2+}N_{2(s)}^{3-} + 6H_2O_{(l)} \rightarrow 3Mg^{2+}(OH^-)_{2(s)} + 2NH_{3(g)}$$

(ii) **Hydrogen**. Under high pressure and in the presence of an iron catalyst, nitrogen and hydrogen will combine together to produce ammonia:

$$N_{2(g)} + 3H_{2(g)} \rightarrow 2NH_{3(g)}$$

This is the basis of the Haber synthesis of ammonia which is discussed in detail in Units 10 and 14.

(iii) **Oxygen**. When nitrogen and oxygen in the air are passed through an electric arc, small quantities of nitrogen monoxide are formed:

$$N_{2(g)} + O_{2(g)} \rightarrow 2NO_{(g)}$$

This is the basis of the 'electric arc' or 'Birkeland and Eyde' process for the manufacture of nitric acid.

6.10 Carbon Dioxide and the Carbon Cycle

Although the quantity of carbon dioxide in the atmosphere is relatively small
(0·03% by volume), it is nevertheless essential either directly or indirectly to the
existence of living things. Green plants remove carbon dioxide from the air in the
process of *photosynthesis*; animal and plant *respiration* and *decay* processes
return carbon dioxide to the air. Together these processes make up nature's
carbon cycle (see Fig. 6.9), and it is because of the delicate balance between them
that the percentage of carbon dioxide in the atmosphere remains remarkably
constant.

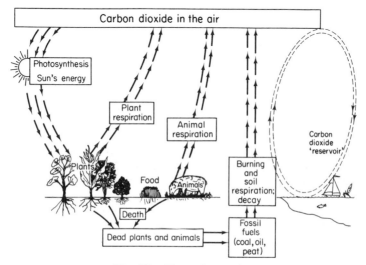

Fig. 6.9   The carbon cycle

The carbon cycle is in two main parts.

(*a*) **Removal of carbon dioxide from the air**. Green plants absorb carbon dioxide
through the underside of their leaves. Carbon dioxide is able to combine with
water in the leaf, producing *carbohydrates* (compounds which contain carbon and
the elements of water, i.e. hydrogen and oxygen in the ratio 2:1) such as
glucose. This process will only occur in the presence of solar energy (sunlight)
and certain biological catalysts. Together these complex reactions which build
up carbohydrates in the leaf are termed **photosynthesis**. They can be summarized
by the equation:

$$6CO_{2(g)} + 6H_2O_{(l)} \rightarrow C_6H_{12}O_{6(aq)} + 6O_{2(g)}$$

The glucose (a sugar) undergoes further reactions to produce more complex plant-
building and food-storage compounds such as cellulose and starch. It will be seen
from the above equation that, *as carbon dioxide is removed from the air, it is
replaced by oxygen.*

(*b*) **Addition of carbon dioxide to the air.** Plants and (to an even greater extent) animals consume carbohydrates and other foodstuffs. The energy-producing chemical reactions between the foodstuff and oxygen from the air result in the formation of carbon dioxide and water, and are together called **respiration**.

When plants and animals die, their bodies decay. Decay means that, by the action of bacteria, complex carbon compounds are broken down into (among other things) carbon dioxide. In this process, as in respiration, *as oxygen is removed from the air, it is replaced by carbon dioxide.*

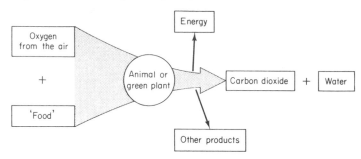

Fig. 6.10   Respiration in plants and animals produces carbon dioxide

Human intervention in the carbon cycle. The combustion of all *carbonaceous fuels* (e.g. oil, coal, wood and petrol) increases the concentration of carbon dioxide in the air. Recent vast increases in such combustion have produced a measurable build-up of carbon dioxide in the atmosphere. This retains heat radiated from the ground and may be causing a slight rise in the average temperature of the earth. If this apparent trend continues there is danger that eventually the polar ice caps will begin to melt, causing the level of the oceans to rise.

6.11   Preparation of Carbon Dioxide

(*a*) **By the action of a dilute acid on a carbonate or a hydrogencarbonate.** When dilute hydrochloric acid is added to any carbonate, a vigorous effervescence occurs as bubbles of carbon dioxide gas are liberated. For example, using calcium carbonate:

$$Ca^{2+}CO_{3(s)}^{2-} + 2H^+Cl_{(aq)}^- \rightarrow Ca^{2+}Cl_{2(aq)}^- + CO_{2(g)} + H_2O_{(l)}$$

The gas can be collected over water if required (see Fig. 6.6 for method of gas collection).

A similar reaction occurs with a hydrogencarbonate, such as sodium hydrogencarbonate:

$$Na^+HCO_{3(s)}^- + H^+Cl_{(aq)}^- \rightarrow Na^+Cl_{(aq)}^- + CO_{2(g)} + H_2O_{(l)}$$

(*b*) **By the action of heat on hydrogencarbonates and certain carbonates.** With the exception of the carbonates of the alkali metals, i.e. sodium, potassium etc., all carbonates and hydrogencarbonates decompose on heating to give carbon dioxide. For example:

$$Ca^{2+}CO_{3(s)}^{2-} \xrightarrow[\text{heating}]{\text{strong}} Ca^{2+}O_{(s)}^{2-} + CO_{2(g)}$$

$$2Na^+HCO_{3(s)}^- \xrightarrow{\text{heat}} Na_2^+CO_{3(s)}^{2-} + CO_{2(g)} + H_2O_{(g)}$$

The action of heat on calcium carbonate in the form of limestone is used commercially to manufacture quicklime (calcium oxide).

(c) **By burning carbonaceous materials.** All natural fuels contain carbon as one of the essential elements; for example, *natural gas* (methane) has a molecular formula CH_4. Combustion of such materials in a plentiful supply of oxygen produces carbon dioxide:

$$CH_{4(g)} + 2O_{2(g)} \rightarrow CO_{2(g)} + 2H_2O_{(g)}$$

(d) **By fermentation processes.** Beer, wines and other alcoholic drinks are made by *fermentation*. This is a complex process in which, by enzyme action, a sugar is converted to ethanol (an alcohol) and carbon dioxide:

$$C_6H_{12}O_{6(aq)} \rightarrow 2C_2H_5OH_{(aq)} + 2CO_{2(g)}$$
$$\text{(a sugar)} \qquad\qquad \text{ethanol}$$

The enzyme *zymase* (a biological catalyst produced by yeast) plays an essential part in this reaction.

6.12 Properties of Carbon Dioxide

(a) **Physical**

Carbon dioxide is a colourless gas with a very faint pungent odour and taste. Its relative molecular mass is 44. It is fairly easy to solidify, and the solid (called 'dry ice') sublimes.

(b) **Chemical**

(i) Carbon dioxide neither burns nor is a supporter of combustion. For this reason, and because it is more dense than air, it is widely used in fire extinguishers. Its ability to extinguish flame can be demonstrated by 'pouring' the gas on to a lighted candle in a large beaker, as shown in Fig. 6.11.

However, burning magnesium will decompose carbon dioxide, and when lowered into a jar of the gas, will continue to burn in the oxygen set free:

$$2Mg_{(s)} + CO_{2(g)} \rightarrow 2Mg^{2+}O_{(s)}^{2-} + C_{(s)}$$

White magnesium oxide is produced and black specks of carbon are seen (confirming that carbon dioxide does indeed contain carbon).

(ii) Carbon dioxide dissolves a little in water (0·335 g in 100 g water at a total

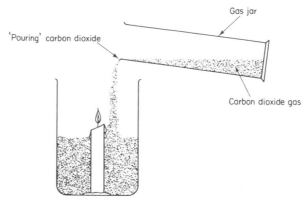

Fig. 6.11   Carbon dioxide is heavier than air and does not support combustion; the candle is therefore extinguished when the gas is poured over it

pressure of 1 atm). The dissolved gas reacts slightly with the water, producing a dilute solution of *carbonic acid*:

$$CO_{2(g)} + H_2O_{(l)} \rightarrow H_2CO_{3(aq)}$$

Carbonic acid is a *weak* acid (the significance of weak and strong acids is discussed in Section 8.8).

If the gas is dissolved under increased pressure, the solubility increases markedly. Effervescent 'fizzy' drinks rely on this property. When the pressure is released by removing the bottle top, the gas can be seen bubbling out of solution.

(iii) Being an acid gas, carbon dioxide will react with and be absorbed by a solution of an alkali. For example, with sodium hydroxide:

$$2Na^+OH^-_{(aq)} + CO_{2(g)} \rightarrow Na_2^+CO_{3(aq)}^{2-} + H_2O_{(l)}$$

This reaction is used to absorb carbon dioxide in the quantitative analysis of a gas mixture.

Lime-water is a colourless solution of calcium hydroxide, a slightly soluble alkali. When carbon dioxide gas is bubbled through this clear solution, a *milky white precipitate* of calcium carbonate is observed:

$$Ca^{2+}(OH^-)_{2(aq)} + CO_{2(g)} \rightarrow Ca^{2+}CO_{3(s)}^{2-} + H_2O_{(l)}$$

This reaction is used to detect carbon dioxide.

If carbon dioxide is passed through lime-water for a long time the original white precipitate dissolves, leaving a colourless solution containing calcium hydrogencarbonate:

$$Ca^{2+}CO_{3(s)}^{2-} + H_2O_{(l)} + CO_{2(g)} \rightarrow Ca^{2+}(HCO_3^-)_{2(aq)}$$

Boiling the clear solution decomposes the calcium hydrogencarbonate, and the white milky precipitate of calcium carbonate returns:

$$Ca^{2+}(HCO_3^-)_{2(aq)} \rightarrow Ca^{2+}CO_{3(s)}^{2-} + H_2O_{(l)} + CO_{2(g)}$$

With this test for carbon dioxide it is easy to demonstrate that exhaled breath contains carbon dioxide. Simply breathing out through a tube (or straw) into a test tube containing lime-water causes the liquid to turn milky.

6.13 Uses of Carbon Dioxide

Carbon dioxide is important in the manufacture of sodium carbonate and sodium hydrogencarbonate by the Solvay (ammonia–soda) process (see Section 14.5). Large quantities of carbon dioxide are used in carbonated beverages, as 'dry ice' in refrigeration, and in fire extinguishers.

(*a*) **Carbon dioxide in the soft-drinks industry.** About 90% of the carbon dioxide manufactured is used in the preparation of carbonated 'fizzy' beverages (see Section 6.12).

(*b*) **Dry ice used as a refrigerant.** Carbon dioxide can be liquefied under high pressure at 20 °C, but when exposed to atmospheric pressure this liquid freezes to the white solid '*dry ice*'. This has a temperature of -78.5 °C. It is used as a refrigerant in shipping fruits, vegetables, dairy products, fish and meat. As a refrigerant it has the advantage over ordinary ice that it (i) sublimes leaving no residue, (ii) maintains a lower temperature, (iii) lasts longer.

(*c*) **Fire extinguishers.** The fact that carbon dioxide is chemically inactive, cheap and easily generated, makes it a useful fire extinguisher. Being considerably denser than air, it settles over the burning material as an occluding (retaining) 'blanket' which prevents more oxygen from reaching the fire. At the same time it dilutes the oxygen in the air to such an extent that combustion is inhibited.

(i) The most convenient type of fire extinguisher consists of a cylinder containing liquid carbon dioxide. When the valve on the cylinder is opened the pressure is released; the liquid turns to a vapour and a snow-white solid. The heavy vapour effectively blankets the fire and lowers the temperature around the burning material. This type of extinguisher is particularly useful in fighting oil fires (where water might spread the fire) and those around electrical apparatus (where water might cause a 'short circuit' and electric shock).

(ii) In the **soda–acid** type of fire extinguisher, shown in Fig. 6.12, separate solutions of sulphuric acid and sodium hydrogencarbonate are mixed when the extinguisher is inverted:

$$2Na^+HCO_{3(aq)}^- + H_2^+SO_{4(aq)}^{2-} \rightarrow Na_2^+SO_{4(aq)}^{2-} + 2H_2O_{(l)} + 2CO_{2(g)}$$

Pressure of carbon dioxide gas produced when the hydrogencarbonate reacts with the acid forces a jet of liquid out of the nozzle which can be directed towards the fire. In this case it is principally the water which puts out the fire and not the carbon dioxide.

(iii) **Foam** fire extinguishers are similar in construction to the soda–acid type. In this case the two reacting solutions are sodium hydrogencarbonate containing a sticky foam-stabilizing substance such as liquorice, and aluminium

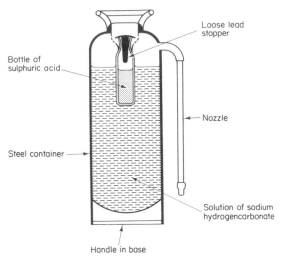

Loose lead
stopper

Bottle of
sulphuric acid

Nozzle

Steel container

Solution of sodium
hydrogencarbonate

Handle in base

Fig. 6.12   The soda–acid type of fire extinguisher

sulphate. The aluminium sulphate acts as an acid and, on inversion of the extinguisher, reacts with the hydrogencarbonate to produce carbon dioxide gas. Gelatinous aluminium hydroxide produced in the reaction helps the liquorice to stabilize the carbon dioxide foam. This type of extinguisher is particularly useful in fighting oil fires.

Baking Soda and Baking Powder
Carbon dioxide makes cakes 'rise' during baking. It is produced within the cake when **baking soda** (sodium hydrogencarbonate) decomposes during the baking process:

$$2Na^+HCO_{3(aq)}^- \xrightarrow{\text{heat}} Na_2^+CO_{3(aq)}^{2-} + H_2O_{(g)} + CO_{2(g)}$$

Baking powder differs from baking soda in that it is a mixture of sodium hydrogencarbonate and a harmless solid acid such as 2,3-dihydroxybutanedioic acid (tartaric acid). These substances do not react with each other while they are dry, but in the moist cake mixture the acid and hydrogencarbonate dissolve and react. Carbon dioxide is produced in tiny bubbles which expand during the cooking process, so making the cake 'rise'.

6.14 Water

Water vapour is always present in the air even over deserts. The amount varies, being greatest at and near the equator. The very small amount of water vapour in the atmosphere at any one time is of vital importance to the world's weather. This water vapour enters the atmosphere by *evaporation* and by *transpiration* (the loss of water vapour from plants and animals). By far the most important of these is evaporation from the oceans. Loss of water vapour from the atmosphere occurs through rain, snow and other forms of *precipitation*. The circulation of water vapour in nature is called the **water cycle** (see Fig. 6.13).

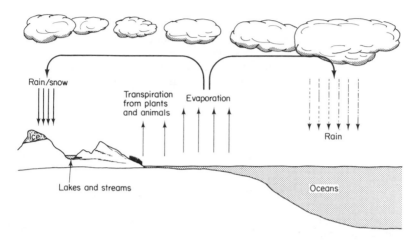

Rain/snow

Transpiration from plants and animals

Evaporation

Ice

Rain

Lakes and streams

Oceans

Fig. 6.13   The water cycle

6.15 Properties of Water

(a) **Physical**

In almost all its physical properties water is unique. It is a liquid; yet compounds similar to it and of higher relative molecular mass are gases, e.g. hydrogen sulphide. This and many other of water's special properties are attributed to the strong *hydrogen bonding* between hydrogen and oxygen atoms of adjacent water molecules (see Section 2.10).

Water has its maximum density at about 4 °C. Thus ice is less dense than water and floats, whereas almost all other substances are denser in the solid state than in the liquid. Also cold water (less than 4 °C) floats on top of warmer water. Freezing therefore takes place from the surface downwards, enabling underwater life to be maintained.

At one atmosphere pressure the boiling point and freezing point of pure water are 100·00 °C and 0·00 °C, respectively. These are used as two *fixed points* in the calibration of thermometers.

(b) **Chemical**

(i) Because of its *polar* nature, water is an excellent solvent for the vast majority of ionic compounds. This particular property is discussed in Unit 5.

(ii) The reaction of water with metals is discussed in Unit 9, and with oxides in Unit 8.

6.16 'Hardness' of Water

(a) Causes of Hardness

When soap forms an insoluble scum and does not easily lather with water, such water is said to be 'hard'. This **hardness** is caused by $Mg^{2+}_{(aq)}$ and $Ca^{2+}_{(aq)}$ ions dissolved in water. Note that only *soluble* compounds of magnesium and calcium can provide these ions: the usual sources are the sulphates, chlorides and hydrogencarbonates.

Whereas the sulphates and chlorides occur in the earth's crust, the hydrogencarbonate is produced when rain-water containing dissolved carbon dioxide passes through calcium carbonate rocks, e.g. chalk and limestone. The dilute solution of carbonic acid (carbon dioxide dissolved in water) reacts with the calcium carbonate to form *soluble* calcium hydrogencarbonate:

$$Ca^{2+}CO^{2-}_{3(s)} + (H_2O_{(l)} + CO_{2(g)}) \rightarrow Ca^{2+}(HCO_3^-)_{2(aq)}$$
$$\text{carbonic acid}$$

Calcium carbonate (chalk etc.) does not *directly* cause hardness since it will not physically dissolve in water.

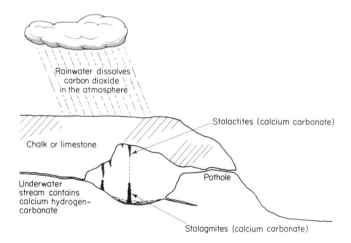

Fig. 6.14   Cave formation in limestone

Countryside in chalk and limestone regions assumes a characteristic appearance, with potholes, springs, and caves containing *stalagmites* and *stalactites*. Both stalagmites and stalactites are formed by slow evaporation and decomposition of calcium hydrogencarbonate solution, producing deposits of calcium carbonate over many years:

$$Ca^{2+}(HCO_3^-)_{2(aq)} \rightarrow Ca^{2+}CO^{2-}_{3(s)} + H_2O_{(l)} + CO_{2(g)}$$

(b) The Nature of Soap

Soap is a name given to a variety of compounds produced when oils and fats are reacted with sodium hydroxide. They are all similar in that they contain a long hydrocarbon chain ending in a carboxylate anion to which is attracted a sodium cation. The hydrocarbon chain is soluble in oils but insoluble in water; the ionic end is soluble in water but insoluble in oils. A typical soap is sodium octadecanoate (sodium stearate) $C_{17}H_{35}COO^-Na^+$:

$$
\underset{\text{hydrocarbon chain}}{H-\overset{\displaystyle H}{\underset{\displaystyle H}{C}}-\overset{\displaystyle H}{\underset{\displaystyle H}{C}}-\overset{\displaystyle H}{\underset{\displaystyle H}{C}}-\overset{\displaystyle H}{\underset{\displaystyle H}{C}}-\overset{\displaystyle H}{\underset{\displaystyle H}{C}}-\overset{\displaystyle H}{\underset{\displaystyle H}{C}}-\overset{\displaystyle H}{\underset{\displaystyle H}{C}}-\overset{\displaystyle H}{\underset{\displaystyle H}{C}}-\overset{\displaystyle H}{\underset{\displaystyle H}{C}}-\overset{\displaystyle H}{\underset{\displaystyle H}{C}}-\overset{\displaystyle H}{\underset{\displaystyle H}{C}}-\overset{\displaystyle H}{\underset{\displaystyle H}{C}}-\overset{\displaystyle H}{\underset{\displaystyle H}{C}}-\overset{\displaystyle H}{\underset{\displaystyle H}{C}}-\overset{\displaystyle H}{\underset{\displaystyle H}{C}}-\overset{\displaystyle H}{\underset{\displaystyle H}{C}}-\overset{\displaystyle H}{\underset{\displaystyle H}{C}}-}\underset{\text{carboxylate anion}}{\overset{\displaystyle O}{\underset{\displaystyle O^-}{C}}\;Na^+}
$$

hydrocarbon chain   carboxylate anion

In water, soap has two functions: (i) it makes the water able to wet material more effectively by lowering the surface tension; (ii) it emulsifies oil and grease. The hydrocarbon chain (the 'tail') dissolves in the grease while the carboxylate–sodium end of the soap molecule (the 'head') remains dissolved in the water (see Fig. 6.15). This grease-removing process is often referred to as 'head-and-tail' detergency. Agitation of the liquid aids dissolution of the grease by the soap.

(c) Effect of Hard Water on Soap

The $Ca^{2+}_{(aq)}$ and $Mg^{2+}_{(aq)}$ ions in hard water react with soap (sodium stearate)

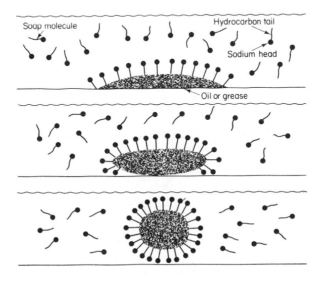

Fig. 6.15   Soap molecules surrounding a grease particle to make it soluble in water

and remove it as an insoluble grey scum (calcium or magnesium stearate). For example:

$$2C_{17}H_{35}COO^-Na^+_{(aq)}+Ca^{2+}_{(aq)} \rightarrow (C_{17}H_{35}COO^-)_2Ca^{2+}_{(s)}+2Na^+_{(aq)}$$
sodium stearate   calcium stearate

Soap is wasted in this way until all the calcium and magnesium ions have been removed. The scum is deposited on fabrics, giving them an unsightly dull appearance. Thus in hard-water districts it is obviously advantageous to remove hardness before washing.

(d) Removal of Hardness
The removal of hardness depends to a certain extent on whether the type of hardness present is 'temporary' or 'permanent'.

(i) **Temporary hardness** is caused by the presence of *calcium* (or magnesium) *hydrogencarbonate* dissolved in the water. It can be removed by boiling the water, when the *insoluble* calcium (or magnesium) carbonate is precipitated:

$$Ca^{2+}(HCO_3^-)_{2(aq)} \rightarrow Ca^{2+}CO^{2-}_{3(s)}+H_2O_{(l)}+CO_{2(g)}$$

This reaction is responsible for the 'furring' of kettles and water pipes.

(ii) **Permanent hardness** is caused by the presence of calcium and magnesium chlorides and sulphates. These compounds do not decompose on heating, and so this type of hardness *cannot* be removed by boiling.

Both temporary and permanent hardness can be removed by the addition of 'washing soda' (sodium carbonate crystals). This reacts with the $Ca^{2+}_{(aq)}$ and $Mg^{2+}_{(aq)}$ ions, precipitating them as the insoluble carbonates:

$$Ca^{2+}_{(aq)}+Na_2^+CO^{2-}_{3(aq)} \rightarrow Ca^{2+}CO^{2-}_{3(s)}+2Na^+_{(aq)}$$

The disadvantage of washing soda as a water softener is that its solution is alkaline and can cause damage to wool and silk.

Hard water can be softened in a continuous process using **ion-exchange resins**. These are compounds which will exchange their own sodium ions for the calcium (or magnesium) ions dissolved in hard water. Thus as sodium ions go into solution, calcium (or magnesium) ions are left on the resin:

$$2Na^+(Resin^-)_{(s)}+Ca^{2+}_{(aq)} \rightarrow Ca^{2+}(Resin^-)_{2(s)}+2Na^+_{(aq)}$$

Tiny beads of the resin are packed in a cylinder and hard water is passed through them, soft water containing sodium ions coming out of the other end. The resin eventually becomes 'spent' when all the sodium ions have been exchanged, but it can be regenerated by passing concentrated sodium chloride solution through it. This increases the concentration of sodium ions to such an extent that they are able to replace the calcium ions on the resin, making it available for use again:

$$Ca^{2+}(Resin^-)_{2(s)}+2Na^+_{(aq)} \rightarrow 2Na^+(Resin^-)_{(s)}+Ca^{2+}_{(aq)}$$

One example of an ion-exchange material is the complex chemical Permutit (sodium aluminium silicate).

Summary of Unit 6

1. **Air** is a *mixture* of colourless, odourless and tasteless gases.
2. Approximately 99% by volume of air is a mixture of *nitrogen* and *oxygen*.
3. Small amounts of the *noble gases* (particularly *argon*) are also present, together with *carbon dioxide* and variable amounts of *water vapour*.
4. Air can be liquefied by compression and cooling. Liquid air is a commercial source of *nitrogen*, *oxygen* and *argon* which are separated by **fractional distillation**.
5. **Oxygen** is the most abundant element in the earth's outer surface.
6. Laboratory methods of preparing oxygen include:
    (a) the action of manganese(IV) oxide on *hydrogen peroxide*
    (b) the action of heat on certain *oxides* and *nitrates*
    (c) the action of heat on *potassium chlorate*(V).
7. Reaction of oxygen with other elements produces **oxides**; this is an oxidation process.
8. **Combustion** is a rapid oxidation process which occurs with the production of heat and light.
9. **Nitrogen** is essential to life. The conversion of stable nitrogen gas into useful nitrogen compounds is called *nitrogen fixation*.
10. The circulation of nitrogen in nature is called the *nitrogen cycle*.
11. Nitrogen is most conveniently obtained by fractional distillation of liquid air, but can be prepared in the laboratory by warming a solution of ammonium nitrite.
12. Nitrogen is an *inert* gas.
13. **Carbon dioxide** occurs in air as a product of *decay*, of *combustion* and of the *respiration* of living things. It is the vital component in nature's *carbon cycle*.
14. Carbon dioxide can be prepared by:
    (a) the action of an acid on a carbonate or hydrogencarbonate
    (b) the action of heat on hydrogencarbonates and certain carbonates
    (c) burning carbonaceous material
    (d) fermentation processes.
15. Carbon dioxide is a *stable gas* and *does not burn*. Its aqueous solution is called *carbonic acid*. It is detected using *lime-water*, a dilute aqueous solution of calcium hydroxide.
16. Carbon dioxide is used in fire extinguishers, as 'dry ice' in refrigeration, in carbonated beverages and as a raising agent in baking processes.
17. The circulation of **water** in nature is called the *water cycle*.
18. Water molecules are *polar* and are joined by *hydrogen bonds*. Hydrogen bonding is responsible for many of the peculiar properties of water.
19. *Calcium* and *magnesium ions* cause hardness in water.
20. The process by which hardness is removed from water is called *water softening*.

Test Yourself on Unit 6

1. Which of the following gases are present in a sample of air?
   (a) hydrogen;
   (b) oxygen;
   (c) carbon dioxide;
   (d) nitrogen;
   (e) argon;
   (f) methane.

2. $200 \, cm^3$ of dry air was shaken with an aqueous alkaline solution of pyrogallol (benzene-1,2,3-triol).
   (a) Name two gases present in the dry air which were removed by the alkaline pyrogallol.
   (b) Name the major constituent of the gas remaining.
   (c) Name one other gas present in the gas remaining.
   (d) What is the volume of gas left after treatment with alkaline pyrogallol?

3. The following equations (not balanced) refer to reactions in which oxygen is produced:
   (i) $H_2O_{2(aq)} \rightarrow H_2O_{(l)} + O_{2(g)}$
   (ii) $Pb_3O_{4(s)} \rightarrow Pb^{2+}O^{2-}_{(s)} + O_{2(g)}$
   (iii) $K^+ClO_{3(s)}^- \rightarrow K^+Cl_{(s)}^- + O_{2(g)}$
   (iv) $K^+NO_{3(s)}^- \rightarrow K^+NO_{2(s)}^- + O_{2(g)}$
   (a) Balance these four equations.
   (b) Name a catalyst which can be used for reactions (i) and (iii).
   (c) Which of these reactions proceeds at room temperature?

4. Separate samples of sulphur and magnesium were burned in oxygen using the following apparatus:

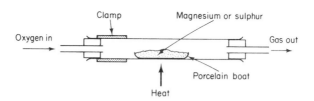

   (a) Name the products formed when these elements burn in oxygen.
   (b) What would you see during these two experiments?

5. (a) Name two ways in which atmospheric nitrogen is 'fixed'.
   (b) Name two artificial fertilizers containing nitrogen.
   (c) Which important nitrogen-containing compound is produced by the Haber process?
   (d) Name the complex process by means of which plants build up carbohydrates from carbon dioxide.

6. Classify as true or false the following statements concerning carbon dioxide:
   (a) It is prepared by the action of an acid on any carbonate or hydrogen-carbonate.
   (b) It forms a white insoluble salt with calcium hydroxide solution.
   (c) It is readily absorbed by both sodium hydroxide solution and concentrated sulphuric acid.
   (d) Solid carbon dioxide ('dry ice') sublimes.

7. 50 cm^3 of tap water from a limestone area required 20 cm^3 of soap solution to form a permanent lather. 50 cm^3 of the same tap water after boiling and cooling required 5 cm^3 of soap solution to form a lather.
   50 cm^3 of tap water after passing through an ion-exchange resin required 0·1 cm^3 of soap solution to form a lather.
   (a) Does the water contain temporary hardness, permanent hardness, or both?
   (b) Why is the volume of soap required to form a lather, less after boiling? Name one substance present in the original tap water which was responsible for this.
   (c) Why does the tap water still require 5 cm^3 of soap solution to produce a lather even after boiling?
   (d) Write an equation showing how rain-water containing dissolved carbon dioxide dissolves limestone.
   (e) Why is only 0·1 cm^3 of soap solution required to produce a permanent lather after passing the tap water through an ion-exchange resin?
   (f) Name two ions responsible for hardness in water.
   (g) If the formula of soap could be represented as Na^+St^-, write an equation to show how soap reacts with the ions present in hard water.

Mark this test out of 40 with the answers provided on page 374.

Unit Seven

Matter and Energy

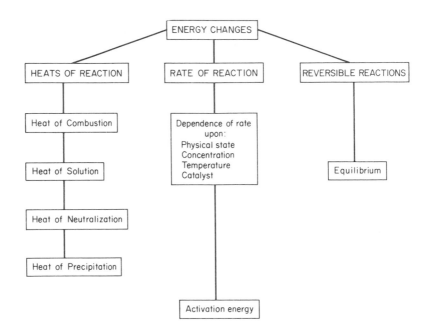

An industrial chemist manufacturing chemicals on a large scale, such as ammonia from nitrogen and hydrogen, has to study factors which influence their formation. Inevitably he or she is concerned with the economics of the process, and in this respect looks closely at energy factors and how they influence speeds of reactions and quantity of product.

Unit 7 introduces us to these physical factors, namely, *energy changes*, *rates of reaction* and *equilibrium*.

7.1 Energy Changes in Chemical Reactions

Fuels are substances which, on burning, produce **energy**. When natural gas (or methane), one of our most familiar forms of household fuel, burns with the oxygen in the air it produces carbon dioxide, water vapour and a large quantity of heat. This heat energy is transferred from the reaction 'system' (i.e. the reactants and products) to the surroundings (i.e. the atmosphere) and the temperature of the surroundings therefore rises. Any reaction in which heat is *given out* by a system to the surroundings is called an **exothermic** reaction. The burning of any fuel is an example of an exothermic reaction.

In contrast, those reactions in which heat is *taken in* from the surroundings by the system are termed **endothermic**. The temperature of the surroundings is lowered in an endothermic reaction.

In a chemical reaction such as the burning of methane, bonds are broken and new bonds formed as the reactants are converted into products. Energy is needed to break any bond and energy is liberated when new bonds are formed. Each different bond is associated with a certain amount of energy. Thus the breaking of some bonds *requires* more energy than others, while the formation of certain bonds *liberates* more energy than others.

A reaction will be exothermic if the energy liberated in bond formation is *greater* than that required for bond breaking, or endothermic if *less*. The combustion of methane

$$CH_{4(g)} + 2O_{2(g)} \rightarrow CO_{2(g)} + 2H_2O_{(g)}$$

is thus exothermic because the energy liberated in the formation of two carbon–oxygen bonds (in carbon dioxide) and four hydrogen–oxygen bonds (in the two water molecules) is greater than that required for the breaking of four carbon–hydrogen bonds (in methane) and two oxygen–oxygen bonds (in the two oxygen molecules).

7.2 Heat of Reaction

The amount of heat evolved or absorbed during a chemical reaction (more specifically, the amount shown in the balanced equation for the reaction) is known as the **heat of reaction**. It is denoted by the symbol ΔH (where Δ, the Greek letter delta, means 'difference') and is measured in joules (J) and kilojoules (kJ).

If a reaction is exothermic, heat is evolved to the surroundings and ΔH is negative. Conversely, in an endothermic reaction ΔH is positive.

To illustrate this convention, let us consider the combustion of methane, which we know from Section 7.1 to be an exothermic reaction. One mole of methane combines with two moles of oxygen to produce one mole of carbon dioxide and two moles of water, and 890·3 kJ of heat is evolved. Thus ΔH is *minus* 890·3 kJ mol^{-1}, and we can write the equation for the reaction as follows:

$$CH_{4(g)} + 2O_{2(g)} \rightarrow CO_{2(g)} + 2H_2O_{(l)}$$

$$\Delta H = -890 \cdot 3 kJ \, mol^{-1}$$

A simple energy diagram for the combustion of methane is shown in Fig. 7.1.

The decomposition of ammonia gas (NH_3) to nitrogen and hydrogen is an example of an *endothermic* reaction. The heat of reaction is therefore positive, and the equation for this reaction can be written as follows:

$$2NH_{3(g)} \rightarrow N_{2(g)} + 3H_{2(g)}$$

$$\Delta H = +92 \cdot 8 \, kJ \, mol^{-1}$$

A simple energy diagram for the decomposition of ammonia is shown in Fig. 7.2.

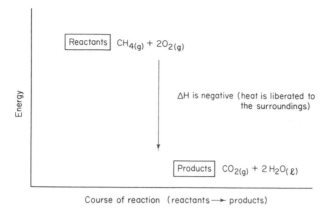

Course of reaction (reactants → products)

Fig. 7.1   Energy diagram to illustrate an exothermic reaction: energy of the products is less than the energy of the reactants, so ΔH is negative

Every chemical change represented by an equation has a heat of reaction. It is often convenient to categorize this heat of reaction according to the *type* of chemical change taking place, e.g. combustion, solution, neutralization or precipitation.

7.3   Heat of Combustion

The heat evolved in the complete combustion of one mole of a particular substance is often referred to as the **heat of combustion**, and is denoted by the symbol $\Delta H_{combustion}$. The heat of combustion of a flammable liquid can be determined by the following method.

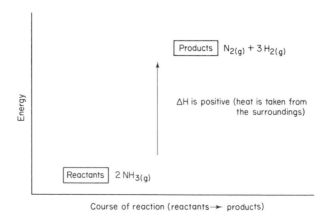

Course of reaction (reactants → products)

Fig. 7.2   Energy diagram to illustrate an endothermic reaction

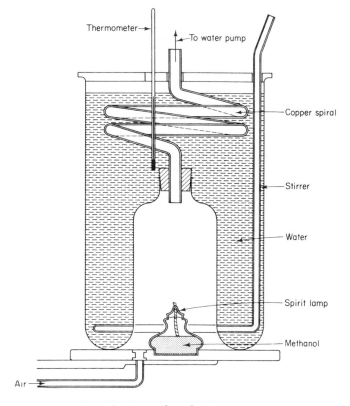

Fig. 7.3   Heat of combustion apparatus

Experiment 7.1   *Determination of the heat of combustion of methanol* (CH_3OH)
The combustion of methanol

$$CH_3OH_{(l)} + 1\tfrac{1}{2}O_{2(g)} \rightarrow CO_{2(g)} + 2H_2O_{(g)}$$

is an exothermic reaction, and ΔH is negative. The experiment consists of burning
a known mass of methanol, using the apparatus shown in Fig. 7.3, and measuring
the consequent rise in temperature in order to determine the quantity of heat
evolved.

First the *heat capacity* of the apparatus is determined, i.e. the heat energy
required to raise the temperature of the apparatus by 1 °C. It can be calculated
by adding the individual heat capacities of the various components of the
apparatus. Thus:

Heat capacity of the glass calorimeter = mass of vessel × specific heat capacity
of glass

$$= 313 \cdot 6 \text{ g} \times 3 \cdot 26 \text{ J g}^{-1}\text{K}^{-1}$$
$$= 1022 \text{ J K}^{-1}$$

Heat capacity of copper spiral = mass of spiral × specific heat capacity of copper
$$= (203 \cdot 0 \times 0 \cdot 38)$$
$$= 77 \text{ J K}^{-1}$$

Heat capacity of brass stirrer = mass of stirrer × specific heat capacity of brass
$$= (28 \cdot 2 \times 1 \cdot 55)$$
$$= 44 \text{ J K}^{-1}$$

Heat capacity of apparatus $= (1022 + 77 + 44) \text{ J K}^{-1}$
$$= 1143 \text{ J K}^{-1}$$

Next, the calorimeter is filled with water so that the upper coil of the copper spiral is covered, and the volume of water recorded. A sensitive thermometer is placed in the water so that the bulb is just above the inner calorimeter vessel.

The methanol is placed in a small spirit burner, which is then weighed. Air is drawn through the copper spiral, using a water pump, and the burner lighted. When a temperature rise of approximately 5 °C is obtained, the burner is extinguished and the maximum temperature recorded. The burner is again weighed to determine the mass of methanol used.

Results
Mass of burner and methanol before experiment = 13·642 g
Mass of burner and methanol after experiment = 12·777 g
∴ mass of methanol used   = 0·865 g

Mass of water   = 546 g
Initial temperature of water   = 23·9 °C
Final temperature of water   = 29·4 °C
∴ temperature rise   = 5·5 °C

Calculation
Heat capacity of calorimeter = 1143 J K^{-1}
Heat capacity of water   = mass of water × specific heat capacity of water
$$= 546 \times 4 \cdot 18$$
$$= 2282 \text{ J K}^{-1}$$

Total heat capacity   = heat capacity of calorimeter + heat capacity of water
$$= (1143 + 2282) \text{ J K}^{-1}$$
$$= 3425 \text{ J K}^{-1}$$

Heat change in reaction   = total heat capacity × rise in temperature
$$= (3425 \times 5 \cdot 5)$$
$$= 18\,840 \text{ J}$$

Hence, in this experiment, 0·865 g of methanol evolved 18 840 J on complete combustion. By simple proportion, 1 g of methanol would evolve

$$\frac{18\,840}{0 \cdot 865} \text{ J}$$

One mole of methanol is 32 g, and the complete combustion of 32 g of methanol would evolve

$$32 \times \frac{18\,840}{0\cdot865} = 697\,000 \text{ J}$$

$$= 697 \text{ kJ}$$

Therefore the heat of combustion of methanol is

$$\Delta H_{\text{combustion}} = -697 \text{ kJ mol}^{-1}$$

(The theoretical value is $-726\cdot3$ kJ mol^{-1}.)

This experiment can be repeated for other alcohols such as ethanol (CH_3CH_2OH), propan-1-ol ($CH_3CH_2CH_2OH$) and butan-1-ol ($CH_3CH_2CH_2CH_2OH$). In a series such as this it is found that the heat of combustion increases as the number of carbon atoms in the molecule increases.

7.4 Heat of Solution

When an *ionic* solid is dissolved in water there is usually, but not always, a change in temperature. This can be explained by taking into account two completely separate energy phenomena which occur during the process of solution.

(*a*) *Energy is absorbed in bond-breaking processes.* This energy is required to break the ionic bonds within the crystal lattice; it is called the **lattice energy**.

(*b*) *Energy is liberated in bond-forming processes.* This energy is evolved when the free ions react with water molecules to become hydrated; it is called **heat of hydration**.

Both phenomena occur simultaneously, and whichever is the greater will determine whether the solution process is exothermic or endothermic, i.e. whether the **heat of solution** is negative or positive.

These effects can be observed when a few grams of the following substances are added to 50 cm^3 of water in a beaker. The solution is stirred with a thermometer and any temperature difference noted.

(i)   Sodium ethanoate (sodium acetate) causes a temperature drop ($a > b$).

(ii) Anhydrous copper(II) sulphate causes a large rise in temperature

$$(b > a).$$

(iii) Sodium chloride causes little change in temperature (a and b similar).

7.5 Heat of Neutralization

Neutralization is the reaction between an acid and a base. Essentially, it is the formation of a bond between $H^+_{(aq)}$ from the acid and $OH^-_{(aq)}$ from the base (see Unit 8):

$$H^+_{(aq)} + OH^-_{(aq)} \rightarrow H_2O_{(l)}$$

Since bond-forming processes liberate energy, all acid–base reactions are exothermic.

Heat of neutralization of an acid by a base is the amount of heat liberated when one mole of hydrogen ions ($H^+_{(aq)}$) from an acid reacts with one mole of hydroxide ions ($OH^-_{(aq)}$) from a base.

Where both the acid and base are strong, it is always found that the heat of neutralization approaches a constant value of 57 kJ liberated. This is to be expected bearing in mind that one mole of any strong monobasic acid (see Section 8.8) produces one mole of hydrogen ions which can be neutralized by one mole of hydroxide ions from a strong base.

$$H^+_{(aq)} + OH^-_{(aq)} \rightarrow H_2O_{(l)}$$

$$\Delta H_{neutralization} = -57 \cdot 3 \text{ kJ mol}^{-1}$$

Where either the acid or the base (or both) is weak, the heat of neutralization is found to be less than 57 kJ. If the acid is weak its solution will not contain many hydrogen ions, and energy will therefore be used to liberate hydrogen ions from the undissociated acid in order to make them available for reaction with $OH^-_{(aq)}$ ions. An example of such a neutralization is the reaction between ethanoic (acetic) acid, a weak acid, and sodium hydroxide, a strong base:

$$CH_3COOH_{(aq)} + Na^+OH^-_{(aq)} \rightarrow CH_3COO^-Na^+_{(aq)} + H_2O_{(l)}$$

$$\Delta H_{neutralization} = -56 \cdot 1 \text{ kJ mol}^{-1}$$

Experiment 7.2   Determination of the heat of neutralization of a strong acid with a strong base

The apparatus required is shown in Fig. 7.4. 100 cm³ of 1 M hydrochloric acid is measured into an expanded-polystyrene cup, and 100 cm³ of 1 M sodium hydroxide solution is measured into a second polystyrene cup. When the two solutions have reached the same temperature the thermometer reading is noted. Then the sodium hydroxide is added to the acid, the mixture is quickly stirred and the highest temperature recorded.

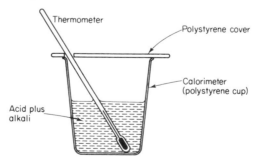

Fig. 7.4   Heat of neutralization apparatus

Results

Initial temperature of both acid and base = 21·2 °C
Final temperature of mixture   = 26·9 °C
∴ rise in temperature   = 5·7 °C

Calculation

The heat capacity of the calorimeter is ignored in this experiment because the mass of the polystyrene cup used as a calorimeter is small compared with that of the solution.

The specific heat capacity of the solution is assumed to be the same as that of pure water, approximately 4·2 joules per gram per degree Celsius.

Heat capacity of solution = mass of solution × specific heat capacity of solution
$$= (200 \times 4·2)$$
$$= 840 \text{ J K}^{-1}$$

Heat change in reaction = heat capacity of solution × rise in temperature
$$= (840 \times 5·7)$$
$$= 4788 \text{ J}$$

Thus the amount of heat liberated when 100 cm³ of 1 M H^+Cl^- neutralizes 100 cm³ of 1 M Na^+OH^- is found to be 4788 J.

Since a 1 M solution contains one mole of substance dissolved in 1000 cm³ or 1 dm³ (see Section 3.7), it follows that 100 cm³ of a 1 M solution contains 0·1 mole of substance, and we can write.

4788 J is liberated in the neutralization of 0·1 mole H^+Cl^- by 0·1 mole Na^+OH^-

∴ 47 880 J is liberated in the neutralization of 1 mole H^+Cl^- by 1 mole Na^+OH^-

Therefore the heat of neutralization of hydrochloric acid by sodium hydroxide is

$$\Delta H_{\text{neutralization}} = -47·9 \text{ kJ mol}^{-1}$$

(The theoretical value is −57·3 kJ mol⁻¹.)

7.6 Heat of Precipitation

When a colourless solution of silver nitrate is added to a colourless solution of sodium chloride a white precipitate of silver chloride is formed. A thermometer held in the solution shows a rise in temperature, indicating an exothermic reaction. The reaction is

$$\underbrace{Ag^+_{(aq)} + NO^-_{3(aq)}}_{\text{in solution}} + \underbrace{Na^+_{(aq)} + Cl^-_{(aq)}}_{\text{in solution}} \rightarrow Ag^+Cl^-_{(s)} + \underbrace{Na^+_{(aq)} + NO^-_{3(aq)}}_{\text{in solution}}$$

The $Na^+_{(aq)}$ and $NO^-_{3(aq)}$ ions take no part in the reaction: they are simply 'spectator' ions. Thus the precipitation can be represented essentially by the following *ionic* equation:

$$Ag^+_{(aq)} + Cl^-_{(aq)} \rightarrow Ag^+Cl^-_{(s)}$$

Like all bond-forming processes, the formation of the ionic bond in $Ag^+Cl^-_{(s)}$

is an exothermic reaction, and the liberation of heat is indicated by the negative value of the heat of precipitation:

$$Ag^+_{(aq)} + Cl^-_{(aq)} \rightarrow Ag^+Cl^-_{(s)}$$

$$\Delta H_{precipitation} = -65 \cdot 7 \text{ kJ mol}^{-1}$$

All precipitation reactions in aqueous solution can be expressed using an ionic equation such as this, and all liberate heat.

7.7 Rate of Chemical Reaction

So far in Unit 7 we have examined heat changes produced by the breaking and forming of chemical bonds of different strengths. Measurable heat changes result from fast reactions, such as the precipitation of silver chloride or the burning of gun-cotton. Other reactions, such as the corrosion of metals, are comparatively slow, and the heat changes associated with them would be difficult to measure. The study of the speeds or **rates** of reactions, which we are about to consider, is a branch of chemistry known as **chemical kinetics**.

As a chemical reaction proceeds, reactants are being used up and new products formed. The **rate of reaction** may be expressed in terms of the *number of moles of reactant converted in unit time* or *the number of moles of product formed in unit time*.

There are four main factors which influence the rate of a chemical reaction. They are:

(a) the physical state of the reactants
(b) the concentration of the reactants
(c) the temperature of the reactants
(d) the presence of a catalyst

Let us consider each of these four effects in turn.

7.8 Rate Dependence on the Physical State of the Reactants

Manufacturers handling extremely fine powders, flour for example, have to take precautions against explosions and hence fires. These occur when the fine powder accidentally rises and mixes with the air. A spontaneous combustion results which is so rapid as to be explosive. An explosion of this type takes place because an airborne fine powder exposes a vast surface area to oxygen molecules in the air. When the flour is bagged there is no danger of explosion because the surface area exposed to oxygen is insignificant. Thus particle size would seem to increase the rate of a chemical reaction. This can be demonstrated in the laboratory by considering the reaction between a solid in different states of subdivision and a liquid, e.g. a solid carbonate and an acid.

Experiment 7.3   Reaction between marble and hydrochloric acid
Marble, a naturally occurring form of calcium carbonate, reacts with hydrochloric acid liberating carbon dioxide:

$$Ca^{2+}CO_3^{2-}{}_{(s)} + 2H^+Cl^-_{(aq)} \rightarrow Ca^{2+}Cl_2^-{}_{(aq)} + H_2O_{(l)} + CO_{2(g)}$$

Two pieces of marble of similar size are selected and one of them is ground to a fine powder. Each sample is placed in a conical flask half filled with dilute hydrochloric acid. The fine powder reacts vigorously and the reaction is completed in a few seconds. In contrast, the single piece of marble shows only slow effervescence and the reaction continues for several minutes.

This is because the fine powder has a greater surface area and exposes more particles for reaction with the liquid. Thus *a reduction in the particle size leads to an increase in the rate of reaction.*

7.9 Rate Dependence on Concentration of the Reactants

Chemical reactions take place when the appropriate atoms, ions or molecules approach and collide with one another. The greater the number of collisions, the more rapid will be the reaction rate. Thus increasing the number of particles per unit volume (i.e. the *concentration*) will cause more frequent collisions and hence an increase in the reaction rate. This picture of how chemical reactions take place is called the **collision theory**.

When a burning spill of wood is plunged into a gas jar of oxygen it burns much more rapidly than it does in the air. This is because the *concentration* of oxygen in the jar is considerably greater than the concentration in the air.

The effect of concentration on the rate of reactions *in solution* can be measured by maintaining one of the reactants at a constant concentration and varying the

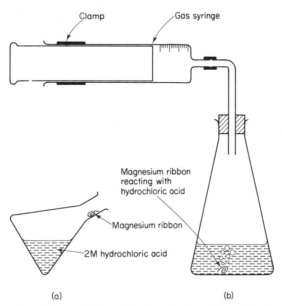

Clamp   Gas syringe

Magnesium ribbon
reacting with
hydrochloric acid

Magnesium ribbon

2M hydrochloric acid

(a)   (b)

Fig. 7.5   Apparatus used in investigating the dependence of rate of reaction on concentration of reactants

concentration of the other. Some measurable change is recorded as the reaction proceeds.

Experiment 7.4   *Reaction between magnesium ribbon and hydrochloric acid.*
In this experiment we measure the volume of hydrogen liberated when different concentrations of hydrochloric acid react with a fixed mass of magnesium ribbon:

$$Mg_{(s)} + 2H^+Cl^-_{(aq)} \rightarrow Mg^{2+}Cl^-_{2(aq)} + H_{2(g)}$$

The apparatus is illustrated in Fig. 7.5.

10 cm³ of 0·50 M hydrochloric acid is placed in the 250 cm³ conical flask and to this is added a suitable length of magnesium ribbon (say 10 cm) so that it rests on the side of the conical flask without touching the acid (see Fig. 7.5a). The flask is then connected to a closed 100 cm³ gas syringe, and tilted so that the magnesium ribbon drops into the acid (see Fig. 7.5b). The volume of hydrogen liberated is recorded at convenient intervals.

The experiment is repeated, this time using 10 cm³ of 0·25 M hydrochloric acid with the same length of magnesium ribbon.

Results
A graph is plotted (Fig. 7.6) to show the **rate curves** (volume of hydrogen against time) for the two different concentrations of magnesium.

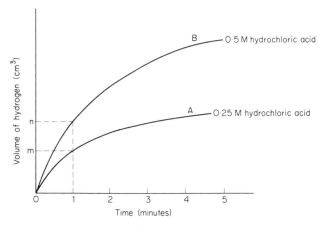

Fig. 7.6   Rate curves for two different concentrations of hydrochloric acid reacting with magnesium

It can be seen from the rate curves that the greater concentration of acid gives the faster reaction rate. For example, *after one minute*

Rate of reaction A = volume/time
$$= m/1$$
$$= m \text{ cm}^3 \text{ per minute}$$

Rate of reaction B $=$ volume/time

$\qquad = n/1$

$\qquad = n \text{ cm}^3$ per minute

As time passes *both* curves become progressively less steep, showing that the rate decreases. This is because the acid is used up as the reaction proceeds and hence its concentration becomes less. Eventually all the acid is used up and, as the volume of hydrogen remains constant, the rate curves flatten to a horizontal straight line.

We can conclude from this experiment that the *rate of the reaction is directly proportional to the concentration of the acid, provided that the magnesium is in excess.*

Experiment 7.5 Reaction between sodium thiosulphate and hydrochloric acid
When hydrochloric acid is added to sodium thiosulphate solution, a precipitate of sulphur slowly forms:

$$Na_2^+ S_2O_{3(aq)}^{2-} + 2H^+Cl_{(aq)}^- \rightarrow S_{(s)} + 2Na^+Cl_{(aq)}^- + H_2O_{(l)} + SO_{2(g)}$$

The rate of this reaction is directly proportional to the concentration of the sodium thiosulphate, provided that the acid is in excess. The time needed for a certain amount of precipitate to form can be used to measure the rate of the reaction. The reaction flask is placed over a piece of paper with a cross on it and the time at which the mark can no longer be seen (owing to the turbidity) is noted. A shorter time means a faster reaction rate and therefore the rate is proportional to $1/\text{time}$ $(1/t)$.

45 cm³ of 0·15 M sodium thiosulphate solution is placed into a clean dry 100 cm³ flask and the flask placed over a cross marked on a paper. 5 cm³ of 2 M hydrochloric acid is added and the time taken for the cross to disappear is noted.

The experiment is repeated using different volumes of sodium thiosulphate

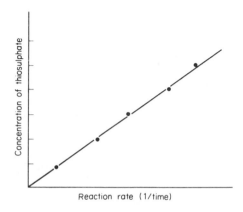

Fig. 7.7 Relationship between rate of reaction and concentration

solution in the same flask. The total volume before the 5 cm³ of acid is added is always kept constant at 45 cm³ with water (for example, 40 cm³ thiosulphate plus 5 cm³ water, or 20 cm³ thiosulphate plus 25 cm³ water).

Results

A graph of concentration of sodium thiosulphate against reaction rate ($1/t$) is plotted (Fig. 7.7). The concentration of sodium thiosulphate in the reaction mixture may be measured as the volume of the original solution or as a fraction of the original concentration.

The straight line graph shows the direct relationship between the rate of reaction and the concentration of sodium thiosulphate.

7.10 Rate Dependence on Temperature

For almost all chemical reactions, the rate of the reaction increases with rise in temperature. Once started, exothermic reactions are self sustaining. The rate increases as the heat produced by the reaction goes to warm the surroundings— including the reactants. In an endothermic reaction once the external heat supply is removed, heat is absorbed from the surroundings, the temperature falls and the reaction rate decreases.

Our *kinetic model* of matter and the *collision theory* explain this variation in rate with temperature. At low temperature the particles will be moving relatively slowly and there would thus be few collisions leading to products. However, as the temperature increases the particles move more rapidly as their kinetic energy increases. The *rate of collision* will be increased and consequently the number of collisions leading to products increases.

When two reactant particles collide, however, they do not necessarily react to form products. Each collision requires a certain amount of energy before reaction can occur. At a particular temperature not all particles have the same kinetic

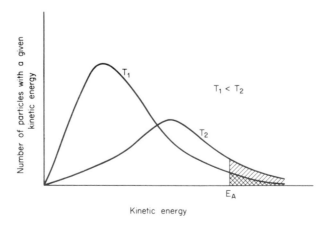

Fig. 7.8   Effect of temperature on the kinetic energy of particles

energy. Those with high energy will be moving rapidly and collisions involving these are likely to be effective in producing reaction. Particles with low kinetic energy will be moving slowly and collisions involving these are not likely to produce reaction.

Raising the temperature increases the kinetic energy of the particles and therefore the number likely to be effective in producing reaction will be increased (see Fig. 7.8).

From Fig. 7.8 it can be seen that increasing the temperature from T_1 to T_2 produces a large increase in the number of particles having kinetic energy greater than the value E_A. This value E_A is the minimum kinetic energy required if two particles are to react to form products, and is called the **activation energy**.

Consider the exothermic reaction:

$$A + B \rightarrow C + D \quad (\Delta H \text{ is negative})$$

Fig. 7.9 illustrates the activation energy E_A for the forward reaction $A + B \rightarrow C + D$.

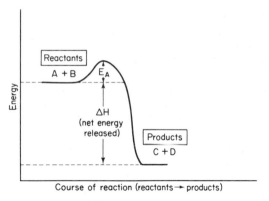

Fig. 7.9   Activation energy E_A is the minimum kinetic energy which the reacting particles must have if they are to form products

In order for the reactants to form products they must have activation energy E_A. The reactants are at a higher energy than are the products. There will therefore be a net *release* of energy indicated by ΔH in Fig. 7.9.

The endothermic reaction $C + D \rightarrow A + B$ would be very slow because the **energy barrier** for this reaction is so large (activation energy is $E_A + \Delta H$) that very few particles would possess this amount of energy.

By using the reaction between fixed concentrations of sodium thiosulphate and hydrochloric acid it can readily be shown that the reaction rate increases with rise in temperature. The time taken for the turbidity to appear is very much shorter at higher temperature than at low ones.

For most chemical reactions, it is found that the rate is approximately doubled for every 10 °C rise in temperature.

7.11 Rate Dependence on Catalysts

A **catalyst** is a substance which alters the rate of a chemical reaction but remains chemically unchanged at the end of the reaction. Those which increase the reaction rate are called **positive catalysts**, while those which slow the reaction are called **negative catalysts** or **inhibitors**.

Although the mechanism of catalysts is often obscure, the ability of certain substances to catalyse reactions used in the production of 'heavy' chemicals is a vital factor in their economic manufacture. Examples of important catalysts in the chemical industry include vanadium(v) oxide (vanadium pentoxide) in the manufacture of sulphuric acid, and finely divided iron in the manufacture of ammonia gas. Indeed the list is almost endless and numerous references to the use of catalysts will be found in later Units.

The effectiveness of a catalyst can be investigated in the laboratory by considering the decomposition of hydrogen peroxide:

$$2H_2O_{2(aq)} \rightarrow 2H_2O_{(l)} + O_{2(g)}$$

Experiment 7.6   The catalytic decomposition of hydrogen peroxide

(*a*) Apparatus similar to that illustrated in Fig. 7.5 is set up. A known volume of dilute hydrogen peroxide solution is measured into the conical flask and to it is added a small weighed amount of manganese(IV) oxide ($Mn^{4+}O_2^{2-}$). The volume of oxygen liberated is recorded at regular intervals by reading the syringe.

The experiment is then repeated using other substances to test their catalytic properties. It is found that of the materials added none is more efficient than manganese(IV) oxide as a catalyst in this decomposition.

(*b*) A known volume of dilute hydrogen peroxide solution is poured into the conical flask and a small weighed amount of manganese(IV) oxide is added. The total volume of oxygen liberated is recorded when reaction ceases. The contents of the flask are filtered using a previously weighed filter paper, then washed with distilled water and dried in a warm oven. The dry manganese(IV) oxide and filter paper are finally weighed in order to obtain the mass of the oxide. *It is found that the mass of manganese*(IV) *oxide remains unchanged.*

The experiment is repeated using different masses of manganese(IV) oxide with the same fixed volume of hydrogen peroxide. In each case the final volume of oxygen liberated is recorded. *This volume is found to be constant and independent of the amount of manganese*(IV) *oxide used.*

The role of a catalyst. It is generally believed that a catalyst assists the reaction by *lowering the activation energy* for the reaction (see Fig. 7.10).

Although the catalyst lowers the activation energy, the heat change for the reaction (ΔH) is unaltered.

7.12 Reversible Reactions

When heating a pan of water, the temperature gradually rises until it reaches 100 °C and the water begins to boil. Boiling occurs with the rapid evolution of

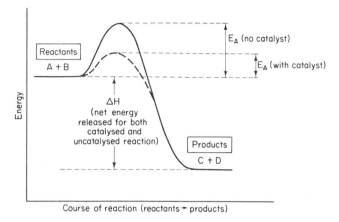

Fig. 7.10 The effect of a catalyst on activation energy

steam, and much of this steam condenses on the cool pan lid into droplets of water which fall back into the pan.

This change of state is plainly reversible, and can be summarized by the equation:

$$H_2O_{(l)} \rightleftharpoons H_2O_{(g)}$$

The sign $\rightleftharpoons$ indicates that the reaction can be made to proceed in either direction simply by altering the conditions. Although not a chemical reaction this is a good example of a reversible process.

When blue copper(II) sulphate crystals are heated they lose their water of crystallization, leaving white anhydrous copper(II) sulphate. If water is added to this white powder (after it has cooled), the blue colour immediately returns and heat is liberated. This indicates that the reverse reaction (hydration) is taking place. This *reversible* reaction, in which the products formed are capable of reacting with each other to re-form the original substance, can be summarized by the equation:

$$Cu^{2+}SO_4^{2-} . 5H_2O_{(s)} \rightleftharpoons Cu^{2+}SO_{4(s)}^{2-} + 5H_2O_{(g)}$$

Most chemical reactions can be considered as being theoretically reversible. Quite often, however, the change is largely in one direction only, and it is then signified by an arrow. For example, the burning of magnesium in oxygen is written:

$$2Mg_{(s)} + O_{2(g)} \rightarrow 2Mg^{2+}O_{(s)}^{2-}$$

to indicate that, to all intents and purposes, the reaction goes to completion.

7.13 Equilibrium

When solid iodine is added to an ethanol–water mixture and stirred, the colourless liquid turns yellowish and the colour gradually deepens to a reddish brown as the iodine dissolves. Eventually a stage is reached when the solution has become saturated and no more iodine seems to dissolve. The system is said to be at **equilibrium**. In order to discover something about the nature of this equilibrium, some solid *radioactive* iodine is added to the saturated solution.

After a while some of the saturated solution is decanted from the solid and tested for radioactivity with a Geiger counter. The test is positive, showing that some of the solid radioactive iodine atoms must have gone into solution—even though the solution was already saturated. Moreover, if some non-radioactive iodine crystals are added to a saturated solution containing radioactive iodine, they become radioactive. Evidently crystals are still dissolving and, at the same time, re-forming despite the fact that the system is in equilibrium.

Thus at equilibrium reaction does not stop. The system is in a *dynamic* (moving) state, continually changing from reactant to product and vice versa. The equilibrium of the iodine system discussed above is indicated by the following reversible reaction, in which the rates of the forward and backward reactions are equal:

$$I_2 \; \rightleftharpoons \; I_2$$
$$\text{solid} \quad \text{dissolved}$$

Effect of Changes in Concentration on the Equilibrium State

When dilute hydrochloric acid is added to a *yellow* solution of potassium chromate(VI), an *orange* solution of potassium dichromate(VI) is produced:

$$2K_2^+ CrO_{4(aq)}^{2-} + 2H^+ Cl_{(aq)}^- \rightarrow K_2^+ Cr_2 O_{7(aq)}^{2-} + 2K^+ Cl_{(aq)}^- + H_2 O_{(l)}$$

or

$$2CrO_{4(aq)}^{2-} + 2H_{(aq)}^+ \rightarrow Cr_2 O_{7(aq)}^{2-} + H_2 O_{(l)}$$
$$\text{yellow} \qquad\qquad \text{orange}$$

Initially the rate of this reaction will be proportional to the concentrations of chromate(VI) ion and hydrogen ion, and it decreases as the concentrations of chromate(VI) ion and hydrogen ion decrease. The reverse reaction can occur as soon as dichromate(VI) ion is formed in solution. The rate of this reverse reaction will be slow at first, but it will increase as the concentration of dichromate(VI) and water increases. *At equilibrium the rates of the two reactions will be the same.*

The equation can thus be written:

$$2CrO_{4(aq)}^{2-} + 2H_{(aq)}^+ \rightleftharpoons Cr_2 O_{7(aq)}^{2-} + H_2 O_{(l)}$$
$$\text{yellow} \qquad\qquad \text{orange}$$

If at equilibrium the concentration of any of these substances is altered, the rate of the forward and backward reactions will no longer be the same. *The equilibrium has been disturbed.*

The addition of hydrogen ion (by adding acid) produces an orange colour, whereas the removal of hydrogen ion (by adding sodium hydroxide) causes the solution to turn yellow. The equilibrium concentrations are affected if the concentrations of reactants and products are altered.

The effect of changing concentrations can be further demonstrated by studying the equilibrium reaction:

$$\text{Fe}^{3+}_{(aq)} \quad + \quad \text{SCN}^{-}_{(aq)} \quad \rightleftharpoons \quad \text{FeSCN}^{2+}_{(aq)}$$

iron(III) ion	thiocyanate ion	monothiocyanato-
(pale brown)	(colourless)	iron(III) ion
		(blood red)

The blood-red colour deepens on the addition of either iron(III) ion, e.g. from iron(III) nitrate, or thiocyanate ion, e.g. from potassium thiocyanate. It becomes paler if the concentration of either (or both) of these reactants is decreased.

Effect of a Catalyst on the Equilibrium State

The addition of a catalyst to a system at equilibrium does not alter the equilibrium state. It simply alters the rate of both the forward and backward reactions so that the equilibrium condition is reached more quickly.

Effect of Temperature on the Equilibrium State

Variation in temperature affects the equilibrium concentrations. For an exothermic reaction, the addition of further heat tends to inhibit the reaction and decrease the concentration of products; for an endothermic reaction, the addition of heat encourages the reaction and increases the concentration of the products.

Summary of Unit 7

1. **Energy changes** accompany all chemical reactions.
2. In **exothermic** reactions *energy is released*; in **endothermic** reactions *energy is absorbed*.
3. The **heat of reaction** ΔH is the amount of heat evolved or absorbed during a chemical reaction. ΔH refers to the amounts shown in the balanced equation for this reaction.
4. **Heat of combustion** is the heat evolved in the complete combustion of one mole of a substance.
5. **Heat of solution** of one mole of an ionic solid is the energy difference between the bond-breaking processes which use up energy and the bond-forming processes which liberate energy.
6. **Heat of neutralization** of an acid or base is the heat evolved when one mole of hydrogen ions (H^+_{aq}) or one mole of hydroxide ions (OH^-_{aq}) is neutralized. For strong acids reacting with strong bases, the value is always about 57.3 kJ mol^{-1}.
7. The **rate of a chemical reaction** indicates how quickly reactants are being used and products are being formed.
8. Reaction rate is influenced by (*a*) the physical state of the reactants, (*b*) the concentration of the reactants, (*c*) the temperature of the reactants and (*d*) the presence of a catalyst.
9. A **reversible reaction** is one in which the products formed are capable of reacting with each other to re-form the original reactants.
10. **Chemical equilibrium** is a state of balance in which the rates of the forward and backward reactions are equal.

11. At a given temperature, a change in concentration of the reactants will alter the equilibrium concentration of the products.
12. The addition of a catalyst to a reaction alters the rate of both the forward and backward reactions but it does not alter the equilibrium state.
13. Variation in temperature affects the equilibrium concentrations.

Test Yourself on Unit 7

1. When solid ammonium chloride dissolves in water, a decrease in temperature results. Is this type of reaction called (a) neutralization, (b) exothermic, (c) endothermic, or (d) double decomposition?

2. Explain why the reaction between anhydrous copper(II) sulphate and water is exothermic.

3. State three means by which the rate of a chemical reaction may be increased.

4. The curves in the graph show the volume of hydrogen evolved during the reaction between zinc and hydrochloric acid at 20 °C. Curve I is obtained when 0·13 g of zinc reacts with 10·0 cm³ of 1 M hydrochloric acid. Curve II is obtained when 0·13 g of zinc reacts with 10·0 cm³ of 1 M hydrochloric acid in the presence of Cu^{2+} ions.

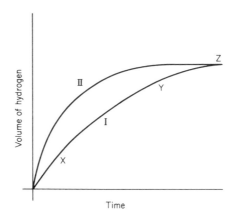

(a) Write a balanced equation for the reaction taking place in the absence of Cu^{2+} ion.
(b) At which of the points X, Y or Z was the rate of reaction greatest?
(c) What can you say about the reaction at point Z?
(d) What fraction of a mole of zinc is 0·13 g?
    (Relative atomic mass of zinc = 65.)
(e) Calculate the volume of 1 M hydrochloric acid required to react exactly with 0·13 g of zinc.
(f) The Cu^{2+} ions increase the initial rate of reaction: True or False?

(g) The Cu^{2+} ions increase the total volume of hydrogen liberated: True or False?

(h) Finely powdered zinc increases the rate of reaction: True or False?

(i) Finely powdered zinc increases the volume of hydrogen liberated: True or False?

(j) Increasing the volume of hydrochloric acid increases the volume of hydrogen produced: True or False?

(k) The presence of Cu^{2+} ions increases the activation energy for the forward reaction: True or False?

(l) Increasing the temperature to 40 °C doubles the volume of hydrogen liberated: True or False?

5. When dilute hydrochloric acid is added to a yellow solution of potassium chromate(VI) an orange solution of potassium dichromate(VI) is produced:

$$K_2^+ CrO_{4(aq)}^{2-} + H^+ Cl_{(aq)}^- \rightleftharpoons K_2^+ Cr_2 O_{7(aq)}^{2-} + K^+ Cl_{(aq)}^- + H_2 O_{(l)}$$

(a) Which of the following sets will balance the equation:

    (i) 1, 1, 2, 2, 1

    (ii) 2, 2, 1, 1, 1

    (iii) 2, 1, 1, 2, 1

    (iv) 2, 2, 1, 2, 1

    (v) 2, 2, 2, 1, 1

(b) What would you observe if dilute sodium hydroxide solution was added to the orange solution? Give a reason for your answer.

Mark this test out of 20 with the answers given on page 375.

Classification of Matter I: Acids, Bases and Salts

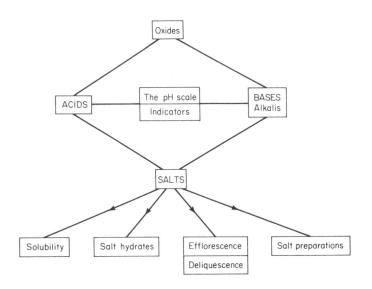

'Heavy' chemicals are those manufactured on a large scale, and prominent among them are *acids* such as sulphuric and nitric, *alkalis* such as sodium hydroxide and ammonia, and the *salts* ammonium sulphate and sodium carbonate. This Unit looks at the chemistry of such compounds, paying particular attention to the properties of acids and bases, the role of the solvent in acidity, volumetric analysis and the preparation of salts.

8.1 Acids and Alkalis

Chemists have known and made attempts to define the term *acid* throughout history. The word 'acid' comes from the Latin *acidus* meaning 'sour', and a sour taste is one of the characteristics that nearly all acids have in common. However, it would be foolish to attempt to classify acids simply by their taste; after all, the deadly poisonous hydrogen cyanide is an acid when dissolved in water.

In contrast to acids, *alkaline* solutions feel slippery or greasy. Alkaline substances were first obtained from the burnt remains of vegetable matter, and the word 'alkali' comes from the Arabic, *al-qaliy*, meaning 'treated ashes'. All the alkalis and a number of insoluble substances can destroy or *neutralize* the characteristic properties of an acid. The collective name given to all these substances, including alkalis, is the word **base**. *A base is a substance which can neutralize an acid.*

8.2 Detecting Acidity and Alkalinity

Solutions of acids and alkalis change the colour of certain dye solutions such as litmus (a mixture of dyes extracted from lichens), red-cabbage juice, flower-petal extracts and many others. Dye solutions which change colour in the presence of an acid or an alkali are called *indicators*.

Different indicators give various colour changes when added to an aqueous solution of an acid or base. *Universal indicator* is a mixture of several indicators and gives a spectrum of colours, each one being characteristic of *the degree of acidity or alkalinity* of the solution to which it is added. Degrees of acidity and alkalinity are conveniently represented by numbers on a scale running from 0 to 14; this is called the *pH scale** (see Table 8.1).

Table 8.1 The pH scale

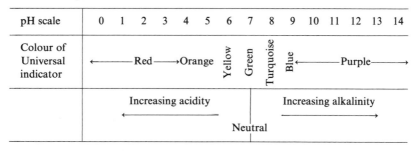

In the range 0 to 7, the lower the pH value the more acid the solution; in the range 7 to 14, the greater the pH value the more alkaline the solution. Thus a solution having a pH value of 2 will be more acid than one having a pH value of 4, and a solution with a pH value of 14 will be *much* more alkaline than one having a pH value of 8. A solution which is neither acid nor alkaline has a *pH value of 7* and is said to be *neutral*: this is the pH value of *pure water*.

In the following experiment the acidity or alkalinity of different oxides dissolved in water is tested with Universal indicator. In particular we test those oxides which were prepared by burning elements in oxygen (see Unit 6).

Experiment 8.1   Acidity and alkalinity of oxides
Gas jars containing samples of gaseous oxides are shaken with a little distilled water, and the resulting solutions are transferred to separate test tubes. Small

* Acidity is closely related to the concentration of hydrogen ion $[H^+_{aq}]$ in the solution. The pH value of a solution is defined as $-\log_{10}[H^+_{aq}]$, i.e. *minus* the logarithm (to base 10) of the hydrogen-ion concentration. For our present purposes the definition is less important than the fact that it offers a useful scale for measuring acidity and alkalinity.

quantities of other oxides are placed in test tubes which are half-filled with distilled water, and shaken.

About 5 drops of Universal indicator are added to each oxide solution and the colour produced compared with standard colours on the Universal indicator bottle. Thus the acidity or alkalinity of each solution is easily determined.

Observations

Oxide	Colour of indicator	Conclusion
Water H_2O	green	neutral
Calcium oxide $Ca^{2+}O^{2-}$	purple	alkaline
Magnesium oxide $Mg^{2+}O^{2-}$	purple	alkaline
Sodium peroxide $Na_2^+O_2^{2-}$	purple	alkaline
Iron(II) di-iron(III) oxide Fe_3O_4	green (does not dissolve)	neutral
Sulphur dioxide SO_2	red	acidic
Carbon dioxide CO_2	orange	weakly acidic
Phosphorus(v) oxide P_4O_{10}	red	acidic

Conclusion

When dissolved in water, the oxides of the non-metals sulphur, carbon and phosphorus produce *acid solutions*, while the oxides of the metals calcium, magnesium and sodium give *alkaline solutions*.

8.3 Classification of Oxides

Experiment 8.1 provides the basis for the classification of oxides. These classes are:

(*a*) **Acidic oxides**. Oxides of non-metals dissolve in water to give an acid solution. For example:

(i) sulphur dioxide gives *sulphurous acid*

$$SO_{2(g)} + H_2O_{(l)} \rightarrow H_2^+ SO_{3(aq)}^{2-},$$

(ii) sulphur trioxide (sulphur(vi) oxide) gives *sulphuric acid*

$$SO_{3(g)} + H_2O_{(l)} \rightarrow H_2^+ SO_{4(aq)}^{2-},$$

(iii) phosphorus(v) oxide (phosphorus pentoxide) gives *phosphoric acid*

$$P_4O_{10(s)} + 6H_2O_{(l)} \rightarrow 4H_3^+ PO_{4(aq)}^{3-}.$$

Thus sulphur dioxide, sulphur trioxide and phosphorus(v) oxide are all acidic oxides.

(b) **Neutral oxides.** These—as their name suggests—give neither an acid nor an alkaline solution when dissolved in water. They are not capable of neutralizing an acid or a base. Examples are water H_2O, carbon monoxide CO, nitrogen monoxide NO, and dinitrogen oxide N_2O.

(c) **Basic oxides.** These are all *oxides of metals*. Only a few of them dissolve in water (giving a strongly alkaline solution of the hydroxide), but they are all capable of neutralizing an acid. For example:

(i) Sodium oxide dissolves in water, producing much heat and giving the alkali sodium hydroxide:

$$Na_2^+O_{(s)}^{2-}+H_2O_{(l)} \rightarrow 2Na^+OH_{(aq)}^-$$

Sodium oxide will also neutralize an acid, e.g. hydrochloric acid:

$$Na_2^+O_{(s)}^{2-}+2H^+Cl_{(aq)}^- \rightarrow 2Na^+Cl_{(aq)}^-+H_2O_{(l)}$$

(ii) Copper(II) oxide does *not* dissolve in water but *is* a *basic oxide* because it neutralizes an acid, e.g. dilute sulphuric acid:

$$Cu^{2+}O_{(s)}^{2-}+H_2^+SO_{4(aq)}^{2-} \rightarrow Cu^{2+}SO_{4(aq)}^{2-}+H_2O_{(l)}$$

In general, all oxides are acidic, basic or neutral; there are some oxides, however, which show additional properties necessitating further classification.

(d) **Amphoteric oxides** display properties which are both acidic and basic. Examples include aluminium oxide ($Al_2^{3+}O_3^{2-}$) and zinc oxide ($Zn^{2+}O^{2-}$). These are discussed in Unit 9.

(e) **Peroxides** liberate hydrogen peroxide (H_2O_2) with an acid. Examples are sodium peroxide ($Na_2^+O_2^{2-}$) and barium peroxide ($Ba^{2+}O_2^{2-}$). When ice-cold dilute sulphuric acid is added to barium peroxide, white barium sulphate is precipitated and a dilute solution of hydrogen peroxide is formed:

$$Ba^{2+}O_{2(s)}^{2-}+H_2^+SO_{4(aq)}^{2-} \rightarrow Ba^{2+}SO_{4(s)}^{2-}+H_2O_{2(aq)}$$

(f) **Compound oxides** behave chemically as a combination of two metal oxides. For example, iron(II) di-iron(III) oxide Fe_3O_4 behaves as if it were $Fe^{2+}O^{2-}.Fe_2^{3+}O_3^{2-}$, and dilead(II) lead(IV) oxide Pb_3O_4 behaves as if it were $(Pb^{2+}O^{2-})_2.Pb^{4+}O_2^{2-}$.

8.4 Properties of Acids and Bases

The properties of acids and bases can be used to provide us with an *operational definition* of these terms.

(a) Acids

(i) All acids *have a sour taste*. It is dangerous to taste unknown acids but we are familiar with the sharp taste of vinegar (ethanoic or acetic acid) and the acids in citrus fruits (citric and ascorbic acids).

(ii) Acidic solutions *change the colour of Universal indicator* to show a *pH value less than 7.*

(iii) They *react with a carbonate or hydrogencarbonate to liberate carbon dioxide.* When dilute hydrochloric acid is added to calcium carbonate or sodium hydrogencarbonate, a rapid effervescence is seen. The gas which is evolved can be identified as carbon dioxide by the fact that it turns lime-water milky.

$$2H^+Cl^-_{(aq)}+Ca^{2+}CO^{2-}_{3(s)} \rightarrow Ca^{2+}Cl^-_{2(aq)}+CO_{2(g)}+H_2O_{(l)}$$

$$H^+Cl^-_{(aq)}+Na^+HCO^-_{3(s)} \rightarrow Na^+Cl^-_{(aq)}+CO_{2(g)}+H_2O_{(l)}$$

(iv) Acids are *neutralized by a base.* If sodium hydroxide is carefully added to a solution of hydrochloric acid containing Universal indicator, the colour changes from red through green to purple, and the solution becomes warm. The green colour represents a stage in the addition when the pH of the mixture is exactly 7 and the solution is neither acid nor alkaline. Such a reaction in which the acidic properties are destroyed is called *neutralization*; the product remaining in solution is a *salt*, in this case sodium chloride:

$$H^+Cl^-_{(aq)}+Na^+OH^-_{(aq)} \rightarrow Na^+Cl^-_{(aq)}+H_2O_{(l)}$$

For any acid/base reaction we may write:

$$ACID+BASE \rightarrow SALT+WATER$$

(b) Bases

Most bases are either *oxides or hydroxides of metals. The only property common to all bases is their ability to neutralize an acid.* For example, insoluble black copper(II) oxide neutralizes sulphuric acid, producing a blue solution of the salt copper(II) sulphate:

$$H_2^+SO^{2-}_{4(aq)}+Cu^{2+}O^{2-}_{(s)} \rightarrow Cu^{2+}SO^{2-}_{4(aq)}+H_2O_{(l)}$$

Similarly the bases sodium hydroxide (Na^+OH^-), magnesium hydroxide ($Mg^{2+}(OH^-)_2$) and calcium oxide ($Ca^{2+}O^{2-}$), will neutralize acids to form a salt and water only.

Soluble hydroxides of metals are also called *alkalis*. Examples include sodium hydroxide (Na^+OH^-), potassium hydroxide (K^+OH^-) and calcium hydroxide ($Ca^{2+}(OH^-)_2$). All alkalis are so classified because they provide $OH^-_{(aq)}$ ions as the only anions in aqueous solution. For example:

$$Na^+OH^-_{(s)} \rightarrow Na^+_{(aq)}+OH^-_{(aq)}$$

Aqueous *ammonia*, although it is not a metal hydroxide, can be classified as an alkali because of the following reaction with water:

$$NH_{3(g)}+H_2O_{(l)} \rightarrow NH^+_{4(aq)}+OH^-_{(aq)}$$
$$\text{ammonium ion}$$

In addition to their ability to neutralize an acid, *alkalis* have the following properties:

(i) they feel greasy or soapy when dissolved in water

(ii) their solution in water changes the colour of Universal indicator to show a pH value greater than 7

(iii) they liberate ammonia when warmed with an ammonium salt:

$$NH_4^+ Cl_{(s)}^- + Na^+ OH_{(aq)}^- \rightarrow Na^+ Cl_{(aq)}^- + NH_{3(g)} + H_2O_{(l)}$$

8.5 Acids and Bases in More Detail

Svante Arrhenius (1859–1927), a Swedish scientist, received the Nobel Prize in Chemistry in 1903 for his work on the dissociation theory of acids, bases and salts. Arrhenius postulated the existence of ions in aqueous solutions. As early as 1884 he suggested that *substances which yield hydrogen ions in aqueous solution are acids*. Hydrogen chloride gas yields hydrogen ions in aqueous solution as follows:

$$HCl_{(g)} \rightarrow H_{(aq)}^+ + Cl_{(aq)}^-$$

The solution is an acid, it is in fact hydrochloric acid. Similarly, nitric acid yields hydrogen ions in solution:

$$HNO_{3(l)} \rightarrow H_{(aq)}^+ + NO_{3(aq)}^-$$

Substances giving hydroxide ions, $OH_{(aq)}^-$, *in aqueous solution are bases* according to Arrhenius' theory. For example, sodium hydroxide and potassium hydroxide yield hydroxide ions as follows:

$$NaOH_{(s)} \rightarrow Na_{(aq)}^+ + OH_{(aq)}^-$$

$$KOH_{(s)} \rightarrow K_{(aq)}^+ + OH_{(aq)}^-$$

A *neutralization reaction* between any acid (a source of $H_{(aq)}^+$) and any alkali (a source of $OH_{(aq)}^-$) is the combination of hydrogen ions and hydroxide ions to form water molecules, and can be written:

$$H_{(aq)}^+ + OH_{(aq)}^- \rightarrow H_2O_{(l)}$$

One of the problems faced by J. N. Brønsted in Denmark and T. M. Lowry in England (1923) was the nature of the hydrogen ion $H_{(aq)}^+$. This problem can be illustrated by looking closely at the acidity of hydrogen chloride. If dry hydrogen chloride gas is dissolved in dry methylbenzene (toluene), the solution shows no acidic properties and is non-conducting. No ions are present in this solution.

When hydrogen chloride is dissolved in water, the solution gives an acid reaction with carbonates and allows an electric current to pass through it with decomposition taking place at the electrodes.

If the solution of hydrogen chloride in methylbenzene is shaken with water, the methylbenzene and water form two layers. The lower aqueous layer displays all the properties of the hydrogen chloride/water solution, showing that some of the hydrogen chloride has passed into the aqueous layer.

Thus the water must play some part in the acidity of hydrochloric acid. We believe that the hydrogen ion is hydrated by a water molecule to form an *oxonium* ion $H_3O_{(aq)}^+$

$$H_{(aq)}^+ + H_2O_{(l)} \rightarrow H_3O_{(aq)}^+$$

and that hydrogen chloride dissolves in water as follows:

$$HCl_{(g)} + H_2O_{(l)} \rightarrow H_3O^+_{(aq)} + Cl^-_{(aq)}$$

This is in line with the Brønsted–Lowry definition of an acid as *any substance consisting of molecules or ions that donate protons*. Similarly, a base may be regarded as *an acceptor of protons*. In the above reaction the hydrogen chloride donates a proton (H^+) to a water molecule and is therefore an acid.

Any compound releasing hydroxide ions must be a base, because the hydroxide ion is capable of accepting a proton to form water:

$$OH^-_{(aq)} + H_3O^+_{(aq)} \rightarrow 2H_2O_{(l)}$$

Similarly, ammonia is a base capable of accepting protons:

$$NH_{3(g)} + H_2O_{(l)} \rightarrow NH^+_{4(aq)} + OH^-_{(aq)}$$

The base (NH_3) accepts a proton from the water.

Although we appreciate that an acid donates a proton to a water molecule it is often convenient to write the hydrated proton simply as $H^+_{(aq)}$.

8.6 Quantitative Reaction between Acids and Bases: Volumetric Analysis

Progressive addition of an alkali to an acid, or vice versa, is called a *titration* and enables us to compare the concentrations of acid and alkali solutions. A titration is carried out in the presence of an *indicator*, such as methyl orange or phenolphthalein, which shows by its change in colour the point at which neither acid nor alkali is present in excess. Table 8.2 illustrates the colours of common indicators in acid and alkaline solution.

Table 8.2 Common indicators

Indicator	Colour in acid	Colour in alkali
Methyl orange	red	yellow
Bromothymol blue	yellow	blue
Litmus	red	blue
Phenolphthalein	colourless	red

Definitions

A *standard* solution is one in which the concentration is known. Concentration of a solution is expressed in terms of *molarity*. Molarity is defined as the number of moles of solute dissolved in one cubic decimetre (1000 cm^3) of solution. Table 8.3 shows the mass of each substance required to make 1 dm^3 (1000 cm^3) of a one-molar solution of some common acids, alkalis and carbonates.

Using these values it is possible to calculate the mass of material required to make any given volume of a standard solution, as in the following examples.

Table 8.3 Molar mass of acids, alkalis and carbonates

Substance	Formula	Mass of 1 mole (in grams)	
Hydrochloric acid	H^+Cl^-	$1 + 35 \cdot 5$	$= 36 \cdot 5$
Sulphuric acid	$H_2^+SO_4^{2-}$	$2 + 32 + (4 \times 16)$	$= 98$
Nitric acid	$H^+NO_3^-$	$1 + 14 + (3 \times 16)$	$= 63$
Sodium hydroxide	Na^+OH^-	$23 + 16 + 1$	$= 40$
Potassium hydroxide	K^+OH^-	$39 + 16 + 1$	$= 56$
Sodium carbonate	$Na_2^+CO_3^{2-}$	$(2 \times 23) + 12 + (3 \times 16)$	$= 106$
Sodium hydrogencarbonate	$Na^+HCO_3^-$	$23 + 1 + 12 + (3 \times 16)$	$= 84$

Example 8.1 What mass of sodium hydroxide is needed to make $1 \, dm^3$ $(1000 \, cm^3)$ of 0·1 M solution?

$1 \, dm^3$ of 1 M sodium hydroxide requires 40 g
hence $1 \, dm^3$ of 0·1 M sodium hydroxide requires 4 g (*Answer*)

Example 8.2 What mass of sulphuric acid is needed to make $250 \, cm^3$ of 1 M solution?

$1 \, dm^3$ $(1000 \, cm^3)$ of 1 M sulphuric acid requires 98 g
hence $250 \, cm^3$ of 1 M sulphuric acid requires 24·5 g (*Answer*)

Example 8.3 What mass of sodium carbonate is needed to make $250 \, cm^3$ of 0·1 M solution?

$1000 \, cm^3$ of 1 M sodium carbonate requires 106 g
hence $250 \, cm^3$ of 1 M sodium carbonate requires 26·5 g
and $250 \, cm^3$ of 0·1 M sodium carbonate requires 2·65 g (*Answer*)

Experiment 8.2 Standardization of hydrochloric acid
'Standardization' of a solution means the determination of its concentration. In this experiment the solution to be standardized is approximately 0·2 M hydrochloric acid, and the method is to titrate a standard solution of sodium carbonate against the acid.

Step 1: preparation of $250 \, cm^3$ of 0·1 M sodium carbonate
As we found in Example 8.3 above, 2·65 g of sodium carbonate is needed to make $250 \, cm^3$ of 0·1 M solution. Exactly 2·65 g of anhydrous sodium carbonate is therefore weighed (in a weighing bottle) and transferred to a clean beaker. The weighing bottle is carefully washed with distilled water, and the washings allowed to drop into the beaker. The solid is dissolved and the solution transferred to a $250 \, cm^3$ graduated flask, using a funnel. Care is taken to ensure that all the solution is transferred from the beaker. The solution is made up to the mark on the flask and shaken well. It is exactly 0·1 M.

Step 2: the titration

A clean, dry 50 cm³ burette is filled with the hydrochloric acid above the 0 cm³ mark. When the tap is opened acid drains from the burette. The level is adjusted so that the bottom of the meniscus is level with the 0 cm³ mark.

Using a clean, dry pipette, 25 cm³ of the sodium carbonate solution is transferred to a clean conical flask. When all the liquid has drained from the pipette the surface of the liquid is touched with the tip of the pipette to ensure that exactly 25 cm³ of solution has been taken. A few drops of *methyl orange* indicator are added, the colour noted, and the flask placed directly below the burette (see Fig. 8.1). A white tile under the flask enables colour changes to be seen more clearly.

Approximately 15 cm³ of acid is allowed to run into the flask, then, shaking

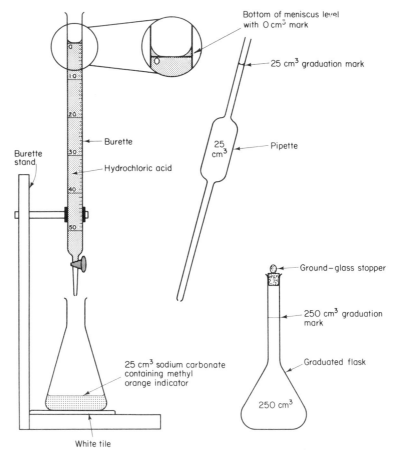

Fig. 8.1   Apparatus for the titration of sodium carbonate with hydrochloric acid

after each addition, acid is run in 1 cm³ at a time until the colour just changes from yellow to pink. The stage at which the colour change occurs is called the *endpoint*.

This first titration is only a trial, but from it we know roughly how much acid is needed to neutralize a 25 cm³ sample of the carbonate solution. Using this result as a guide, at least two further titrations are carried out on separate 25 cm³ portions of the carbonate solution; the volumes of acid used in two consecutive titrations should agree within 0·1 cm³.

Results

	First burette reading	*Second burette reading*	*Volume of acid used*
	cm³	cm³	cm³
Trial	0·0	25·0	25·0
Accurate	0·0	24·5	24·5
Accurate	24·5	49·0	24·5

Thus 24·5 cm³ of hydrochloric acid requires 25·0 cm³ of 0·1 M sodium carbonate solution for neutralization. The equation for the reaction is:

$$Na_2^+CO_{3(aq)}^{2-}+2H^+Cl_{(aq)}^- \rightarrow CO_{2(g)}+2Na^+Cl_{(aq)}^-+H_2O_{(l)}$$

Calculation

The molarity of the acid is calculated from the following formula (which is applicable to many other types of volumetric exercise and should be committed to memory):

$$\frac{\text{number of moles of Substance 1 in the balanced equation}}{\text{number of moles of Substance 2 in the balanced equation}}$$

$$= \frac{(\text{volume of Solution 1 } (V_1)) \times (\text{molarity of Solution 1 } (M_1))}{(\text{volume of Solution 2 } (V_2)) \times (\text{molarity of Solution 2 } (M_2))}$$

or $\quad \dfrac{\text{moles of 1 in balanced equation}}{\text{moles of 2 in balanced equation}} = \dfrac{V_1 \times M_1}{V_2 \times M_2}$

If Substance 1 is hydrochloric acid and Substance 2 is sodium carbonate, substitution in the formula gives:

$$\frac{2}{1} = \frac{24\cdot5 \times \text{molarity of acid}}{25\cdot0 \times 0\cdot1}$$

$$\text{molarity of acid} = \frac{2 \times 25\cdot0 \times 0\cdot1}{24\cdot5}$$

$$= 0\cdot204$$

Thus the hydrochloric acid solution has been standardized: its molarity is found to be 0·204 M.

8.7 Calculations Involving Acid/Alkali Reactions

The method described above for the titration of hydrochloric acid with sodium carbonate can be used for any other acid/alkali reaction. Reactions involving hydrochloric acid or nitric acid or sulphuric acid with sodium hydroxide or potassium hydroxide can use either phenolphthalein or methyl orange as indicator, whereas acid/carbonate and acid/hydrogencarbonate reactions always use methyl orange or some similar related indicator. The indicator chosen for any particular titration depends on its range of acidity and alkalinity.

Example 8.4 Calculate the volume of 0·1 M hydrochloric acid that is neutralized by 20 cm^3 of 0·2 M sodium hydroxide solution.

Calculation
The formula for use in all examples such as this is:

$$\frac{\text{(volume of acid)} \times \text{(molarity of acid)}}{\text{(volume of alkali)} \times \text{(molarity of alkali)}} = \frac{\text{number of moles of acid in the balanced equation}}{\text{number of moles of alkali in the balanced equation}}$$

The equation for the reaction between sodium hydroxide and hydrochloric acid is:

$$Na^+OH^-_{(aq)} + H^+Cl^-_{(aq)} \rightarrow Na^+Cl^-_{(aq)} + H_2O_{(l)}$$

showing that one mole of acid reacts with one mole of alkali. Substituting in the formula:

$$\frac{\text{volume of acid} \times 0·1}{20 \times 0·2} = \frac{1}{1}$$

$$\text{volume of acid} = \frac{1 \times 20 \times 0·2}{1 \times 0·1}$$

$$= 40 \text{ cm}^3 \ (Answer)$$

Example 8.5 Calculate the molarity of a sulphuric acid solution, 24 cm^3 of which is neutralized by 18 cm^3 of 0·04 M potassium hydroxide solution.

Calculation
The equation for this reaction is:

$$H_2^+SO_{4(aq)}^{2-} + 2K^+OH^-_{(aq)} \rightarrow K_2^+SO_{4(aq)}^{2-} + 2H_2O_{(l)}$$

showing that one mole of acid reacts with *two* moles of alkali. Substituting in the formula:

$$\frac{24 \times \text{molarity of acid}}{18 \times 0·04} = \frac{1}{2}$$

$$\text{molarity of acid} = \frac{18 \times 0{\cdot}04 \times 1}{2 \times 24}$$

$$= 0{\cdot}015 \text{ M } (Answer)$$

Example 8.6 Calculate the molarity of a solution of sodium hydroxide containing $0{\cdot}1$ g in 50 cm^3 solution.

Calculation

A molar (1 M) solution of sodium hydroxide contains 40 g in 1 dm^3. The given solution contains $0{\cdot}1$ g sodium hydroxide in 50 cm^3,

i.e. $\dfrac{1000}{50} \times 0{\cdot}1$ g in 1000 cm^3

i.e. 2 g in 1 dm^3

The given solution is only $\frac{2}{40}$ as concentrated as a molar solution: its molarity is therefore $\frac{2}{40} = 0{\cdot}05$ M (*Answer*)

8.8 Weak and Strong Acids

The terms 'weak' and 'strong' when applied to electrolytes indicate the degree to which they *dissociate into ions* when dissolved in water. A *strong electrolyte* will give many ions in aqueous solutions, a *weak electrolyte* few. These terms must not be confused with 'concentrated', indicating a large mass of dissolved material in a given volume of solution, and 'dilute' meaning a small mass of dissolved material in a given volume.

As mentioned in Unit 5, ethanoic (acetic) acid solution conducts an electric current with difficulty. This indicates that *ions* must be present, but not in the same amount as are present in a solution of, say, nitric acid. Thus the reaction

$$CH_3COOH_{(aq)} + H_2O_{(l)} \rightarrow CH_3COO^-_{(aq)} + H_3O^+_{(aq)}$$

occurs to a far less extent than

$$H^+NO^-_{3(aq)} + H_2O_{(l)} \rightarrow NO^-_{3(aq)} + H_3O^+_{(aq)}$$

Nitric acid is a *strong* acid, fully dissociated into oxonium ions $H_3O^+_{(aq)}$ and nitrate ions $NO^-_{3(aq)}$ in aqueous solution, whereas ethanoic (acetic) acid is a *weak* acid with few oxonium ions in aqueous solution. Much of the ethanoic acid remains in solution as undissociated CH_3COOH molecules.

Because of their different degrees of dissociation, equally concentrated solutions of ethanoic acid and nitric acid will differ in their pH values. Tests with Universal indicator show that the pH value of $0{\cdot}1$ M nitric acid is 1, and that of $0{\cdot}1$ M ethanoic acid is approximately 3. (The lower the pH value, the stronger the acid.)

Many weak acids occur naturally. They are often organic compounds containing carbon, hydrogen and oxygen. For example:

methanoic acid (formic acid) HCOOH is present in the 'sting' of ants;

ethanoic acid (acetic acid) CH_3COOH occurs in vinegar;

butanoic acid (butyric acid) C_3H_7COOH is responsible for the odour of rancid butter and 'strong' cheese;

ethanedioic acid (oxalic acid) $(COOH)_2$ occurs as a potassium salt in the leaves of rhubarb.

Basicity of an Acid

The number of replaceable hydrogen ions in one molecule of an acid is called its *basicity*. Thus the basicity of hydrochloric acid (H^+Cl^-) is 1, that of sulphuric acid $(H_2^+SO_4^{2-})$ is 2, while that of phosphoric acid $(H_3^+PO_4^{3-})$ is 3. Hydrochloric acid is described as a 'monobasic' acid, sulphuric acid is a 'dibasic' acid, and phosphoric acid is a 'tribasic' acid.

8.9 Salts

The 'common salt' used in everyday life as a seasoning is only one of a whole group of chemical compounds, the collective name of which is *salts*.

In general a salt is formed when the hydrogen ion of an acid is replaced wholly or partly by some other cation, e.g. the metal ion Cu^{2+} or the ammonium ion NH_4^+.

Thus when pieces of magnesium ribbon are added to dilute hydrochloric acid $(H^+Cl^-_{(aq)})$, the metal dissolves and bubbles of colourless hydrogen gas are given off. The magnesium replaces the hydrogen ions in the acid, forming the salt magnesium chloride in solution:

$$Mg_{(s)} + 2H^+Cl^-_{(aq)} \rightarrow Mg^{2+}Cl^-_{2(aq)} + H_{2(g)}$$

Normal Salts and Acid Salts

When *all* the hydrogen ions of an acid are replaced by other cations, the product is a *normal salt*; but if one or more hydrogen ions *remain* associated with the anion, the product is an *acid salt*. The term 'acid salt' does not necessarily imply a pH less than 7.

Consider dibasic carbonic acid $H_2^+CO_3^{2-}$ with two replaceable hydrogen ions. It can form *normal salts* in which *both* hydrogen ions are replaced, e.g. sodium carbonate $Na_2^+CO_3^{2-}$ or calcium carbonate $Ca^{2+}CO_3^{2-}$. However when only *one* of the hydrogen ions is replaced an *acid salt* is produced, e.g. sodium hydrogencarbonate $Na^+HCO_3^-$ or calcium hydrogencarbonate $Ca^{2+}(HCO_3^-)_2$. Note that in the acid salt the remaining hydrogen forms part of the hydrogencarbonate anion. Table 8.4 lists the different types of salts which can be formed from a number of the common acids.

8.10 Solubility of Salts

The oceans contain about 2·7% by mass of dissolved sodium chloride, and enormous quantities of this salt are extracted by solar evaporation of sea water. As the water evaporates the solution eventually becomes *saturated*, and solid

Table 8.4 Salts derived from common acids

Acid	Formula	Salt	Type	Example
Nitric	$H^+NO_3^-$	Nitrates	normal	$K^+NO_3^-$; $Ba^{2+}(NO_3^-)_2$
Hydrochloric	H^+Cl^-	Chlorides	normal	Na^+Cl^- ; $Cu^{2+}Cl_2^-$
Hydrobromic	H^+Br^-	Bromides	normal	K^+Br^- ; $Pb^{2+}Br_2^-$
Hydriodic	H^+I^-	Iodides	normal	$NH_4^+I^-$; $Pb^{2+}I_2^-$
Ethanoic (acetic)	CH_3COOH	Ethanoates (acetates)	normal	$CH_3COO^-Na^+$; $(CH_3COO^-)_2Cu^{2+}$
Sulphuric	$H_2^+SO_4^{2-}$	Sulphates	normal	$Ag_2^+SO_4^{2-}$; $Mg^{2+}SO_4^{2-}$
		Hydrogen-sulphates	acid	$Na^+HSO_4^-$
Carbonic	$H_2^+CO_3^{2-}$	Carbonates	normal	$Na_2^+CO_3^{2-}$; $Ca^{2+}CO_3^{2-}$
		Hydrogen-carbonates	acid	$Na^+HCO_3^-$; $Ca^{2+}(HCO_3^-)_2$
Phosphoric	$H_3^+PO_4^{3-}$	Phosphates	normal	$Ag_3^+PO_4^{3-}$; $Fe^{3+}PO_4^{3-}$
		Dihydrogen-phosphates	acid	$Na^+H_2PO_4^-$
		Monohydrogen-phosphates	acid	$Na_2^+HPO_4^{2-}$

sodium chloride begins to form as crystals. The term *saturated* is used to indicate a stage when the solvent contains as much solute as it can dissolve at a particular temperature; at this stage dissolved solute and crystals of undissolved solute can exist together.

Sodium chloride is quite soluble in water but lead(II) chloride is only sparingly soluble. Thus the degree of *solubility* varies from one salt to another, and we define solubility as *the number of grams of a solute necessary to saturate a fixed amount of solvent* (usually 100 g) *at a given temperature in the presence of undissolved solute*.

Solubility Curves

A solubility curve shows the variation of solubility with temperature. In general the solubility of a salt increases with increasing temperature, as illustrated in Fig. 8.2.

Experiment 8.3 *Determination of the solubility of potassium chloride, and construction of a solubility curve.*

A weighed amount (10 g) of potassium chloride is placed in a boiling tube. A small measured volume of distilled water is added from a burette and the tube warmed in a water bath to 80 °C. Distilled water is added—with constant

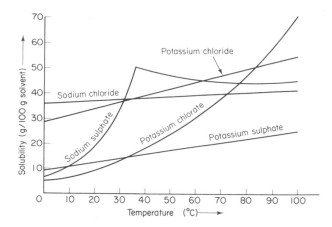

Fig. 8.2   Solubility curves for five typical salts

stirring—until all the solid potassium chloride dissolves, the temperature being maintained at 80 °C. The tube is removed from the water bath and allowed to cool. During cooling the solution is stirred continuously with a thermometer and the temperature at which the first crystals appear is recorded, see Fig. 8.3. The solution is reheated and cooled until consistent readings for the crystalliza-tion point are obtained. Thus the mass of water (1 cm³ of water can be considered as having a mass of 1 g approximately) which is saturated by 10 g of potassium chloride is known at the recorded temperature and hence the solubility can be calculated at this temperature.

A small measured quantity of water is then added from the burette and the new, lower, temperature at which crystals appear is recorded. This procedure is repeated several times and a graph is plotted from the solubilities at the various temperatures: such a graph is called a *solubility curve*. The solubilities can be determined at each temperature, as illustrated by the following example.

Results

Mass of potassium chloride used   = 10·00 g
Volume of water added from the burette   = 20·0 cm³
∴ mass of water   = 20·0 g
Average temperature at which crystals appear = 80 °C

Calculation

20·0 g of water at 80 °C is saturated by 10 g of potassium chloride

1·0 g of water at 80 °C is saturated by $\frac{10}{20} \times 1\cdot0$ g of potassium chloride
100·0 g of water at 80 °C is saturated by $\frac{10}{20} \times 100 = 50$ g of potassium chloride

Thus the solubility of potassium chloride at 80 °C is 50 g. This result provides one point for the construction of the solubility curve of potassium chloride in

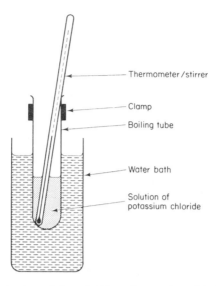

Thermometer/stirrer

Clamp

Boiling tube

Water bath

Solution of
potassium chloride

*Fig. 8.3   Determination of the solubility of potassium chloride at different
temperatures*

water. Other points for *different* temperatures can be obtained in a similar manner. A typical curve for potassium chloride is illustrated in Fig. 8.2.

At room temperature it is convenient to classify the solubility of many salts as follows:

(*a*) All the common salts of sodium, potassium, and ammonium (NH_4^+) are *soluble*.

(*b*) All nitrates are *soluble*.

(*c*) All the common chlorides, *except* those of silver, lead(II) and mercury(I), are *soluble*.

(*d*) All the common sulphates, *except* those of lead(II), barium and calcium, are *soluble*.

(*e*) All carbonates, *except* those of the alkali metals and ammonium, are *insoluble*.

8.11 Hydrated Salts

The evaporation of solutions of many salts eventually produces crystals having a definite shape and colour. Dry crystals of many of these salts liberate water vapour when heated. Such water is chemically combined with the cation (and in some cases the anion) and is incorporated in the structure of the crystal. This 'combined' water is called **water of crystallization**, and salts which contain it are said to be **hydrated**. Those salts which do not contain water of crystallization are said to be **anhydrous**: they may or may not be crystalline.

Some commonly occurring hydrated salts are listed in Table 8.5.

Table 8.5 Examples of hydrated salts

Name	Formula	Common name
Copper(II) sulphate-5-water	$Cu^{2+}SO_4^{2-}.5H_2O$	blue vitriol
Magnesium sulphate-7-water	$Mg^{2+}SO_4^{2-}.7H_2O$	Epsom salt
Sodium carbonate-10-water	$Na_2^+CO_3^{2-}.10H_2O$	washing soda
Sodium sulphate-10-water	$Na_2^+SO_4^{2-}.10H_2O$	Glauber's salt
Iron(II) sulphate-7-water	$Fe^{2+}SO_4^{2-}.7H_2O$	green vitriol
Calcium sulphate-2-water	$Ca^{2+}SO_4^{2-}.2H_2O$	gypsum

The presence of water of crystallization in a salt can be detected by heating a sample of the dry crystals in a boiling tube, as shown in Fig. 8.4.

On heating dry crystals of copper(II) sulphate the deep blue colour gradually fades and a colourless liquid is collected in the water-cooled test tube. The boiling point of this liquid (100 °C) shows it to be water. (Confirmation that a liquid is water can be obtained by measuring its *boiling point, freezing point* and *density* under specified conditions.)

When water is added to cold, white, anhydrous copper(II) sulphate the solid turns blue and much heat is produced. This is a test which indicates that a liquid contains water, but *not* necessarily that it is *pure* water.

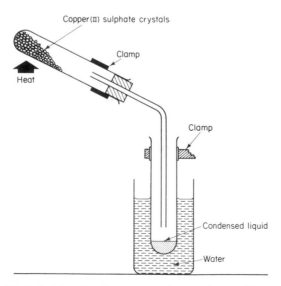

Fig. 8.4   Method of showing the presence of water of crystallization in a salt

Experiment 8.4   *Determination of the number of moles of water of crystallization associated with one mole of magnesium sulphate.*

An evaporating basin is weighed and about 5 g of magnesium sulphate crystals is added. The basin and contents are reweighed and then heated gently until no visible change can be observed. After cooling, the basin and contents are weighed. The basin is reheated for a short time, cooled and reweighed. This cycle of operations is repeated until two successive weighings are the same, showing that all the water of crystallization has been driven off. This process is termed 'heating to constant weight'.

Results

Mass of evaporating basin	= 36·00 g
Mass of evaporating basin + magnesium sulphate crystals	= 41·24 g
Mass of evaporating basin + anhydrous magnesium sulphate after heating to constant weight	= 38·56 g
Hence: Mass of magnesium sulphate crystals	= 5·24 g
Mass of anhydrous magnesium sulphate	= 2·56 g
Mass of water	= 2·68 g

Calculation

Since one mole of magnesium sulphate ($Mg^{2+}SO_4^{2-}$) has a mass of 120·5 g, the number of moles of magnesium sulphate present is

$$\frac{2 \cdot 56}{120 \cdot 5} = 0 \cdot 0213$$

Since one mole of water has a mass of 18 g, the number of moles of water present is

$$\frac{2 \cdot 68}{18} = 0 \cdot 149$$

Hence 0·0213 moles of magnesium sulphate is combined with 0·149 moles of water of crystallization. One mole of magnesium sulphate is therefore combined with

$$\frac{0 \cdot 149}{0 \cdot 0213} \text{ moles of water of crystallization}$$

$$= 6 \cdot 995 = 7 \text{ to the nearest whole number}$$

Thus the formula of magnesium sulphate crystals is $Mg^{2+}SO_4^{2-} \cdot 7H_2O$.

8.12 Deliquescence and Efflorescence

Substances which absorb water from the atmosphere and dissolve in it forming a solution are said to be **deliquescent** (see Section 6.2).

In contrast, some hydrated salts lose their water of crystallization simply on standing in a dry atmosphere at room temperature. Thus translucent crystals of sodium carbonate-10-water (washing soda) crumble to a white powder on standing. In this case nine of the ten molecules of water of crystallization are spontaneously lost to the air:

$$Na_2^+CO_3^{2-}.10H_2O_{(s)} \rightarrow Na_2^+CO_3^{2-}.H_2O_{(s)}+9H_2O_{(g)}$$

Such loss of water of crystallization is called **efflorescence** and the hydrated salt which loses the water is said to be efflorescent.

8.13 Preparation of Salts

Salts can be prepared in a variety of ways, and the choice of method is often dictated in practice by the solubility characteristics of the salt and of the reactants. This section illustrates the preparation of some simple salts and the isolation of a pure sample of the salt from solution.

(a) **Preparation of Salts by the Action of a Dilute Acid on a Carbonate, Basic Oxide, or a Metal**

The carbonates, basic oxides or metals used in these preparations are all insoluble in water. When reaction with the dilute acid is complete (warming if necessary) the excess oxide, metal or carbonate can be filtered leaving only a pure sample of the salt in solution. Suitable reactions include:

$$Cu^{2+}O^{2-}_{(s)} + H_2^+SO_{4(aq)}^{2-} \xrightarrow{\text{heat}} Cu^{2+}SO_{4(aq)}^{2-} + H_2O_{(l)}$$

$$Mg^{2+}CO_{3(s)}^{2-}+H_2^+SO_{4(aq)}^{2-} \rightarrow Mg^{2+}SO_{4(aq)}^{2-}+H_2O_{(l)}+CO_{2(g)}$$

$$Cu^{2+}CO_{3(s)}^{2-}+H_2^+SO_{4(aq)}^{2-} \rightarrow Cu^{2+}SO_{4(aq)}^{2-}+H_2O_{(l)}+CO_{2(g)}$$

$$Mg_{(s)}+H_2^+SO_{4(aq)}^{2-} \rightarrow Mg^{2+}SO_{4(aq)}^{2-}+H_{2(g)}$$

$$Fe_{(s)}+H_2^+SO_{4(aq)}^{2-} \rightarrow Fe^{2+}SO_{4(aq)}^{2-}+H_{2(g)}$$

Experiment 8.5   Preparation of copper(II) sulphate crystals from copper(II) oxide and dilute sulphuric acid

Approximately 50 cm^3 of 2 M sulphuric acid is poured into a beaker or conical flask and heated almost to boiling on a tripod and gauze. Copper(II) oxide is added carefully, a little at a time, until no more will react even after further heating:

$$Cu^{2+}O_{(s)}^{2-} + H_2^+SO_{4(aq)}^{2-} \rightarrow Cu^{2+}SO_{4(aq)}^{2-}+H_2O_{(l)}$$

The excess black solid copper(II) oxide is filtered from the hot solution.

The blue filtrate containing copper(II) sulphate is collected in an evaporating basin and a few drops of 2 M sulphuric acid are added to maintain a clear bright-blue solution. The solution is then heated to *crystallization point*. This point is detected by dipping a glass rod into the solution and transferring a drop of the liquid on to a white tile: when small crystals form on cooling, the whole solution is set aside to crystallize. The blue crystals of copper(II) sulphate are filtered and then dried between filter papers.

$$Cu^{2+}SO_{4(aq)}^{2-}+5H_2O_{(l)} \rightarrow Cu^{2+}SO_4^{2-}.5H_2O_{(s)}$$

The following preparations are similar to Experiment 8.5 (above) except that heat is not required.

(i) Magnesium sulphate crystals from magnesium carbonate and dilute sulphuric acid

(ii) Copper(II) sulphate crystals from copper(II) carbonate and dilute sulphuric acid

(iii) Magnesium sulphate crystals from magnesium metal and dilute sulphuric acid

(iv) Iron(II) sulphate crystals from iron filings and dilute sulphuric acid

(b) Preparation of Salts by the Action of an Acid on an Alkali

Unlike the previous method, both reactants are soluble in water. A titration (see Experiment 8.2) is required to ensure that the sample of salt contains neither excess acid nor excess alkali. The method is illustrated in the two following experiments.

Experiment 8.6   Preparation of sodium sulphate

Two clean dry burettes are required. One is filled with approximately 2 M sodium hydroxide solution, and the other with approximately 2 M sulphuric acid. Exactly 25 cm³ of the acid is run into a clean conical flask, and a few drops of phenolphthalein indicator are added. The solution remains colourless. The reading on the second burette is noted and alkali is added to the acid a little at a time, shaking between each addition, until one drop of alkali turns the solution pink. The reading on the second burette is again noted, and the experiment is repeated until two consecutive readings agree to within 0·1 cm³. The added sodium hydroxide has now neutralized the acid:

$$H_2^+SO_{4(aq)}^{2-}+2Na^+OH_{(aq)}^- \rightarrow Na_2^+SO_4^{2-}+2H_2O_{(l)}$$

Suppose the volume of sodium hydroxide required to neutralize the 25 cm³ of acid is 48·6 cm³. These volumes of acid (25·0 cm³) and alkali (48·6 cm³) are then measured from the burettes into an evaporating basin, this time *without any indicator*. The solution is evaporated to crystallization point (see Experiment 8·5) and the crystals of sodium sulphate-10-water filtered from solution and dried between filter papers.

$$Na_2^+SO_{4(aq)}^{2-}+10H_2O_{(l)} \rightarrow Na_2^+SO_4^{2-}.10H_2O_{(s)}$$

Experiment 8.7   Preparation of sodium hydrogensulphate

The two burettes are filled with the *same* solutions of sodium hydroxide and sulphuric acid that were used in the previous experiment. Exactly 50 cm³ of acid is mixed with 48·6 cm³ of the sodium hydroxide in a crystallizing dish and set on one side. Again no indicator is used to contaminate the salt. Doubling the quantity of acid used ensures that the alkali added is sufficient to neutralize exactly *half* the hydrogen ions from the sulphuric acid, leaving the acid salt:

$$2H_2^+SO_{4(aq)}^{2-}+2Na^+OH_{(aq)}^- \rightarrow 2Na^+HSO_{4(aq)}^-+2H_2O_{(l)}$$

After several days the crystals of sodium hydrogensulphate are filtered and dried between filter papers.

$$Na^+HSO_{4(aq)}^- + H_2O_{(l)} \rightarrow Na^+HSO_4^- . H_2O_{(s)}$$

Care of burettes. Burettes must be carefully rinsed with distilled water after use and left to dry with the tap removed. Sodium hydroxide solution tends to dissolve grease from taps and also to absorb carbon dioxide (from the air) forming sodium carbonate. This can cause the tap to become firmly cemented to its socket; moreover, the formation of solid in a restricted space can break the burette.

(c) Preparation of Insoluble Salts by Precipitation

Insoluble salts can be prepared by 'metathesis' or **double decomposition**. In this type of reaction the insoluble salt is precipitated by mixing two solutions, one containing the cation and the other the anion of the insoluble salt. For example, yellow insoluble lead(II) iodide is precipitated when solutions of lead(II) nitrate (containing the cation Pb^{2+}) and potassium iodide (containing the anion I^-) are mixed:

$$Pb^{2+}(NO_3^-)_{2(aq)} + 2K^+I_{(aq)}^- \rightarrow Pb^{2+}I_{2(s)}^- + 2K^+NO_{3(aq)}^-$$

The yellow precipitate of lead(II) iodide is centrifuged or filtered, washed with distilled water and dried.

Other insoluble salts such as silver chloride, lead(II) chloride and barium sulphate are prepared by this method.

If one of the materials available is insoluble it must be converted into a solution containing the appropriate ion. Thus in the preparation of lead(II) chloride from lead(II) carbonate, the *insoluble* lead(II) carbonate must first be converted into *soluble* lead(II) nitrate by the action of excess nitric acid:

$$Pb^{2+}CO_{3(s)}^{2-} + 2H^+NO_{3(aq)}^- \rightarrow Pb^{2+}(NO_3^-)_{2(aq)} + CO_{2(g)} + H_2O_{(l)}$$

White insoluble lead(II) chloride can then be precipitated by the addition of a solution containing chloride ions. For example, using dilute hydrochloric acid ($H^+Cl_{(aq)}^-$):

$$Pb^{2+}(NO_3^-)_{2(aq)} + 2H^+Cl_{(aq)}^- \rightarrow Pb^{2+}Cl_{2(s)}^- + 2H^+NO_{3(aq)}^-$$

(d) Preparation of Salts by Direct Combination

A few salts such as iron(II) sulphide, iron(III) chloride and copper(II) sulphide may be obtained by direct combination of the elements.

For example, when an intimate mixture of iron filings and sulphur is heated, a strongly exothermic reaction occurs. The mixture glows and black solid iron(II) sulphide is produced:

$$Fe_{(s)} + S_{(s)} \rightarrow Fe^{2+}S_{(s)}^{2-}$$

Summary of Unit 8

1. An **acid** yields hydrogen ions in aqueous solution; more generally, it is a substance which *donates* protons to other substances. Other properties of acids include:
    (*a*) sour taste
    (*b*) reaction with a carbonate or hydrogencarbonate to liberate carbon dioxide
    (*c*) ability to change the colour of certain dye solutions (indicators)
    (*d*) neutralization by a base.
2. A **base** is the oxide or hydroxide of a metal. It will neutralize an acid.
3. An **alkali** is a *soluble* base. It yields hydroxide ions in aqueous solution; more generally, it is a substance which *accepts* protons from other substances. Other properties of alkalis include:
    (*a*) 'soapy' feel
    (*b*) ability to change the colour of indicators
    (*c*) liberation of ammonia from ammonium salts.
4. The **pH scale** is a measure of the degree of acidity or alkalinity of a solution. The scale ranges from 0 to 14, a pH value of 7 representing a *neutral solution*. In the range 0 to 7, the lower the pH value the more acid the solution; in the range 7 to 14, the greater the pH value the more alkaline the solution.
5. **Oxides** are compounds of an element with oxygen. They can be classified as *acidic, basic, neutral* or *amphoteric*. Other classes include *peroxides* and *compound oxides*.
6. **Titration** is the progressive addition of one solution (e.g. an acid) to another (e.g. an alkali) until reaction between them is complete.
7. A **standard solution** is one in which the concentration of solute is known.
8. A **one-molar solution** contains one mole of solute dissolved in 1 dm^3 of solution.
9. The molarity of a solution can be calculated from the following formula:

$$\frac{(\text{volume of Solution 1}) \times (\text{molarity of Solution 1})}{(\text{volume of Solution 2}) \times (\text{molarity of Solution 2})} = \frac{\text{number of moles of Substance 1 in the balanced equation}}{\text{number of moles of Substance 2 in the balanced equation}}$$

10. A *strong* acid yields many $H^+_{(aq)}$ ions (actually as oxonium ions H_3O^+) in aqueous solution, whereas a *weak* acid yields few.
11. A **concentrated solution** is one which contains a large mass of solute in a given volume of solution; a **dilute solution** contains a small mass of solute in a given volume of solution.
12. A **salt** is formed when the hydrogen ion of an acid is replaced wholly (a *normal salt*) or partly (an *acid salt*) by some other cation.
13. A **saturated solution** contains as much solute as the solvent will dissolve at a particular temperature and in the presence of crystals of undissolved solute.
14. **Solubility** is defined as the number of grams of solute necessary to saturate a fixed amount of solvent at a given temperature.

15. **Hydrated salts** contain a definite molecular proportion of water, chemically combined in a crystal of the salt, e.g. $Cu^{2+}SO_4^{2-}.5H_2O$. Such combined water is called *water of crystallization*.
16. A **deliquescent** substance absorbs water from the atmosphere and dissolves in it forming a solution.
17. An **efflorescent salt** loses all or part of its water of crystallization when exposed to the air.
18. Salts may be prepared by the following methods:
    (a) action of a dilute acid on a carbonate, basic oxide or a metal
    (b) neutralization of an acid by an alkali
    (c) precipitation   (d) direct combination.

Test Yourself on Unit 8

1. A chemist labelled three shelves *acids*, *bases* and *salts*. On which shelf would you place the following chemicals?
   (a) Na^+OH^-
   (b) $Na^+NO_3^-$
   (c) H^+Cl^-
   (d) $Ca^{2+}O^{2-}$
   (e) $Cu^{2+}SO_4^{2-}$

2. The pH value of a solution of nitric acid is always:
   (a) 1,   (b) 7,   (c) 14,   (d) less than 7,   (e) greater than 7.

3. Distilled water has a pH of:
   (a) 1,   (b) 7,   (c) 14,   (d) less than 7,   (e) greater than 7.

4. State whether the following statements are true or false:
   (a) A base is always soluble in water.
   (b) Sodium hydroxide is both a base and an alkali.
   (c) Hydrogen is released whenever an acid and metal are mixed.
   (d) All indicators give a red solution with an acid.
   (e) A gardener neutralizes an alkaline soil by the addition of 'lime' (calcium hydroxide).

5. On standing in air, (i) sodium carbonate-10-water (washing soda) crystals crumble to a white powder, whereas (ii) sodium hydroxide pellets gradually turn into a colourless liquid.
Select the most appropriate name for each of these two different processes (i) and (ii) from the following list:
   (a) effervescence,
   (b) efflorescence,
   (c) hydration,
   (d) decrepitation,
   (e) deliquescence,
   (f) diffusion.

6. What volume of 0·100 M sodium hydroxide solution:
(*a*) contains 4 g of sodium hydroxide?
(*b*) neutralizes 25 cm³ of 0·05 M hydrochloric acid solution?
(*c*)neutralizes 25 cm³ of 0·05 M sulphuric acid solution? Write the equation for this reaction.
(*d*) reacts exactly with 0·5 mole of hydrogen ion from an acid?
(Relative atomic masses: Na = 23, O = 16 and H = 1)

7. In the following directions for the laboratory preparation of pure dry copper(II) sulphate crystals certain words and phrases are underlined. Correct these directions by replacing the underlined words and phrases.

Approximately 50 cm³ of <u>concentrated hydrochloric acid</u> is warmed with a <u>little</u> copper(II) oxide. When no more of the oxide reacts the mixture is <u>poured</u> into an evaporating basin and the <u>residue</u> evaporated to <u>dryness</u>. After cooling, the crystals are filtered, washed with a <u>large volume of water</u> and dried <u>over a bunsen flame</u>.

8. The following graph shows the solubility curves for potassium nitrate, sodium nitrate, potassium chloride and sodium chloride.

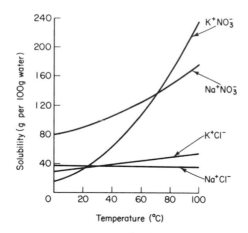

Potassium nitrate can be prepared by mixing hot saturated solutions of potassium chloride and sodium nitrate. Use the solubility curves to answer the following questions:
(*a*) Which salt crystallizes first from solution at 80°C?
(*b*) Which salt crystallizes first from solution at 10°C?
(*c*) At which temperature are the solubilities of potassium nitrate and sodium nitrate the same?
(*d*) Which is the most soluble salt at 40°C?
(*e*) If a saturated solution of sodium nitrate, at 80°C, containing 150 g sodium nitrate in 100 g water was cooled to 0°C, how much sodium nitrate would crystallize?

9. Insert the following oxides in their appropriate position in the table below:
(a) carbon monoxide, (b) aluminium oxide, (c) dilead(II) lead(IV) oxide, Pb_3O_4,
(d) phosphorus(V) oxide, (e) sodium peroxide $N_2^+O_2^{2-}$, (f) copper(II) oxide.

Classification	Oxide
Acidic	
Basic	
Neutral	
Amphoteric	
Peroxide	
Compound oxide	

Mark this test out of 40 with the answers given on page 376.

Classification of Matter II: The Metals

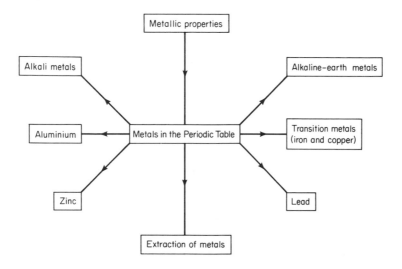

The elements in the periodic table can be classified in many different ways, e.g. solid, liquid and gas. However, one of the most convenient classifications has four distinct categories: *metals*, *non-metals*, the *noble gases* and *hydrogen* (which is in a class by itself). This Unit begins by looking at those properties which can be used to classify elements as *metals* and then goes on to consider some of the more important metals. The chemistry of metals is developed to show property patterns and hence relationships in families or groups within the periodic table.

9.1 What are Metals?

Aluminium, iron, tin, copper and lead are very much part of the world we live in. For most people recognition of these substances as metals lies in their hardness and in their shiny lustrous surface. They can be hammered into shape (*malleability*) and drawn into wires (*ductility*). These are only a few of the physical properties which the chemist uses to classify elements as metals. In order to explain these physical properties we need to look again at the structure of metals. A useful model of a metal is one in which the atoms exist as spherical, positive ions (cations) arranged in a regular three-dimensional network or 'crystal lattice' (see Unit 2). The electrons present in the metal are distributed in such a way that

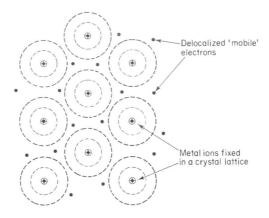

Delocalized 'mobile' electrons

Metal ions fixed in a crystal lattice

Fig. 9.1   The metallic bond

the groups of positive metal ions in the crystal lattice are surrounded by a 'sea' of mobile electrons (see Fig. 9.1). These mobile electrons constitute the *metallic bond*.

9.2   Metallic Properties and the Metallic Bond

(*a*) **Conductivity**

High *electrical conductivity* in metals results from the ease of electron movement from one place in the metal crystal to another (see Section 5.1).

High *thermal conductivity* in metals is largely due to the movement of their mobile electrons. By contrast, heat conduction in ionic and covalent solids is accomplished by energy changes between particles during the slight thermal vibrations of the ions or atoms in their localized positions.

(*b*) **Metallic Lustre**

Metals are crystalline and their flat crystal surfaces are able to reflect light. This may also be explained, in part, by the surface mobile electrons absorbing and re-emitting light energy.

(*c*) **Malleability and Ductility**

In contrast to ionic and covalent crystals, which are *brittle*, metals have high malleability and ductility because it is relatively easy for metal atoms to be moved about within the lattice without destroying the bonding.

9.3   Metals in the Periodic Table

Recognition of a metal by its general physical properties is not always easy, especially for those elements near the centre of the periodic table. Let us therefore consider the *chemical* properties of the metals and how they fit into the periodic classification of the elements.

(*a*) Metals are elements which readily *lose electrons* to form cations. Magnesium, for example, loses two electrons to become the magnesium cation:

$$Mg \rightarrow Mg^{2+} + 2e$$

(*b*) Metals form *basic oxides* (see Unit 8). Copper, for example, forms copper(II) oxide which will neutralize an acid.

(*c*) The halides (chlorides, bromides etc.) of the metals have an ionic structure; when dissolved in water their solution conducts an electric current and is decomposed by it.

Elements exhibiting these metallic properties are found on the *left-hand side* of the periodic table (see Fig. 2.4). Typical metallic families are the *alkali metals* (Group 1) and the *alkaline-earth metals* (Group 2), while some of the more common and important metals occur in the *transition elements*.

Study of a typical family such as the alkali metals shows that the elements become more metallic in character as the atomic number increases. Thus potassium (At. No. 19) is more metallic than sodium (At. No. 11), which in turn is more metallic than lithium (At. No. 3).

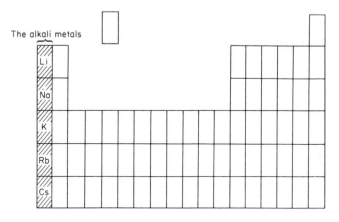

Fig. 9.2   Position of the alkali metals in the periodic table

9.4 The Alkali Metals (Group 1)

The position of the alkali metals in the periodic table is shown in Fig. 9.2, and their names, symbols, electronic configurations and important physical properties are listed in Table 9.1.

Table 9.1 The alkali metals

	Lithium (Li)	Sodium (Na)	Potassium (K)	Rubidium (Rb)	Caesium (Cs)
Electronic configuration	2.1	2.8.1	2.8.8.1	2.8.18.8.1	2.8.18.18.8.1
Atomic number	3	11	19	37	55
Melting point (K)	452	371	337	312	302
Boiling point (K)	1590	1165	1047	961	960
Density (kg m^{-3})	534	970	860	1530	1870

The alkali metals are soft solids and all are easily cut with a knife. When freshly cut, the surface shows a metallic silvery lustre which quickly tarnishes on exposure to the air. They are all good conductors of heat and electricity.

9.5 Chemical Properties of the Alkali Metals

The chemistry of the alkali metals is the chemistry of their ions. All of these metals readily lose their single outer electron to form positive ions with unit charge and the corresponding noble-gas structure. For example:

$$Na \rightarrow Na^+ + e$$

The more easily the atoms lose this outer electron, the more reactive they are. Reactivity increases with increasing atomic number for two reasons: (a) the number of electron shells increases, shielding the outer negative electron from the attraction of the positive nucleus; (b) the distance of the outer electron from the nucleus increases, causing a decrease in the electrostatic attraction. Thus potassium is more reactive than sodium which in turn is more reactive than lithium.

Reaction of the Alkali Metals with Water
The gradation in chemical activity with increasing atomic number can be illustrated by the reactions of the alkali metals with water.

Experiment 9.1   *Reaction between sodium and water*
A small piece of sodium is cut from a stick of the metal and freed from the paraffin oil under which it is stored by pressing it between filter paper (*avoid contact with the skin*). After the white surface coating has been scraped off, exposing the silvery metal, the piece is transferred to a trough containing water and a few drops of phenolphthalein indicator. The sodium melts into a silvery globule which darts about on the surface of the water, becoming smaller and smaller until it disappears. A bright-red solution in the trough indicates that an alkali (sodium hydroxide) has been produced:

$$2Na_{(s)} + 2H_2O_{(l)} \rightarrow 2Na^+OH^-_{(aq)} + H_{2(g)}$$

If a piece of sodium is placed in a sodium spoon and then plunged *beneath* the surface of water in a trough, bubbles of a colourless gas rise from the metal and can be collected in a test tube filled with water (see Fig. 9.3).

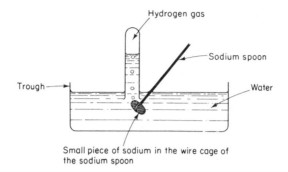

Fig. 9.3   Sodium reacts with water to produce hydrogen

On ignition the gas burns with a pale-blue flame. The gas is hydrogen.

These experiments are then repeated with freshly cut pieces of lithium and potassium. The lithium sample moves around the surface of the water, gradually dissolving. Reaction is less vigorous than with sodium. On the other hand, pieces of potassium melt and skate around on the surface of the water very quickly; so much heat is generated that the hydrogen evolved is ignited and burns with a lilac flame characteristic of potassium. This reaction is much more vigorous than that of sodium or lithium.

Caution: care must be taken when pieces of sodium, lithium or potassium are added to water. If possible cover the trough with a sheet of glass to minimize the risk of splashing. Small pieces of unused metal must be disposed of by dissolving them in a beaker containing excess methylated spirit placed in a fume cupboard.

Reaction of the Alkali Metals with Air

The alkali metals exhibit a distinct gradation in reactivity when exposed to air or oxygen. The surface of freshly cut lithium quickly tarnishes in air, but the metal needs to be warmed before it will react—quite vigorously—to produce lithium oxide:

$$2Li_{(s)} + \tfrac{1}{2}O_{2(g)} \rightarrow Li_2^+ O_{(s)}^{2-}$$

Sodium forms mainly the *peroxide*:

$$2Na_{(s)} + O_{2(g)} \rightarrow Na_2^+ O_{2(s)}^{2-}$$

Potassium, rubidium and caesium form the rather unusual ionic *superoxide* in which the orange-coloured O_2^- ion is present:

$$K_{(s)} + O_{2(g)} \rightarrow K^+ O_{2(s)}^-$$

9.6 Uses of the Alkali Metals

Sodium plays an important part in the manufacture of tetraethyl-lead(IV) which is used as an 'anti-knock' additive in petrol. Sodium is also used as a coolant and heat-transfer agent in some nuclear reactors. Rubidium and caesium are used in photoelectric cells. The photoelectric effect is observed when light of high energy strikes certain metals (particularly caesium) causing electrons to be emitted by the metallic surface.

9.7 Detection of the Alkali Metals

Chemical detection of most metal ions depends on their ability to form a precipitate (often coloured) with a suitable anion. However, the alkali-metal ions form few insoluble compounds and their presence is most easily detected by the colour

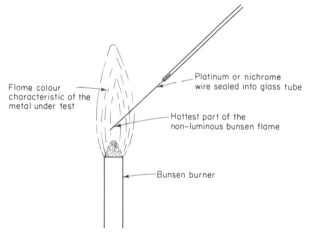

Fig. 9.4   Flame test for metals

Table 9.2 Characteristic flame colours

Metal	Flame coloration
Lithium	crimson-red
Sodium	yellow
Potassium	lilac (pink when viewed through blue cobalt glass)
Rubidium	red
Caesium	blue
Barium	apple-green
Strontium	red
Calcium	brick-red
Copper	blue-green

they impart to a bunsen flame. A platinum or nichrome wire is dipped in a solution of the alkali-metal salt with concentrated hydrochloric acid and then held in a hot, non-luminous bunsen flame (see Fig. 9.4).

The colour produced in the flame by the vaporized metal atom is characteristic of the metal. Certain other metals also give characteristic flame colours and are consequently included with the alkali metals in Table 9.2.

9.8 Important Compounds of the Alkali Metals

The similarities in chemical properties between the corresponding compounds of the various alkali metals are such that we can summarize the chemistry of the whole group by studying only one metal. We choose to consider sodium compounds simply because they are the most common.

(a) **Sodium chloride**, Na^+Cl^-
Sodium chloride occurs extensively as deposits of *rock salt* and in *sea water* (approximately 3%). Rock salt may be obtained by mining, but more often water is pumped down into the deposit and the salt is brought to the surface as brine.

In warm regions 'salt' is produced by solar evaporation of sea water or taken from deposits around the dried-up fringes of salt lakes.

Because of its widespread distribution and availability, sodium chloride finds many uses (see Fig. 9.5).

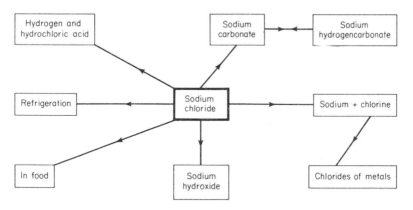

Fig. 9.5   Some of the uses of sodium chloride

(b) **Sodium hydroxide**, Na^+OH^- (caustic soda)
Sodium hydroxide is a white solid (melting point 591 K) which is sold in the form of flakes, pellets or sticks. It is a *caustic alkali* (caustic = burning) which should be handled with care. Most of it is manufactured by the electrolysis of sodium chloride solution in a *mercury cell* (described in Section 14.3).

When the solid is exposed to air it *deliquesces*, i.e. takes up moisture and dissolves in it. A solution of sodium hydroxide readily absorbs carbon dioxide, forming sodium carbonate:

$$2Na^+OH^-_{(s)} + CO_{2(g)} \rightarrow Na_2^+CO_{3(aq)}^{2-} + H_2O_{(l)}$$

Thus sodium hydroxide left standing in a moist atmosphere acquires a white coating of solid sodium carbonate as the solution produced by deliquescence absorbs carbon dioxide.

Other notable properties of sodium hydroxide include the following:

(i) It is a strong alkali capable of neutralizing acids to produce salts. For example:

$$Na^+OH^-_{(aq)} + H^+Cl^-_{(aq)} \rightarrow Na^+Cl^-_{(aq)} + H_2O_{(l)}$$

(ii) The hydroxides of many metals can be precipitated from solution using sodium hydroxide solution. For example:

$$2Na^+OH^-_{(aq)} + Cu^{2+}SO_{4(aq)}^{2-} \rightarrow Cu^{2+}(OH^-)_{2(s)} + Na_2^+SO_{4(aq)}^{2-}$$

or

$$2OH^-_{(aq)} + Cu^{2+}_{(aq)} \rightarrow Cu^{2+}(OH^-)_{2(s)}$$

(iii) Sodium hydroxide, like any strong alkali, will liberate *ammonia* gas from ammonium salts on warming. This is a test for an ammonium salt.

$$Na^+OH^-_{(aq)} + NH_4^+Cl^-_{(s)} \rightarrow NH_{3(g)} + H_2O_{(l)} + Na^+Cl^-_{(aq)}$$

or

$$OH^-_{(aq)} + NH_{4(s)}^+ \rightarrow NH_{3(g)} + H_2O_{(l)}$$

(c) **Sodium carbonate** $(Na^{2+}CO_3^{2-})$ and **sodium hydrogencarbonate** $(Na^+HCO_3^-)$

Sodium carbonate-10-water $Na_2^+CO_3^{2-}.10H_2O$ is called *washing soda*. It is manufactured by the *Solvay* or *ammonia–soda* process, one of the most economically efficient industrial processes (see Section 14.5).

Washing soda is *efflorescent*, losing nine of its ten molecules of water of crystallization on standing in dry air:

$$Na_2^+CO_3^{2-}.10H_2O_{(s)} \rightarrow Na_2^+CO_3^{2-}.H_2O_{(s)} + 9H_2O_{(g)}$$

(i) *Action of heat*

On heating the carbonate, both the mono- and decahydrates give the anhydrous salt. Further heating has no effect: *the carbonate is not decomposed*. In contrast, sodium hydrogencarbonate decomposes on heating to give the carbonate, carbon dioxide and water vapour:

$$2Na^+HCO_{3(s)}^- \xrightarrow{\text{heat}} Na_2^+CO_{3(s)}^{2-} + H_2O_{(g)} + CO_{2(g)}$$

(ii) *Action of acids*

Both the carbonate and hydrogencarbonate of sodium and the other alkali metals liberate carbon dioxide on addition of an acid. For example:

$$Na_2^+CO_{3(s)}^{2-} + 2H^+Cl^-_{(aq)} \rightarrow 2Na^+Cl^-_{(aq)} + CO_{2(g)} + H_2O_{(l)}$$

$$Na^+HCO_{3(s)}^- + H^+NO_{3(aq)}^- \rightarrow Na^+NO_{3(aq)}^- + CO_{2(g)} + H_2O_{(l)}$$

(iii) *Action of water*
Solutions in water of both sodium hydrogencarbonate and (particularly) sodium carbonate give an alkaline reaction with Universal indicator (pH is greater than 7).

(*d*) **Sodium sulphate, $Na_2^+SO_4^{2-}$, and sodium hydrogensulphate, $Na^+HSO_4^-$**
The laboratory preparation of both these compounds is described in Unit 8. Sodium sulphate-10-water is commonly known as *Glauber's salt* after the 17th-century German chemist, Johann Glauber.

A solution of sodium hydrogensulphate is strongly acidic and exhibits some of the properties of sulphuric acid.

(*e*) **Sodium nitrate, $Na^+NO_3^-$**
Sodium nitrate occurs naturally as *Chile saltpetre*. It decomposes on heating into sodium nitrite and oxygen:

$$2Na^+NO_{3(s)}^- \xrightarrow{\text{heat}} 2Na^+NO_{2(s)}^- + O_{2(g)}$$

Potassium nitrate, commonly called saltpetre or nitre, is used extensively in the production of fireworks and explosives.

9.9 The Alkaline-earth Metals (Group 2)

The position of the alkaline-earth metals in the periodic table is shown in Fig. 9.6, and the names, symbols, electronic configurations and important physical properties are listed in Table 9.3.

The Group 2 metals look more like metals than the elements of Group 1 because of their silvery appearance and greater hardness.

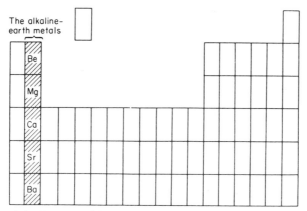

Fig. 9.6   Position of the alkaline-earth metals in the periodic table

Table 9.3 The alkaline-earth metals

	Beryllium (Be)	Magnesium (Mg)	Calcium (Ca)	Strontium (Sr)	Barium (Ba)
Electronic configuration	2.2	2.8.2	2.8.8.2	2.8.18.8.2	2.8.18.18.8.2
Atomic number	4	12	20	38	56
Melting point (K)	1550	924	1120	1042	1000
Boiling point (K)	3243	1380	1760	1657	1910
Density ($kg\,m^{-3}$)	1800	1741	1540	2600	3600

They tarnish on exposure to air, but not as rapidly as the Group 1 metals. All these elements are excellent conductors of heat and electricity.

9.10 Chemical Properties of the Alkaline-earth Metals

The elements in Group 2 can all easily attain a noble-gas structure by losing their two outer electrons. Thus the chemistry of these metals is essentially the chemistry of their dipositive ions (Mg^{2+}, Ca^{2+} etc.). As with the Group 1 metals the reactivity increases with increasing atomic number. The properties of magnesium and calcium are typical of the group as a whole, and for this reason these two metals will be studied in some detail.

(i) *Reaction with water*
Unlike the alkali metals, magnesium does not react readily with cold water. However, if it is heated it will react with steam (see Fig. 9.7):

$$Mg_{(s)} + H_2O_{(g)} \rightarrow Mg^{2+}O^{2-}_{(s)} + H_{2(g)}$$

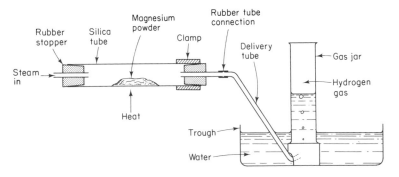

Fig. 9.7   Experiment to demonstrate the action of magnesium on steam

In contrast, calcium when freshly cut sinks in cold water and reacts steadily, liberating a colourless gas (hydrogen) which can be collected as shown in Fig. 9.8:

$$Ca_{(s)} + 2H_2O_{(l)} \rightarrow Ca^{2+}(OH^-)_{2(aq)} + H_{2(g)}$$

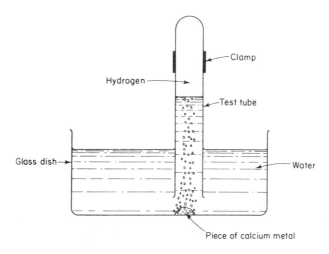

Fig. 9.8   *Calcium reacts with water to produce hydrogen*

(ii) *Reaction with acids*

Both calcium and magnesium react with dilute mineral acids, but the reaction of calcium is violent and should not be attempted in the laboratory. Hydrogen is liberated from most dilute acids. For example:

$$Mg_{(s)} + 2H^+Cl^-_{(aq)} \rightarrow Mg^{2+}Cl^-_{2(aq)} + H_{2(g)}$$

With *nitric acid* and magnesium, hydrogen is liberated only when the acid is very dilute (less than 5%). Magnesium is one of the few metals that will liberate hydrogen from nitric acid. With most other metals nitric acid acts as an oxidizing agent and the products include oxides of nitrogen. Copper, for example, reacts with moderately concentrated nitric acid to produce nitrogen monoxide and a solution of copper(II) nitrate:

$$3Cu_{(s)} + 8H^+NO^-_{3(aq)} \rightarrow 3Cu^{2+}(NO_3^-)_{2(aq)} + 2NO_{(g)} + 4H_2O_{(l)}$$

9.11 Uses of the Alkaline-earth Metals

Magnesium is the only Group 2 element which finds important industrial use, particularly in light, strong, corrosion-resistant alloys. Beryllium is used in high-strength light alloys.

9.12 Important Compounds of the Alkaline-earth Metals

(a) **Calcium carbonate**, $Ca^{2+}CO_3^{2-}$

Calcium carbonate is one of the most abundant compounds in the earth's crust. It is found in a number of different forms.

(i) *In limestone*

Limestone occurs in layers as a sedimentary rock, having been formed from deposits of shells of minute marine animals millions of years ago.

(ii) *In marble*

Marble is limestone that has been changed by heat and pressure so that it is hard and capable of taking a high polish.

(iii) *In chalk*

Chalk has a similar origin to limestone but is softer and more porous.

(iv) *Calcite*

Calcite is a pure crystalline variety of calcium carbonate. Transparent colourless crystals are called *Iceland spar*.

Like all carbonates, calcium carbonate reacts with acids liberating carbon dioxide. For example:

$$Ca^{2+}CO_{3(s)}^{2-}+2H^+Cl_{(aq)}^- \rightarrow Ca^{2+}Cl_{2(aq)}^- + H_2O_{(l)} + CO_{2(g)}$$

The reaction with sulphuric acid is rapidly inhibited by a layer of almost insoluble calcium sulphate.

At high temperatures calcium carbonate is converted into white calcium oxide (quicklime):

$$Ca^{2+}CO_{3(s)}^{2-} \rightarrow Ca^{2+}O_{(s)}^{2-}+CO_{2(g)}$$

A lump of calcium carbonate (limestone) glows when hot but otherwise appears unchanged. After cooling, the addition of a few drops of water causes the lump to swell, crack, crumble and become so hot that steam is liberated. This chemical change is called *slaking of quicklime*:

$$Ca^{2+}O_{(s)}^{2-}+H_2O_{(l)} \rightarrow Ca^{2+}(OH^-)_{2(s)}$$

Calcium carbonate is used as a raw material for the production of many other compounds, and its industrial importance is discussed in Unit 14.

(b) **Calcium hydroxide**, $Ca^{2+}(OH^-)_2$ (slaked lime)

Calcium hydroxide is a white solid, slightly soluble in water: the clear solution is called *lime-water*. Lime-water is used to test for carbon dioxide (see Section 6.12) and has the properties of an alkali. A suspension of slaked lime (calcium hydroxide) in water is called *milk of lime*.

(c) **Calcium sulphate**, $Ca^{2+}SO_4^{2-}$

Calcium sulphate occurs naturally as the mineral *gypsum* ($Ca^{2+}SO_4^{2-}$.

$2H_2O$) and as *anhydrite* ($Ca^{2+}SO_4^{2-}$). Because of their slight solubility both of these minerals produce permanent hardness in water (see Section 6.16).

When gypsum is heated it loses part of its water of crystallization, forming a white powder known as *plaster of Paris* (calcium sulphate-$\frac{1}{2}$-water):

$$2Ca^{2+}SO_4^{2-}.2H_2O_{(s)} \rightarrow (Ca^{2+}SO_4^{2-})_2.H_2O_{(s)} + 3H_2O_{(g)}$$

When mixed with water plaster of Paris is rapidly reconverted to gypsum, the mass setting to a solid. The setting process is accompanied by slight expansion, which makes plaster of Paris particularly useful for taking casts and for supporting broken bones.

9.13 Aluminium

Aluminium is the most important metal in Group 3. It is the most abundant metal in the earth's crust, occurring in clays, feldspar, mica and many other natural deposits.

The electronic configuration and important physical properties of aluminium are listed in Table 9.4.

Table 9.4 Physical properties of aluminium

	Electronic configuration	Atomic number	Melting point (K)	Boiling point (K)	Density (kg m^{-3})
Aluminium (Al)	2.8.3	13	933	2740	2700

Aluminium is a silvery white metal which quickly becomes dulled with a thin layer of oxide. This oxide film forms an inert protective layer on the metal surface and makes the aluminium very resistant to corrosive action. The thickness of the film can be increased artificially by an electrolytic process producing 'anodized' aluminium.

9.14 Chemical Properties of Aluminium

We have seen that the chemistry of the Group 1 and Group 2 metals is essentially that of their ions. The chemistry of aluminium, however, is not dominated by the Al^{3+} ion because the removal of *three* electrons ($Al \rightarrow Al^{3+} + 3e$) to form a noble-gas structure requires a great deal of energy.

(a) Reaction with Air

Aluminium is resistant to further oxidation and corrosion by the air once the protective oxide film has been formed. Removal of this oxide film renders the metal reactive. Salt water attacks the oxide film allowing the aluminium to become corroded; ordinary aluminium is therefore not used for marine purposes.

(b) Reaction with Acids

Once again the oxide film minimizes chemical attack by acids on the metal. The protective film is strengthened by oxidizing acids such as sulphuric and, particularly, nitric acid: hence no reaction occurs with them. Moderately concentrated hydrochloric acid will dissolve the metal, forming the chloride and liberating hydrogen:

$$2Al_{(s)} + 6H^+Cl^-_{(aq)} \rightarrow 2Al^{3+}Cl^-_{3(aq)} + 3H_{2(g)}$$

(c) Reaction with Alkalis

In marked contrast to the Group 1 and 2 metals, aluminium dissolves in sodium or potassium hydroxide solution forming the *tetrahydroxo-aluminate* ion (see Section 9.16a).. Once started the exothermic reaction is vigorous and hydrogen is liberated:

$$2Al_{(s)} + 2Na^+OH^-_{(aq)} + 6H_2O_{(l)} \rightarrow 2\underset{\substack{\text{sodium tetrahydroxo-}\\\text{aluminate}}}{Na^+Al(OH)^-_{4(aq)}} + 3H_{2(g)}$$

(d) The Thermite Reaction

When aluminium powder is mixed with iron(III) oxide in a fire-clay crucible, and the mixture ignited with a magnesium fuse (see Fig. 9.9), an extremely exothermic reaction takes place producing molten iron:

$$2Al_{(s)} + Fe_2^{3+}O_{3(s)}^{2-} \rightarrow Al_2^{3+}O_{3(s)}^{2-} + 2Fe_{(l)}$$

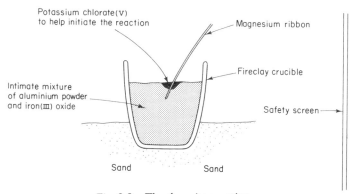

Fig. 9.9   The thermite reaction

This reaction can be used for on-the-spot welding. (*Caution*: the thermite reaction can be very dangerous owing to the production of molten iron and showers of sparks. Observers should maintain a safe distance behind a safety screen.)

In general any reaction between aluminium and the oxide of a less active metal may be called a thermite reaction; it will be similarly vigorous and highly exothermic.

9.15 Uses of Aluminium

Because of its metallic nature aluminium is a good conductor of heat and electricity. Its lightness in comparison to copper makes it a useful alternative for electric power lines. It is also used for making cooking utensils which are light, durable and corrosion-resistant. Aluminium is used extensively with other metals to form tough light-weight alloys. For example, *duralumin* (aluminium, copper and magnesium) and *magnalium* (aluminium and magnesium) are both widely used in aircraft manufacture.

Aluminium foil is used in cooking, in packaging and for milk-bottle tops.

The metal finds widespread use in the building industry for window frames, panels, etc.

9.16 Compounds of Aluminium

(*a*) **Aluminium hydroxide**, $Al^{3+}(OH^-)_3$

When sodium or potassium hydroxide is added to a solution of aluminium salt, a white gelatinous precipitate of aluminium hydroxide is formed:

$$Al_2^{3+}(SO_4^{2-})_{3(aq)} + 6Na^+OH_{(aq)}^- \rightarrow 2Al^{3+}(OH^-)_{3(s)} + 3Na_2^+SO_{4(aq)}^{2-}$$

Aluminium hydroxide is insoluble in water, but if an excess of sodium hydroxide is added the precipitate dissolves forming sodium tetrahydroxo-aluminate:

$$Al^{3+}(OH^-)_{3(s)} + Na^+OH_{(aq)}^- \rightarrow Na^+Al(OH)_{4(aq)}^-$$

This reaction is unusual in that the aluminium hydroxide shows *acidic properties* in its reaction with the caustic alkali sodium hydroxide.

On the other hand, aluminium hydroxide dissolves in hydrochloric acid showing its expected *basic properties*:

$$Al^{3+}(OH^-)_{3(s)} + 3H^+Cl_{(aq)}^- \rightarrow Al^{3+}Cl_{3(aq)}^- + 3H_2O_{(l)}$$

Hydroxides (or oxides) which show *both* acidic and basic properties are said to be *amphoteric*.

(*b*) **Aluminium oxide**, $Al_2^{3+}O_3^{2-}$ (alumina)

Aluminium oxide can be prepared in the laboratory by heating the hydroxide:

$$2Al^{3+}(OH^-)_{3(s)} \xrightarrow{\text{heat}} Al_2^{3+}O_{3(s)}^{2-} + 3H_2O_{(g)}$$

When freshly prepared it reacts with dilute acids and caustic alkalis, showing its *amphoteric* nature.

The finely divided powder is inert and is used as the absorbing medium in column chromatography.

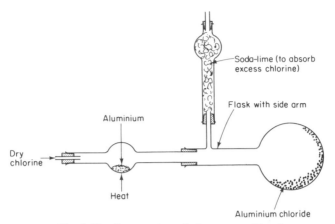

Fig. 9.10   Preparation of aluminium chloride

(c) **Aluminium chloride**, $Al^{3+}Cl_3^-$ or Al_2Cl_6

The chloride can be prepared as shown in Fig. 9.10 by passing dry chlorine over heated aluminium:

$$2Al_{(s)} + 3Cl_{2(g)} \rightarrow 2Al^{3+}Cl_{3(s)}^- \rightarrow Al_2Cl_{6(s)}$$

This reaction is strongly exothermic.

Aluminium chloride sublimes and may be collected as its vapour cools. Physical measurements indicate that the pure anhydrous chloride exists as Al_2Cl_6 double molecules which dissociate on heating to give $Al^{3+}Cl_3^-$. In aqueous solution the Al^{3+} ion is highly hydrated and a solution of the chloride shows an acid reaction.

(d) **Aluminium sulphate-18-water**, $Al_2^{3+}(SO_4^{2-})_3.18H_2O$, and **the alums** $M^+Al^{3+}(SO_4^{2-})_2.12H_2O$

Aluminium sulphate crystallizes from solution with eighteen molecules of water of crystallization. When a solution containing aluminium sulphate and potassium sulphate in equimolar quantities is allowed to stand, the *double* salt aluminium potassium sulphate-12-water, $K^+Al^{3+}(SO_4^{2-})_2.12H_2O$, known as *potash alum*, is obtained. Like all double salts, potash alum in solution gives reactions which indicate the presence of all the constituent ions: in this case the potassium ion K^+, the aluminium ion Al^{3+} and the sulphate ion SO_4^{2-}.

It is possible to prepare a series of 'alums', each of which contains the unipositive ion K^+ or NH_4^+ combined with the tripositive ion Al^{3+}, Fe^{3+} or Cr^{3+}. For example, *chrome alum* is chromium(III) potassium sulphate-12-water, $K^+Cr^{3+}(SO_4^{2-})_2.12H_2O$; *ammonium alum* is aluminium ammonium sulphate-12-water, $NH_4^+Al^{3+}(SO_4^{2-})_2.12H_2O$.

The alums are *isomorphous*, i.e. they are identical in crystalline form. Thus it is possible for a crystal of purple chrome alum suspended in a saturated solution of potash alum to acquire a colourless overgrowth of the latter having an identical octahedral form.

9.17 The Transition Metals

The location of the transition metals within the periodic table is indicated in Fig. 9.11.

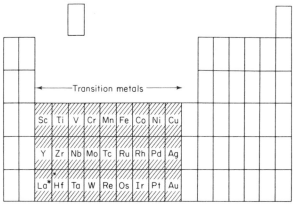

Fig. 9.11   *Position of the transition metals in the periodic table*

The *first transition series* of elements (from scandium Sc to copper Cu, inclusive) contains some of the most abundant as well as the most industrially important metals. These elements all have two electrons in their outermost shell and an incomplete penultimate shell of similar energy; hence they display several features in common. For example, they have

(a) high melting points (a consequence of strong metallic bonding)
(b) pronounced malleability and ductility
(c) coloured ions (in most cases)
(d) the ability to catalyse certain reactions
(e) variable oxidation states, e.g. iron(II) and iron(III), associated with variable positive charge on the ions, e.g. Fe^{2+} and Fe^{3+}

Iron and copper are typical of the transition metals and have been selected for special study because of their enormous practical importance. Table 9.5 lists the electronic configurations and physical properties of these two metals.

Table 9.5 Physical properties of iron and copper

	Atomic number	Electronic configuration	Density $(kg\ m^{-3})$	Melting point (K)	Boiling point (K)
Iron(Fe)	26	2.8.14.2	7870	1808	3300
Copper(Cu)	29	2.8.17.2	8930	1356	2868

9.18 Iron

Iron is the fourth most abundant element in the earth's crust and the second most abundant metal. Minerals containing iron which can be profitably extracted are called *iron ores*. These include *haematite* (Fe_2O_3) and *magnetite* (Fe_3O_4). Because of its abundance and relative ease of extraction, iron (including its alloys) is the most widely used of all the metals.

The extraction of iron is carried out in a *blast furnace* (see Unit 14) and during this 'smelting' process the molten iron absorbs carbon, silicon, sulphur, manganese and phosphorus. The percentage of these elements will vary slightly and determine the quality of the iron. Although some of the molten iron from the blast furnace may be run into moulds of sand to make *pig iron*, most of it is used directly to make *steel*.

9.19 Properties of Iron

Pure iron is a silvery white metal which melts at 1808 K. It is usually coated with a layer of the oxide Fe_3O_4 which dulls the surface lustre.

(a) Rusting

Iron does not readily corrode when exposed to the action of either dry air or pure, air-free water. In the presence of *both* air and water, however, corrosion occurs resulting in the formation of **rust**. This rust is essentially a hydrated form of iron(III) oxide, $Fe_2^{3+}O_3^{2-} . xH_2O$.

To show that the formation of rust requires both air and water, some dry iron nails are placed in three test tubes as illustrated in Fig. 9.12.

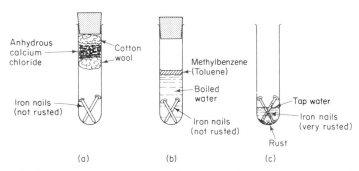

Fig. 9.12   Experiment to show that both air and water are necessary for rusting

There is no water in tube (*a*) and no dissolved air in tube (*b*). Rusting occurs after a few days only in tube (*c*) where the iron is attacked by both dissolved air and water. No rusting takes place in tubes (*a*) or (*b*).

The rusting of iron is an electrochemical process. Impurities and scratches on the iron surface collect water containing dissolved air. A tiny electrical cell is set up, and the iron dissolves as part of an oxidation–reduction reaction within the cell. This solution of iron is eventually converted by the oxygen in the air into hydrated iron(III) oxide, which we call rust. Rusting is accelerated by the presence of an electrolyte (e.g. sodium chloride from sea water or from roads that have been 'salted' to prevent icing) which increases the conductivity of the solution in the tiny cell.

Prevention of rusting. A considerable amount of time and money is spent on protecting iron from corrosion, particularly in the motor-car industry, the building and engineering trades, indeed wherever iron is exposed to the atmosphere. The common methods of protection include:

(i) *Painting.* A coating of paint protects iron by excluding air and moisture.

(ii) *Galvanizing.* In this process the iron is dipped into molten zinc, thus covering the surface of the iron with a thin layer of zinc which does not readily corrode.

(iii) *Plating.* Iron is covered with a thin layer of a metal which is resistant to corrosion. Metals such as chromium and nickel are deposited on the iron by an electrolytic process. Tin-plate, from which tin cans are made, is produced by dipping sheets of iron or steel in molten tin.

(iv) *Phosphatizing.* A thin, tenacious layer of impervious iron(II) phosphate can be produced on the surface of iron by reaction with a suitable phosphate solution.

(b) Reaction with Steam

Iron at red heat decomposes steam, producing hydrogen and the 'compound oxide' iron(II) di-iron(III) oxide. This reaction is reversible:

$$3Fe_{(s)} + 4H_2O_{(g)} \rightleftharpoons Fe_3O_{4(s)} + 4H_{2(g)}$$

(c) Reaction with Acids

Iron dissolves in most *dilute* acids and in concentrated hydrochloric acid, liberating hydrogen and producing iron(II) salts. For example:

$$Fe_{(s)} + 2H^+Cl^-_{(aq)} \rightarrow Fe^{2+}Cl^-_{2(aq)} + H_{2(g)}$$

$$Fe_{(s)} + H_2^+SO_{4(aq)}^{2-} \rightarrow Fe^{2+}SO_{4(aq)}^{2-} + H_{2(g)}$$

With dilute nitric acid a complex reaction takes place due to the strong oxidizing nature of the acid. Concentrated nitric acid renders iron passive by forming an impervious surface film of iron(II) di-iron(II) oxide, Fe_3O_4.

9.20 Compounds of Iron

Having two oxidation states, iron(II) Fe^{2+} and iron(III) Fe^{3+}, iron is capable of forming two types of compound. Whether an atom loses two electrons to become an Fe^{2+} ion, or loses three electrons to become an Fe^{3+} ion depends on the nature of the reacting substance.

The two oxidation states of iron are related as follows:

$$\text{iron(II)} \underset{\text{reduction}}{\overset{\text{oxidation}}{\rightleftarrows}} \text{iron(III)}$$

In an *oxidation* reaction, iron(II) *loses* an electron:

$$Fe^{2+} \xrightarrow{\text{oxidation}} Fe^{3+} + e$$

In a *reduction* reaction, iron(III) *gains* an electron:

$$Fe^{3+} + e \xrightarrow{\text{reduction}} Fe^{2+}$$

(a) Oxides

Iron(II) *oxide*, $Fe^{2+}O^{2-}$, is stable only at high temperatures.

Iron(III) *oxide*, $Fe_2^{3+}O_3^{2-}$, is prepared as a red powder when green iron(II) sulphate crystals are heated:

$$2Fe^{2+}SO_4^{2-}.7H_2O_{(s)} \rightarrow Fe_2^{3+}O_{3(s)}^{2-} + SO_{2(g)} + SO_{3(g)} + 14H_2O_{(g)}$$

This oxide is used for polishing metals and glass ('jeweller's rouge') and as a red pigment.

As well as the oxides corresponding to iron(II) and iron(III), there is a third oxide, *iron*(II) *di-iron*(III) *oxide*, whose formula can be written as either $Fe^{2+}O^{2-}.Fe_2^{3+}O_3^{2-}$ or Fe_3O_4. This is in fact the most readily prepared of the three oxides, and is formed when steam is passed over heated iron or when iron burns in air:

$$3Fe_{(s)} + 2O_{2(g)} \rightarrow Fe_3O_{4(s)}$$

(b) Hydroxides

Iron(II) *hydroxide*, $Fe^{2+}(OH^-)_2$, is obtained as a pale-green precipitate when sodium hydroxide solution or aqueous ammonia is added to a solution of an iron(II) salt. For example:

$$Fe^{2+}SO_{4(aq)}^{2-} + 2Na^+OH_{(aq)}^- \rightarrow Fe^{2+}(OH^-)_{2(s)} + Na_2^+SO_{4(aq)}^{2-}$$

The hydroxide dissolves in acids forming a solution of a salt. For example:

$$Fe^{2+}(OH^-)_{2(s)} + H_2^+SO_{4(aq)}^{2-} \rightarrow Fe^{2+}SO_{4(aq)}^{2-} + 2H_2O_{(l)}$$

This solution shows the characteristic green colour of the hydrated $Fe_{(aq)}^{2+}$ ion.

Iron(III) *hydroxide*, $Fe^{3+}(OH^-)_3$, is obtained as a 'foxy' red-brown precipitate when sodium hydroxide or aqueous ammonia is added to a solution of an iron(III) salt. For example:

$$Fe_2^{3+}(SO_4^{2-})_{3(aq)} + 6Na^+OH_{(aq)}^- \rightarrow 2Fe^{3+}(OH^-)_{3(s)} + 3Na_2^+SO_{4(aq)}^{2-}$$

The precipitate dissolves readily in acids forming a solution showing the characteristic yellow-brown colour of the hydrated $Fe_{(aq)}^{3+}$ ion.

(c) **Chlorides**

Iron(II) *chloride*, $Fe^{2+}Cl_2^-$, is formed as a white solid when hydrogen chloride gas is passed over heated iron:

$$Fe_{(s)} + 2HCl_{(g)} \rightarrow Fe^{2+}Cl_{2(s)}^- + H_{2(g)}$$

The apparatus used for this preparation is shown in Fig. 9.10.

When iron is added to dilute or concentrated hydrochloric acid, a solution of iron(II) chloride is produced:

$$Fe_{(s)} + 2H^+Cl_{(aq)}^- \rightarrow Fe^{2+}Cl_{2(aq)}^- + H_{2(g)}$$

Iron(II) chloride-6-water can be crystallized from this solution, but it is difficult to obtain the anhydrous salt.

Iron(III) *chloride*, $Fe^{3+}Cl_3^-$, is formed as a red-black solid when chlorine is passed over heated iron:

$$2Fe_{(s)} + 3Cl_{2(g)} \rightarrow 2Fe^{3+}Cl_{3(s)}^-$$

Once the reaction has begun the heat produced is sufficient to sustain the reaction without further external heating. The chloride sublimes and can be collected in a cooled receiver like that shown in Fig. 9.10.

When iron(III) oxide dissolves in hot concentrated hydrochloric acid a deep yellow-brown solution containing iron(III) chloride is produced:

$$Fe_2^{3+}O_{3(s)}^{2-} + 6H^+Cl_{(aq)}^- \rightarrow 2Fe^{3+}Cl_{3(aq)}^- + 3H_2O_{(l)}$$

It is again difficult to obtain the anhydrous salt from this solution.

(d) **Sulphates**

Iron(II) *sulphate*, $Fe^{2+}SO_4^{2-}$, is produced when dilute sulphuric acid acts on iron:

$$Fe_{(s)} + H_2^+SO_{4(aq)}^{2-} \rightarrow Fe^{2+}SO_{4(aq)}^{2-} + H_{2(g)}$$

Hydrogen is evolved and the solution obtained has the pale-green colour characteristic of the hydrated Fe^{2+} ion. Crystallization from this solution yields the heptahydrate $Fe^{2+}SO_4^{2-} \cdot 7H_2O$.

Iron(III) *sulphate*, $Fe_2^{3+}(SO_4^{2-})_3$, can be prepared by the oxidation of iron(II) sulphate in sulphuric acid using a suitable oxidizing agent. For example, using hydrogen peroxide:

$$2Fe^{2+}SO_{4(aq)}^{2-} + H_2^+SO_{4(aq)}^{2-} + H_2O_{2(aq)} \rightarrow Fe_2^{3+}(SO_4^{2-})_{3(aq)} + 2H_2O_{(l)}$$

9.21 Oxidation and Reduction of Iron Salts

The conversion of iron(II) to iron(III) is an *oxidation* process. Thus oxidation of any iron(II) salt with an appropriate oxidizing agent produces an iron(III) salt. Suitable oxidizing agents are hydrogen peroxide, concentrated nitric acid, chlorine and even concentrated sulphuric acid. For example, with chlorine:

$$2Fe^{2+}Cl_{2(aq)}^- + Cl_{2(g)} \rightarrow 2Fe^{3+}Cl_{3(aq)}^-$$

The conversion of iron(III) to iron(II) is a *reduction* process. Suitable reducing agents include hydrogen sulphide and metals as they ionize, e.g. zinc powder and dilute sulphuric acid.

When hydrogen sulphide is passed into iron(III) chloride solution a precipitate of sulphur is observed:

$$2Fe^{3+}Cl_{3(aq)}^- + H_2S_{(g)} \rightarrow 2Fe^{2+}Cl_{2(aq)}^- + 2H^+Cl_{(aq)}^- + S_{(s)}$$

On filtering this pale-yellow precipitate, the characteristic green solution of iron(II) chloride remains.

Metals towards the top of the electrochemical series (see 5.9) can dissolve in acid or aqueous solution to yield electrons which reduce iron(III) to iron(II). For example, if zinc powder is added to an acidified solution of iron(III) sulphate, iron(II) sulphate is obtained.

$$Zn_{(s)} \rightarrow Zn_{(aq)}^{2+} + 2e^-$$

$$2Fe_{(aq)}^{3+} + 2e^- \rightarrow 2Fe_{(aq)}^{2+}$$

Adding: $$Zn_{(s)} + 2Fe_{(aq)}^{3+} \rightarrow Zn_{(aq)}^{2+} + 2Fe_{(aq)}^{2+}$$

9.22 Tests for Iron(II) and Iron(III) Ions

(*a*). When sodium hydroxide or aqueous ammonia is added to a solution of an iron salt, a green gelatinous precipitate indicates iron(II), whereas a 'foxy' red-brown gelatinous precipitate indicates iron(III) (see Section 9.20*b*).

(*b*) The addition of potassium thiocyanate to a solution of an iron salt produces a blood-red coloration if iron(III) ions are present; iron(II) ions do not react. The test is extremely sensitive and trace amounts of iron(III) ions can often be detected in solutions containing mainly iron(II) ions.

(*c*) When potassium hexacyanoferrate(II), $K_4^+Fe(CN)_6^{4-}$, is added to a solution of an iron salt, a deep-blue precipitate ('Prussian blue') indicates the presence of an iron(III) salt. With iron(II) ions a white precipitate forms which rapidly turns blue.

9.23 Copper

Although copper does not occur abundantly in nature, its consumption among the metals is only exceeded by iron and aluminium. Most of the world's copper is obtained from the sulphide ore *chalcopyrite* (copper pyrites) $CuFeS_2$. The sulphide ore is heated in a *reverberatory furnace* to produce 'blister copper', a crude form of the metal. A reverberatory furnace (Latin *reverberare* = to beat back) is one in which the flame is reflected down on to the hearth from the vaulted roof of the heating chamber; thus the ore does not come into contact with the fuel.

The impure copper is refined by making it the anode of an electrolytic cell which has thin strips of pure copper as the cathode and copper(II) sulphate as the electrolyte (see Fig. 9.13).

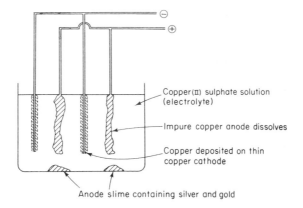

Copper(II) sulphate solution (electrolyte)

Impure copper anode dissolves

Copper deposited on thin copper cathode

Anode slime containing silver and gold

Fig. 9.13   Electrolytic refining of copper

During electrolysis the anode dissolves and pure copper is deposited on the cathode (see Section 5.10 for details). The impurities (including valuable amounts of silver and gold) from the crude copper collect as a sludge below the anode.

9.24 Properties of Copper

Pure copper is the only *red* metal. On exposure to the atmosphere it becomes coated with a protective patina of green basic copper(II) carbonate. The metal is not attacked by water (or steam). Copper will not dissolve in acids to liberate hydrogen, but it will dissolve in oxidizing acids, such as nitric and sulphuric, as indicated in the following reactions.

(*a*) With 50% nitric acid, nitrogen monoxide is produced as the copper dissolves to form a blue solution containing the hydrated copper(II) ion:

$$3Cu_{(s)} + 8H^+NO_{3(aq)}^- \rightarrow 3Cu^{2+}(NO_3^-)_{2(aq)} + 2NO_{(g)} + 4H_2O_{(l)}$$

(*b*) With concentrated nitric acid, brown nitrogen dioxide is formed:

$$Cu_{(s)} + 4HNO_{3(l)} \rightarrow Cu^{2+}(NO_3^-)_{2(aq)} + 2NO_{2(g)} + 2H_2O_{(l)}$$

(*c*) With hot concentrated sulphuric acid, colourless sulphur dioxide gas is liberated:

$$Cu_{(s)} + 2H_2SO_{4(l)} \rightarrow Cu^{2+}SO_{4(aq)}^{2-} + SO_{2(g)} + 2H_2O_{(l)}$$

9.25 Uses of Copper

Because of its excellent electrical conductivity, copper finds widespread use in electrical engineering, from large-scale generators to minute electronic equipment.

Pure copper is relatively soft, but its strength can be improved by alloying with other metals. The chief alloys are *brass* (copper with up to 40% zinc) and *bronze* (copper with up to 10% tin). Bronzes with small quantities of phosphorus are called *phosphor bronze* and are particularly valuable as springs for delicate instruments.

Copper is used as a roofing material, in water piping, in radiators, in under-floor heating and in many other applications throughout the building industry. Copper and its alloys find a use in almost every walk of life.

9.26 Compounds of Copper

Copper has two oxidation states: copper(I) Cu^+ and copper(II) Cu^{2+}. However, the hydrated copper(I) ion is unstable and decomposes in solution into copper and the copper(II) ion. For this reason the only copper(I) compound we need to consider is copper(I) oxide.

(a) Oxides

Copper(I) oxide, $Cu_2^+O^{2-}$, is obtained as an orange-red precipitate in the reduction of an alkaline solution of copper(II) sulphate. This reaction is used as a test for a reducing agent such as glucose ($C_6H_{12}O_6$).

When this oxide is warmed with dilute sulphuric acid, a blue solution of copper(II) sulphate is produced together with a dark-red precipitate of copper:

$$Cu_2^+O_{(s)}^{2-} + H_2^+SO_{4(aq)}^{2-} \rightarrow Cu^{2+}SO_{4(aq)}^{2-} + Cu_{(s)} + H_2O_{(l)}$$

Copper(II) oxide, $Cu^{2+}O^{2-}$, is obtained as a black solid when the nitrate, carbonate or hydroxide of copper(II) is heated:

$$2Cu^{2+}(NO_3^-)_{2(s)} \rightarrow 2Cu^{2+}O_{(s)}^{2-} + 4NO_{2(g)} + O_{2(g)}$$

$$Cu^{2+}CO_{3(s)}^{2-} \rightarrow Cu^{2+}O_{(s)}^{2-} + CO_{2(g)}$$

$$Cu^{2+}(OH^-)_{2(s)} \rightarrow Cu^{2+}O_{(s)}^{2-} + H_2O_{(g)}$$

When copper itself is heated in air its surface becomes blackened with a coating of copper(II) oxide:

$$2Cu_{(s)} + O_{2(g)} \rightarrow 2Cu^{2+}O_{(s)}^{2-}$$

Copper(II) oxide is a basic oxide, insoluble in water but dissolving in dilute acids to form copper(II) salts. Hydrogen will reduce the heated oxide to copper:

$$Cu^{2+}O_{(s)}^{2-} + H_{2(g)} \rightarrow Cu_{(s)} + H_2O_{(g)}$$

(b) Copper(II) hydroxide, $Cu^{2+}(OH^-)_2$

A blue-green gelatinous precipitate of copper(II) hydroxide is obtained when a solution of a caustic alkali (e.g. sodium hydroxide) is added to a solution of a copper(II) salt such as copper(II) sulphate:

$$Cu^{2+}SO_{4(aq)}^{2-} + 2Na^+OH_{(aq)}^- \rightarrow Cu^{2+}(OH^-)_{2(s)} + Na_2^+SO_{4(aq)}^{2-}$$

If aqueous ammonia is added to a solution of a copper(II) salt, the initial result is the precipitation of copper(II) hydroxide:

$$Cu^{2+}SO_{4(aq)}^{2-} + 2NH_{3(aq)} + 2H_2O_{(l)} \rightarrow Cu^{2+}(OH^-)_{2(s)} + (NH_4^+)_2SO_{4(aq)}^{2-}$$

Further addition of aqueous ammonia causes the blue-green precipitate of copper(II) hydroxide to dissolve, forming a deep royal-blue solution containing the tetra-ammine-copper(II) ion:

$$Cu^{2+}(OH^-)_{2(s)}+4NH_{3(aq)} \rightarrow Cu(NH_3)_4^{2+}(OH^-)_{2(aq)}$$

At one time this solution was used to dissolve cellulose in the production of 'artificial silk'.

(c) **Copper(II) sulphate-5-water**, $Cu^{2+}SO_4^{2-}.5H_2O$
The preparation of copper(II) sulphate crystals is described in Section 8.13. This blue crystalline salt containing five molecules of water of crystallization is probably the most familiar of the copper(II) salts and has many uses. It is used in electroplating, in dyeing and printing textiles, as a wood preservative and as a fungicide in Bordeaux mixture.

(d) **Copper(II) nitrate-3-water**, $Cu^{2+}(NO_3^-)_2.3H_2O$
This salt is prepared in solution by the action of moderately concentrated nitric acid on copper:

$$3Cu_{(s)}+8H^+NO_{3(aq)}^- \rightarrow 3Cu^{2+}(NO_3^-)_{2(aq)}+2NO_{(g)}+4H_2O_{(l)}$$

It is very deliquescent, and isolation of the anhydrous salt is difficult.
On strong heating, copper(II) nitrate decomposes forming black copper(II) oxide, brown nitrogen dioxide and oxygen:

$$2Cu^{2+}(NO_3^-)_{2(s)} \rightarrow 2Cu^{2+}O_{(s)}^{2-}+4NO_{2(g)}+O_{2(g)}$$

(e) **Copper(II) carbonate**
A basic copper(II) carbonate occurs naturally in the mineral *malachite*, $Cu^{2+}CO_3^{2-}.Cu^{2+}(OH^-)_2$. On heating, it forms copper(II) oxide and liberates carbon dioxide:

$$Cu^{2+}CO_3^{2-}.Cu^{2+}(OH^-)_{2(s)} \rightarrow 2Cu^{2+}O_{(s)}^{2-}+CO_{2(g)}+H_2O_{(g)}$$

9.27 Zinc

Zinc is the first non-transition metal to occur after copper (copper is the last metal in the first transition series). Its major physical characteristics are listed in Table 9.6.

Table 9.6 Physical properties of zinc

	Atomic number	Electronic configuration	Density (kg m⁻³)	Melting point (K)	Boiling point (K)
Zinc (Zn)	30	2.8.18.2	7140	693	1180

Zinc occurs naturally as *zinc blende*, $Zn^{2+}S^{2-}$, and to a lesser extent as *calamine*, $Zn^{2+}CO_3^{2-}$. These ores are first roasted in air to produce the oxide:

$$2Zn^{2+}S_{(s)}^{2-} + 3O_{2(g)} \rightarrow 2Zn^{2+}O_{(s)}^{2-} + 2SO_{2(g)}$$

$$Zn^{2+}CO_{3(s)}^{2-} \rightarrow Zn^{2+}O_{(s)}^{2-} + CO_{2(g)}$$

Then the oxide is reduced to the metal by heating with powdered coke:

$$Zn^{2+}O_{(s)}^{2-} + C_{(s)} \rightarrow Zn_{(g)} + CO_{(g)}$$

Zinc vapour is condensed and the liquid run into moulds.

(a) Properties

Pure zinc is a silvery white metal; on exposure to the air it becomes coated with a thin protective layer of the oxide and the basic carbonate. The metal does not react with water, but heated zinc will displace hydrogen from steam:

$$Zn_{(s)} + H_2O_{(g)} \rightarrow Zn^{2+}O_{(s)}^{2-} + H_{2(g)}$$

Most mineral acids react with the metal to liberate hydrogen, although the reaction is much slower than might be expected. Impure zinc dissolves quickly in both dilute hydrochloric and sulphuric acids. For example:

$$Zn_{(s)} + 2H^+Cl_{(aq)}^- \rightarrow Zn^{2+}Cl_{2(aq)}^- + H_{2(g)}$$

Reaction with nitric acid is complicated because of the strong oxidizing nature of this acid. Depending on the conditions, nitrogen dioxide, nitrogen monoxide and dinitrogen oxide (nitrous oxide) are formed together with zinc nitrate.

Zinc dissolves in caustic alkali solutions forming the tetrahydroxozincate ion and liberating hydrogen. For example, with sodium hydroxide:

$$Zn_{(s)} + 2Na^+OH_{(aq)}^- + 2H_2O_{(l)} \rightarrow Na_2^+Zn(OH)_{4(aq)}^{2-} + H_{2(g)}$$

(b) Uses

Large quantities of the metal are used in galvanizing iron to protect it from corrosion.

Zinc is also important in alloys such as *brass* (zinc and copper), and as the negative electrode in dry batteries.

9.28 Compounds of Zinc

In contrast to the transition metals, zinc has only one oxidation state (Zn^{2+}) and thus forms only one series of compounds. The common zinc compounds show few anomalous properties.

(a) Zinc oxide, $Zn^{2+}O^{2-}$

Zinc oxide can be prepared in the usual way by the action of heat on the hydroxide, carbonate or nitrate. For example:

$$2Zn^{2+}(NO_3^-)_{2(s)} \rightarrow 2Zn^{2+}O_{(s)}^{2-} + 4NO_{2(g)} + O_{2(g)}$$

The oxide is a white solid which changes to pale yellow on being heated and back to white again on cooling.

It is an *amphoteric oxide*, dissolving in both acid and alkali:

$$Zn^{2+}O^{2-}_{(s)} + 2H^+Cl^-_{(aq)} \rightarrow Zn^{2+}Cl^-_{2(aq)} + H_2O_{(l)}$$

$$Zn^{2+}O^{2-}_{(s)} + 2Na^+OH^-_{(aq)} + H_2O_{(l)} \rightarrow Na^+_2Zn(OH)^{2-}_{4(aq)}$$

Zinc oxide is used in antiseptic ointments and as a white pigment in paints.

(*b*) **Zinc hydroxide, $Zn^{2+}(OH^-)_2$**
The hydroxide is formed as a white gelatinous precipitate when an alkali is added to a solution of a zinc salt. For example:

$$2Na^+OH^-_{(aq)} + Zn^{2+}SO^{2-}_{4(aq)} \rightarrow Zn^{2+}(OH^-)_{2(s)} + Na^+_2SO^{2-}_{4(aq)}$$

If the sodium hydroxide is added in excess, the precipitate dissolves:

$$Zn^{2+}(OH^-)_{2(s)} + 2Na^+OH^-_{(aq)} \rightarrow Na^+_2Zn(OH)^{2-}_{4(aq)}$$

Zinc hydroxide is also soluble in aqueous ammonia owing to the formation of the soluble colourless tetra-ammine-zinc ion, $Zn(NH_3)^{2+}_4$.

(*c*) **Zinc sulphate-7-water, $Zn^{2+}SO^{2-}_4.7H_2O$**
The sulphate is a white crystalline solid formed by dissolving the hydroxide, oxide, carbonate or metal in dilute sulphuric acid and evaporating the resulting solution to crystallization point.

(*d*) **Zinc nitrate-6-water, $Zn^{2+}(NO_3^-)_2.6H_2O$**
This very deliquescent hydrated salt is decomposed on heating into zinc oxide, nitrogen dioxide and oxygen (see zinc oxide above).

9.29 Lead

Lead has been known for at least 5000 years and by Roman times was in common use for making water pipes etc. The chief source of lead is the sulphide ore *galena* $Pb^{2+}S^{2-}$, which is often found with zinc blende $Zn^{2+}S^{2-}$.

Galena is first roasted in air to form the oxide:

$$2Pb^{2+}S^{2-}_{(s)} + 3O_{2(g)} \rightarrow 2Pb^{2+}O^{2-}_{(s)} + 2SO_{2(g)}$$

The oxide is then reduced with coke in a small blast furnace, molten lead being 'tapped' from the base of the furnace:

$$Pb^{2+}O^{2-}_{(s)} + C_{(s)} \rightarrow Pb_{(l)} + CO_{(g)}$$

The major physical characteristics of lead are listed in Table 9.7.

Table 9.7 Physical properties of lead

	Atomic number	Electronic configuration	Density $(kg\ m^{-3})$	Melting point (K)	Boiling point (K)
Lead (Pb)	82	2.8.18.32.18.4	11340	600	2017

(a) Properties

Pure lead is a soft silvery grey metal which becomes coated with a whitish protective film of basic carbonate on exposure to damp air. It slowly dissolves in
'soft' water, but in 'hard' water a protective coating of the sulphate or carbonate is formed on the inside of lead pipes preventing dissolution.

Lead is low in the electrochemical series and does not readily dissolve in dilute
sulphuric or hydrochloric acids: reaction is inhibited by the insoluble lead(II)
chloride and lead(II) sulphate which forms on the surface of the metal.

However, lead does dissolve in nitric acid, dilute or concentrated, forming
the nitrate. With dilute acid the reaction is

$$3Pb_{(s)} + 8H^+NO_{3(aq)}^- \rightarrow 3Pb^{2+}(NO_3^-)_{2(aq)} + 2NO_{(g)} + 4H_2O_{(l)}$$

With concentrated acid the reaction is

$$Pb_{(s)} + 4HNO_{3(l)} \rightarrow Pb^{2+}(NO_3^-)_{2(aq)} + 2NO_{2(g)} + 2H_2O_{(l)}$$

(b) Uses

The compound tetraethyl-lead(IV), $Pb(C_2H_5)_4$, is widely used as an 'anti-knock'
additive in petrol to inhibit pre-ignition.

Large quantities of lead are used in making the plates of lead accumulators
(car batteries).

Lead is an efficient absorber of X-rays, gamma rays and other harmful radiation
and is therefore used for screening purposes wherever radioactive materials are
handled.

9.30 Compounds of Lead

In most of its common compounds lead shows a Pb^{2+} oxidation state, but with
four electrons in its outer shell the possibility of a Pb^{4+} oxidation state exists.
This latter state is not common and the only example of it that we need to consider
is the oxide $Pb^{4+}O_2^{2-}$.

(a) Oxides

Lead(II) *oxide*, $Pb^{2+}O^{2-}$, is yellow in colour. It can be prepared by heating the
carbonate or nitrate:

$$2Pb^{2+}(NO_3^-)_{2(s)} \rightarrow 2Pb^{2+}O_{(s)}^{2-} + 4NO_{2(g)} + O_{2(g)}$$

To some extent this is an amphoteric oxide, but its basic properties are more
pronounced than its acidic properties.

Lead(IV) *oxide*, $Pb^{4+}O_2^{2-}$, is a dark-brown solid which can be prepared by the
action of dilute nitric acid on 'red lead' (see below). It decomposes on heating
to lead(II) oxide and oxygen:

$$2Pb^{4+}O_{2(s)}^{2-} \rightarrow 2Pb^{2+}O_{(s)}^{2-} + O_{2(g)}$$

As well as the oxides corresponding to lead(II) and lead(IV), there is a third
common oxide, *dilead*(II) *lead*(IV) *oxide*, whose formula can be written as either

$Pb^{4+}O_2^{2-} \cdot 2Pb^{2+}O^{2-}$ or Pb_3O_4. This compound is often referred to as 'red lead' because of its bright scarlet colour. It can be obtained by maintaining yellow lead(II) oxide at a temperature of 400–450 °C in air. At higher temperatures the 'red lead' decomposes into lead(II) oxide again:

$$6Pb^{2+}O^{2-}+O_2 \underset{>450\,°C}{\overset{400-450\,°C}{\rightleftharpoons}} 2Pb_3O_4$$

Dilead(II) lead(IV) oxide behaves as a mixture of lead(II) and lead(IV) oxides. When it is warmed with dilute nitric acid the red colour quickly disappears and insoluble dark-brown *lead(IV) oxide* remains:

$$Pb^{4+}O_2^{2-} \cdot 2Pb^{2+}O_{(s)}^{2-}+4H^+NO_{3(aq)}^- \rightarrow Pb^{4+}O_{2(s)}^{2-}+2Pb^{2+}(NO_3^-)_{2(aq)}+2H_2O_{(l)}$$

All of the lead oxides can be readily reduced to the metal when heated in hydrogen, carbon monoxide or with carbon. Essentially the reaction is the reduction of lead(II) oxide.

$$Pb^{2+}O_{(s)}^{2-}+H_{2(g)} \rightarrow Pb_{(s)}+H_2O_{(g)}$$

$$Pb^{2+}O_{(s)}^{2-}+CO_{(g)} \rightarrow Pb_{(s)}+CO_{2(g)}$$

$$Pb^{2+}O_{(s)}^{2-}+C_{(s)} \rightarrow Pb_{(s)}+CO_{(g)}$$

One of the characteristic tests for lead is to heat a little of the suspected lead compound in a depression on a charcoal block using a blowpipe. If lead is present it will be reduced by the charcoal (carbon) and a silvery globule of the metal will result.

(b) Salts

Most lead(II) compounds, including the sulphate, halides, carbonate and sulphide, are insoluble in cold water and can be prepared by double-decomposition reactions involving precipitation. In order to prepare these compounds it is necessary to obtain a soluble lead(II) salt such as the nitrate and add a solution containing the appropriate anion.

Lead(II) nitrate, $Pb^{2+}(NO_3^-)_2$, is obtained by dissolving the metal, carbonate or lead(II) oxide in warm dilute nitric acid. For example:

$$Pb^{2+}O_{(s)}^{2-}+2H^+NO_3^- \rightarrow Pb^{2+}(NO_3^-)_{2(aq)}+H_2O_{(l)}$$

The nitrate crystallizes from solution and has no water of crystallization.

Lead(II) hydroxide, $Pb^{2+}(OH^-)_2$, is obtained as a white precipitate when sodium hydroxide solution is added to lead(II) nitrate solution:

$$Pb^{2+}(NO_3^-)_{2(aq)}+2Na^+OH_{(aq)}^- \rightarrow Pb^{2+}(OH^-)_{2(s)}+2Na^+NO_{3(aq)}^-$$

This precipitate will dissolve in excess sodium hydroxide.

Lead(II) chloride, $Pb^{2+}Cl_2^-$, can be prepared by adding a solution containing Cl^- ions to lead(II) nitrate solution. For example:

$$Pb^{2+}(NO_3^-)_{2(aq)}+2Na^+Cl_{(aq)}^- \rightarrow Pb^{2+}Cl_{2(s)}^-+2Na^+NO_{3(aq)}^-$$

The white precipitate of lead(II) chloride is soluble in hot water.

Lead(II) iodide, $Pb^{2+}I_2^-$, is obtained as a bright yellow precipitate when potassium iodide is added to a solution of a lead(II) salt. For example:

$$Pb^{2+}(NO_3^-)_{2(aq)} + 2K^+I_{(aq)}^- \rightarrow Pb^{2+}I_{2(s)}^- + 2K^+NO_{3(aq)}^-$$

If the precipitate is redissolved by heating, bright yellow 'spangles' appear as the solution cools. This reaction can test for the presence of Pb^{2+} ions in solution.

Lead(II) sulphate, $Pb^{2+}SO_4^{2-}$, is obtained as a white precipitate when any solution containing the sulphate ion $SO_{4(aq)}^{2-}$ is added to lead(II) nitrate solution. For example, using dilute sulphuric acid:

$$Pb^{2+}(NO_3^-)_{2(aq)} + H_2^+SO_{4(aq)}^{2-} \rightarrow Pb^{2+}SO_{4(s)}^{2-} + 2H^+NO_{3(aq)}^-$$

9.31 Principles Underlying the Extraction of Metals

The position of a metal in the electrochemical series (see Table 9.8) is a useful indication of its reactivity. Oxidation potential (the reverse of reduction potential) measures the tendency of the metal to lose electrons and go into solution as positive hydrated ions. Thus potassium near the top of the series is much more likely to lose an electron

$$K_{(s)} \rightarrow K_{(aq)}^+ + e$$

than is silver which lies near the bottom of the series:

$$Ag_{(s)} \rightarrow Ag_{(aq)}^+ + e$$

The extraction of metals from their ores is essentially the reverse of this process, i.e. it is a reduction process, and thus we can predict that silver and metals low in the electrochemical series are obtainable from their ores much more easily than potassium and metals high in the series. A metal's position in the electrochemical series therefore has a strong bearing on the choice of method for its extraction.

Extraction processes can be classified into three main types:

(*a*) *Heating the ore* (for metals low in the electrochemical series)
Metals low in the electrochemical series include silver, mercury and gold. The tendency for these metals to become ions is so small that they sometimes occur 'native' as the free metal or can be extracted simply by heating the ore in air. For example, mercury is obtained by heating the sulphide ore *cinnabar*:

$$Hg^{2+}S_{(s)}^{2-} + O_{2(g)} \rightarrow Hg_{(l)} + SO_{2(g)}$$

(*b*) *Reduction of the oxide*
(i) *With carbon or carbon monoxide* (for copper, lead, iron and zinc)
Ores of metals higher than silver in the series do not decompose directly to the free metal simply on heating. If the ore is an oxide it must be reduced by heating with a reducing agent such as carbon or carbon monoxide:

$$Fe_2^{3+}O_{3(s)}^{2-} + 3CO_{(g)} \rightarrow 2Fe_{(l)} + 3CO_{2(g)}$$

Ores other than oxides, particularly sulphides, have to be roasted in air to produce the oxide which can then be reduced. Thus:

$$2Zn^{2+}S^{2-}_{(s)}+3O_{2(g)} \rightarrow 2Zn^{2+}O^{2-}_{(s)}+2SO_{2(g)}$$

then
$$Zn^{2+}O^{2-}_{(s)}+C_{(s)} \rightarrow Zn_{(l)}+CO_{(g)}$$

(ii) *With a metal higher in the electrochemical series*
The *thermite* reaction (described in Section 9.14) is useful for the extraction of small quantities of carbon-free metals such as chromium, manganese and titanium. This method is expensive and would thus be used only for obtaining in a very pure state special-purpose metals whose oxides are not readily reduced by carbon. For example:

$$2Al_{(s)}+Cr^{3+}_2O^{2-}_{3(s)} \rightarrow 2Cr_{(l)}+Al^{3+}_2O^{2-}_{3(s)}.$$

(c) *Electrolysis of the fused ore* (for metals high in the electrochemical series)
Electrolytic processes, in which electrical energy is used to convert metal ions into atoms, are some of the chemist's most powerful devices for winning reactive metals from their ores. They form the basis for the industrial manufacture of sodium and aluminium (see Unit 14), potassium, magnesium and calcium.

In addition, electrolytic processes can be used in the purification of many metals, e.g. copper (see Sections 5.10 and 9.23). In this process the impure metal acts as anode and a small amount of pure metal is used as the cathode. The metal is transferred from the anode to the cathode during electrolysis.

The relationship of the above methods of metal extraction to the electro-chemical series is summarized in Table 9.8.

Table 9.8 Extraction of metals

	Electrochemical series (based on reduction potentials)		Method of extraction
↑ Decreasing ease of metal extraction →	Potassium,	K	Electrolysis of fused ore
	Calcium,	Ca	
	Sodium,	Na	
	Magnesium,	Mg	
	Aluminium,	Al	
	Manganese,	Mn	Conversion of ore to oxide
	Zinc,	Zn	followed by reduction with
	Chromium,	Cr	carbon, carbon monoxide
	Iron,	Fe	or aluminium
	Tin,	Sn	
	Lead,	Pb	
	Copper,	Cu	
	Silver,	Ag	
	Mercury,	Hg	Heating the ore
	Gold,	Au	

Summary of Unit 9

1. **Metals** have a lattice structure composed of positive ions surrounded by a sea of mobile electrons. This structure accounts for their high *electrical* and *thermal conductivity, metallic lustre, malleability* and *ductility*.
2. Metals readily lose electrons to form *cations*. They also form basic oxides.
3. The **alkali metals** each have a single electron in their outermost energy level. Their chemistry is the chemistry of their ions (M^+). Reactivity increases with increasing atomic number, as illustrated by their reactions with water and air.
4. Alkali metals impart a characteristic colour to a bunsen flame.
5. Important compounds of the alkali metals include *sodium chloride, sodium hydroxide, potassium hydroxide, sodium carbonate* and *sodium hydrogen-carbonate*.
6. The **alkaline-earth metals** can all easily attain a noble-gas structure by losing their two outer electrons to give stable ions (M^{2+}). Important compounds include *calcium carbonate, calcium hydroxide, calcium sulphate*.
7. **Aluminium** is the most abundant metal in the earth's crust: it occurs in clays, bauxite, etc. Its reactivity is reduced by an inert protective oxide layer which makes the metal resistant to attack.
8. The oxide and hydroxide of aluminium are **amphoteric**.
9. The *first transition series* of elements includes **iron** and **copper**. Metals in this series have two electrons in their outermost shell and an incomplete penultimate shell of similar energy.
10. Iron *rusts* in moist air. This oxidation process is prevented by coating the metal with a protective film.
11. Iron forms two series of compounds; it shows an *oxidation state* of either $+2$ (iron(II)) or $+3$ (iron(III))

$$\text{iron(II)} \underset{\text{reduction}}{\overset{\text{oxidation}}{\rightleftharpoons}} \text{iron(III)}$$

12. Compounds of iron containing the hydrated iron(II) ion are green, whereas those containing the hydrated iron(III) ion are brownish.
13. Reaction with sodium hydroxide or potassium thiocyanate or potassium hexacyanoferrate(II) can be used to distinguish between solutions containing iron(II) and iron(III) ions.
14. **Copper** is purified *electrolytically*.
15. Copper is not attacked by *water* and will *not* liberate hydrogen from dilute acids.
16. The two oxidation states of copper are copper(I) and the more common copper(II).
17. Hydrated copper(II) ions are blue.
18. **Zinc** is extracted from its sulphide ore zinc blende by *roasting* to the oxide followed by *reduction with carbon*.
19. Zinc resembles the alkaline-earth metals, and its only oxidation state is zinc(II).

20. Zinc oxide and zinc hydroxide are *amphoteric*, dissolving in both acid and alkali.

21. **Lead** is extracted by roasting to convert its ores to the oxide which is then reduced to the metal with carbon (coke).

22. Being *low* in the electrochemical series, lead is less reactive than zinc or iron.

23. Lead can exist in two oxidation states: lead(II) and the less common lead(IV).

24. The three common oxides of lead are lead(II) oxide (yellow), lead(IV) oxide (dark brown) and dilead(II) lead(IV) oxide (red).

25. Most lead salts are *insoluble* and can be prepared by appropriate double-decomposition reactions involving precipitation.

26. Metals are extracted from their ores according to their position in the electrochemical series.

> (*a*) Reactive metals high in the series require a high-energy process such as electrolysis.
>
> (*b*) Unreactive metals low in the series are either found naturally or are recovered simply by heating the ore.
>
> (*c*) Other metals are usually extracted by heating the ore to convert it to the oxide and then reducing this with carbon, carbon monoxide or aluminium.

Test Yourself on Unit 9

1. From the following list of metals: sodium, calcium, aluminium, iron, magnesium and copper, which one has

(*a*) no reaction with dilute hydrochloric acid?

(*b*) ions with oxidation states of $+3$ and $+2$?

(*c*) a violent reaction with water?

(*d*) a brick-red coloration in a bunsen flame?

(*e*) a brilliant white flame as it burns?

(*f*) a reaction with sodium hydroxide solution liberating hydrogen?

(*g*) a red-black solid chloride which readily sublimes?

(*h*) a deliquescent hydroxide?

(*i*) a black oxide which dissolves readily in dilute acids to give a blue solution?

(*j*) the greatest possibility of being found native?

2. Identify the following metals from the information given:

(*a*) Metal A can be cut with a knife. When a small piece of A is dropped on to water it melts into a silvery ball, burns with a lilac flame and rushes around the surface of the water, gradually disappearing.

(*b*) Metal B has a yellow oxide, a dark-brown oxide and a red oxide.

(*c*) Metal C is the only red metal. It dissolves readily in nitric acid to form a blue solution.

(*d*) Metal D has an amphoteric oxide which is pale yellow when hot and white when cold.

(*e*) Metal E is the alkali metal which is the least reactive with water.

3. The electronic configurations of four metals W, X, Y and Z are

W: 2.8.8.1
X: 2.8.14.2
Y: 2.8.8.2
Z: 2.8.1

Which metal(s) would you expect to
   (a) be a transition metal?
   (b) form a unipositive ion?
   (c) show more than one oxidation state in its compounds?
   (d) be in the same group in the periodic table?
   (e) form a colourless dipositive ion?

4. Compound A is a green crystalline salt of iron which gives the following results when tested:
   (a) Addition of barium chloride solution to a solution of A results in the formation of a white precipitate B, which is insoluble in dilute hydrochloric acid.
   (b) On heating A, water vapour and two oxides of sulphur (C and D) are liberated, leaving a red-brown residue E.
   (c) E dissolves in warm concentrated hydrochloric acid to give a yellow solution F.
   (d) With hydrogen sulphide, the solution F yields a yellowish-white precipitate G, which when filtered leaves a green filtrate H.
Identify the substances A to H. Name and give the colour of the precipitates formed when excess aqueous ammonia is added to (i) a solution of A, and (ii) a solution of A treated with concentrated nitric acid.

5. Are the following statements true or false?
   (a) Elements high in the electrochemical series (e.g. potassium) are extracted from their ores by an electrolytic process.
   (b) Being lower in the electrochemical series, lead is less reactive than magnesium.
   (c) Copper liberates hydrogen from dilute hydrochloric acid.
   (d) Metals readily gain electrons to form cations.
   (e) All ammonium salts liberate ammonia gas on warming with a dilute acid.
   (f) Both sodium carbonate and calcium carbonate give carbon dioxide with a dilute acid and are decomposed to the oxide on heating.
   (g) Aluminium metal reduces iron(III) oxide to iron in the thermite reaction.
   (h) An amphoteric oxide shows both basic and acidic properties.

Mark this test out of 40 with the answers given on page 278.

Classification of Matter III: The Non-metals

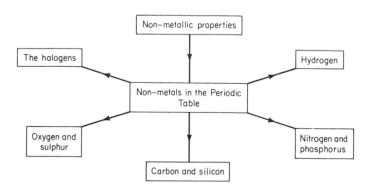

This Unit looks at the properties of some of the more important non-metals and their compounds. The properties and group relationships are developed for the halogens, and the properties of non-metals from other groups are studied in some detail.

10.1 What are Non-metals?

The properties of a small group of elements (about one-sixth of the total number of elements) located in the upper right-hand corner of the periodic table (see Fig. 10.1) are quite unlike those discussed in Unit 9.

These elements are termed *the non-metals*. In general they have melting and boiling points below 500 °C (in fact most are gases), and in the solid or liquid state do not conduct electricity. Can the marked difference in properties between the metals and non-metals be explained in terms of electronic structure?

In Unit 9 the physical properties of metals were explained by a model which envisaged positive metal ions held in a 'sea' of delocalized mobile electrons (see Fig. 9.1). Metallic properties are therefore associated with the metal in bulk and not with individual isolated atoms or molecules. In contrast, non-metallic properties point to the existence of small, isolated, individual units. *Most non-metals exist as small separate molecules with very weak attractive forces* (known as van der Waals' forces) *between them*. There are no delocalized mobile electrons to hold the individual units together, and non-metals therefore lack those properties (such as lustre and electrical conductivity) which are associated with mobile electrons.

Chlorine, a non-metal, is composed of Cl_2 molecules. The two atoms in each chlorine molecule are held together by a strong covalent bond, but the attractive

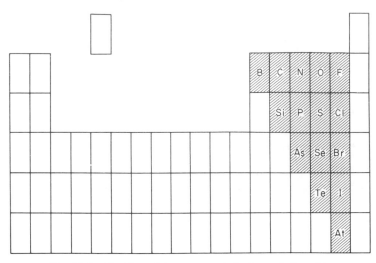

Fig. 10.1   Position of the non-metals in the periodic table

forces *between* molecules are merely the weak van der Waals' forces (see Fig. 10.2). It is because the forces of attraction between chlorine molecules are so weak that, under standard conditions, the element exists as a gas.

10.2 Non-metals with a Giant Structure

Not all non-metallic elements have the physical properties (e.g. low boiling and melting points) which characterize chlorine gas. Some, notably carbon with four

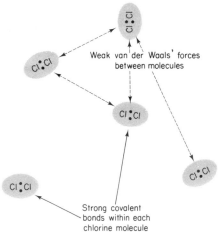

Fig. 10.2   Structure of chlorine gas, showing the weak forces of attraction between neighbouring molecules

electrons in its outer shell, are able to form giant molecules composed of many individual atoms joined together by strong covalent bonds. Carbon forms two such giant structures: diamond (shown in Fig. 2.16) and graphite (shown in Fig. 2.17). Such structures have a marked influence on the physical properties of these non-metals. Thus diamond is very hard with a high melting point, while graphite is soft (the layers slide over each other) and is an electrical conductor due to the mobile electrons between the layers.

10.3 Important Chemical Properties of Non-metals

(a) In contrast to the metals, which form *cations*, non-metals and groups of non-metals (e.g. carbonate CO_3^{2-}) tend to form *anions*. Thus chlorine forms the chloride ion (see Section 10.6) by gaining an electron:

$$Cl + e \rightarrow Cl^-$$

(b) They form *acidic* (occasionally neutral) *oxides* (see Unit 8). Thus sulphur forms sulphur dioxide which will dissolve in water to form sulphurous acid.

(c) The halides of the non-metals have a covalent structure and tend to react with, rather than dissolve in, water. Thus phosphorus pentachloride reacts with water to form phosphoric acid:

$$PCl_{5(s)} + 4H_2O_{(l)} \rightarrow H_3^+PO_{4(aq)}^{3-} + 5H^+Cl_{(aq)}^-$$

Typical non-metal families are Group 7 (the halogens), Group 6 and Group 5. Study of a typical non-metal family such as the halogens illustrates the gradation in physical and chemical properties within the group. It is found that non-metallic character *decreases* as the atomic number increases.

10.4 The Halogens (Group 7)

The names, symbols, electronic configuration and important physical properties of the halogen elements are listed in Table 10.1.

Table 10.1 The halogens

	Fluorine (F)	*Chlorine* (Cl)	*Bromine* (Br)	*Iodine* (I)
Electronic configuration	2.7	2.8.7	2.8.18.7	2.8.18.18.7
Atomic number	9	17	35	53
Melting point (K)	54	172	266	387
Boiling point (K)	85	239	332	457
Density (kg m^{-3})	1·7	3·2	3100	4940

At room temperature fluorine and chlorine are poisonous gases with irritating choking smells. Fluorine is pale yellow, and chlorine is greenish-yellow. Bromine is a dark-red volatile liquid with a poisonous, choking vapour; iodine is a dark shiny solid.

All four halogens exist in the form of diatomic molecules F_2, Cl_2, Br_2 and I_2, with strong covalent bonds holding the two atoms together and only weak van der Waals' forces between individual molecules. Being covalently bonded, they dissolve readily in covalent solvents: chlorine produces a colourless solution in tetrachloromethane (carbon tetrachloride), bromine gives a red solution, and iodine a purple solution.

10.5 Preparation of the Halogens

(a) Fluorine
Fluorine will accept one electron into its outer energy level more readily than any other element and is thus *the most powerful oxidizing agent known*.
The reaction

$$F + e \rightarrow F^-$$

occurs with great ease. In contrast the reaction

$$F^- \rightarrow F + e$$

occurs with extreme difficulty, and it is thus virtually impossible to obtain fluorine from fluoride ion by chemical means. The only method available is an electrolytic one in which fluorine is generated under anhydrous conditions from a solution of potassium hydrogendifluoride in anhydrous liquid hydrogen fluoride. Fluorine is discharged at the *anode*.

  Anode reaction
$$2F^- \rightarrow F_2 + 2e$$
  Cathode reaction
$$2H^+ + 2e \rightarrow H_2$$

(b) Chlorine, Bromine and Iodine
The laboratory preparation of these three halogens is much easier than that of fluorine because they are less powerful oxidizing agents. The halide ion decreases in stability down the group, i.e. the iodide ion is the most readily reducible.

In general all these halogens can be prepared by warming the corresponding halide (chloride, bromide or iodide) with manganese(IV) oxide and concentrated sulphuric acid:

$$Mn^{4+}O^{2-}_{2(s)} + 3H_2SO_{4(l)} + 2Na^+X^-_{(s)} \rightarrow Mn^{2+}SO^{2-}_{4(aq)} + 2Na^+HSO^-_{4(aq)} + 2H_2O_{(l)} + X_{2(g)}$$

where X_2 represents a halogen and X^- its corresponding halide.

If the halogen is chlorine, it can be dried by passing it through concentrated sulphuric acid and collected by downward delivery (as shown on the right-hand side of Fig. 10.3).

For bromine, Br_2, the vapour is condensed and collected as the liquid in a cooled receiver; for iodine, I_2, the vapour sublimes and can be collected as a solid on a cooled surface.

The common laboratory method for the preparation of chlorine (Fig. 10.3) involves the direct oxidation of hydrochloric acid with the oxidizing agent potassium manganate(VII) (potassium permanganate):

$$2K^+MnO_{4(s)}^- + 16H^+Cl_{(aq)}^- \rightarrow 2K^+Cl_{(aq)}^- + 2Mn^{2+}Cl_{2(aq)}^- + 8H_2O_{(l)} + 5Cl_{2(g)}$$

As an alternative, manganese(IV) oxide can be used as the oxidizing agent but this reaction requires heat:

$$Mn^{4+}O_{2(s)}^{2-} + 4H^+Cl_{(aq)}^- \rightarrow Mn^{2+}Cl_{2\ (aq)}^- + 2H_2O_{(l)} + Cl_{2(g)}$$

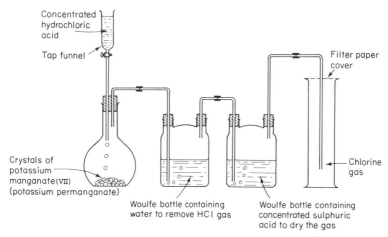

Fig. 10.3   Laboratory preparation and collection of dry chlorine gas

A small sample of chlorine can be conveniently prepared by adding an acid to 'bleaching powder'. Using hydrochloric acid and writing the active constituent of bleaching powder as $Ca^{2+}(ClO^-)_2$ the reaction is

$$Ca^{2+}(ClO^-)_{2(s)} + 4H^+Cl_{(aq)}^- \rightarrow Ca^{2+}Cl_{2(aq)}^- + 2H_2O_{(l)} + 2Cl_{2(g)}$$

10.6 Chemical Properties of the Halogens

All these elements have seven electrons in their outer energy level and readily *gain* one electron to achieve a stable octet of electrons. Hence there is a strong tendency to form anions, as mentioned earlier:

$$Cl + e \rightarrow Cl^-$$

The more easily the atom gains this outer electron, the more reactive it is. Reactivity thus decreases with increasing atomic number. This is because the *smaller* the atom the greater the attraction of the positive nucleus for the negative electron. Thus fluorine (atomic number 9) is much more reactive than chlorine (atomic number 17), and chlorine in turn is more reactive than bromine (atomic number 35). Iodine (atomic number 53) is the least reactive.

Fluorine is the most reactive non-metal and combines directly with almost all other elements; yet the compounds of fluorine are often extremely stable. This is well illustrated by PTFE or poly(tetrafluoroethylene), a material which is very resistant to chemical attack and is used for coating 'non-stick' cooking utensils.

The chemistry of fluorine is discussed separately from that of the other halogens because of its extreme reactivity. Chlorine, bromine and iodine show a definite gradation in chemical properties, and so the chemistry of all three can be illustrated by looking at the element chlorine in some detail with only minor references to bromine and iodine.

(*a*) Reactions with Metals

Chlorine, bromine and iodine react readily with most metals. This is because of the strong tendency of the halogen to gain an electron and the strong tendency of metals to lose electrons. Thus heated sodium reacts vigorously with chlorine gas producing the white ionic solid, sodium chloride:

$$2Na_{(s)} + Cl_{2(g)} \rightarrow 2Na^+Cl^-_{(s)}$$

Metals having more than one oxidation number tend to form the 'higher' chloride when reacting directly with chlorine. For example, iron reacts with chlorine to give iron(III) chloride and *not* iron(II) chloride:

$$2Fe_{(s)} + 3Cl_{2(g)} \rightarrow 2Fe^{3+}Cl^-_{3(s)}$$

(*b*) Reactions with Water

Chlorine and bromine are moderately soluble in water. Chlorine dissolves to give a solution called 'chlorine water' which contains both hydrochloric acid and chloric(I) acid (hypochlorous acid):

$$Cl_{2(g)} + H_2O_{(l)} \rightleftharpoons H^+Cl^-_{(aq)} + H^+ClO^-_{(aq)}$$

The chlorate(I) ion (hypochlorite ion) ClO^- is a constituent of liquid bleach and kills bacteria. For this latter reason small amounts of chlorine are dissolved in swimming-pool water.

On standing in sunlight 'chlorine water' liberates oxygen:

$$2H^+ClO^-_{(aq)} \rightarrow 2H^+Cl^-_{(aq)} + O_{2(g)}$$

Iodine is only sparingly soluble in water.

(*c*) Reactions with Alkalis

Chlorine is absorbed by a *cold dilute* solution of sodium hydroxide to form sodium chloride and sodium *chlorate*(I) (sodium hypochlorite):

$$2Na^+OH^-_{(aq)} + Cl_{2(g)} \rightarrow Na^+Cl^-_{(aq)} + Na^+ClO^-_{(aq)} + H_2O_{(l)}$$

In general the reaction of chlorine with any cold, dilute alkali can be written

$$2OH^-_{(aq)} + Cl_{2(g)} \rightarrow Cl^-_{(aq)} + ClO^-_{(aq)} + H_2O_{(l)}$$

With *hot, concentrated* sodium hydroxide, chlorine reacts to give sodium chloride and sodium *chlorate*(v) (sodium chlorate):

$$6Na^+OH^-_{(aq)}+3Cl_{2(g)} \rightarrow 5Na^+Cl^-_{(aq)}+Na^+ClO^-_{3(aq)}+3H_2O_{(l)}$$

In general the reaction with any hot, concentrated alkali can be written

$$6OH^-_{(aq)}+3Cl_{2(g)} \rightarrow 5Cl^-_{(aq)}+ClO^-_{3(aq)}+3H_2O_{(l)}$$

Bromine reacts with alkalis to give the bromide and *bromate*(v); iodine similarly gives the iodide and iodate(v).

When chlorine is passed over cold, solid calcium hydroxide (slaked lime) the product is a mixture called *bleaching powder*:

$$3Ca^{2+}(OH^-)_{2(s)}+2Cl_{2(g)} \rightarrow Ca^{2+}(ClO^-)_{2(s)}+Ca^{2+}Cl^-_2 . Ca^{2+}(OH^-)_2 . H_2O_{(s)}$$
$$+H_2O_{(l)}$$

(d) Oxidizing Properties of the Halogens

All the halogens are oxidizing agents because of their ability to accept electrons. As this ability decreases with increasing atomic number, fluorine is the strongest oxidizing agent and iodine is the weakest.

This ease of oxidation can be readily demonstrated for chlorine, bromine and iodine by the following series of **displacement reactions**.

Experiment 10.1   Determination of the relative oxidizing powers of chlorine, bromine and iodine

Separate solutions of each halogen in tetrachloromethane (carbon tetrachloride) are prepared by (i) bubbling chlorine gas through the liquid, (ii) adding a drop of bromine liquid and (iii) adding a small crystal of iodine. The chlorine solution is colourless; the bromine solution is red, and the iodine solution is purple. A little of each of these solutions is separately shaken with aqueous solutions of potassium chloride, potassium bromide and potassium iodide. The more dense tetrachloromethane (carbon tetrachloride) is allowed to settle as a separate layer and the colour of this lower layer is recorded.

The results can be tabulated as follows:

	Potassium chloride solution	Potassium bromide solution	Potassium iodide solution
Chlorine in CCl_4 (colourless)	no change	colourless to red (see Eqn. 1 below)	colourless to purple (see Eqn. 2 below)
Bromine in CCl_4 (red)	no change	no change	red to purple (see Eqn. 3 below)
Iodine in CCl_4 (purple)	no change	no change	no change

Conclusion

Chlorine oxidizes bromide ion to bromine, and iodide ion to iodine:

$$Cl_{2(CCl_4)} + 2Br^-_{(aq)} \rightarrow Br_{2(CCl_4)} + 2Cl^-_{(aq)} \tag{1}$$
$$\text{colourless} \qquad\qquad\qquad \text{red}$$

$$Cl_{2(CCl_4)} + 2I^-_{(aq)} \rightarrow I_{2(CCl_4)} + 2Cl^-_{(aq)} \tag{2}$$
$$\text{colourless} \qquad\qquad\qquad \text{purple}$$

Bromine oxidizes iodide ion to iodine:

$$Br_{2(CCl_4)} + 2I^-_{(aq)} \rightarrow I_{2(CCl_4)} + 2Br^-_{(aq)} \tag{3}$$
$$\text{red} \qquad\qquad\qquad \text{purple}$$

Iodine oxidizes neither bromide ion nor chloride ion.

Thus the order of oxidizing ability is *chlorine > bromine > iodine*.

(e) Other Oxidizing Properties of Chlorine

(i) Chlorine has a great affinity for hydrogen, either free or when it is combined with carbon in hydrocarbons. In bright sunlight hydrogen and chlorine mixtures explode. A jet of hydrogen will burn in chlorine, and vice versa. The product in either case is hydrogen chloride:

$$H_{2(g)} + Cl_{2(g)} \rightarrow 2HCl_{(g)}$$

A burning waxed taper (wax is a hydrocarbon mixture) burns with a red smoky flame when lowered into a gas jar containing chlorine.

Cotton wool soaked in warm turpentine (another hydrocarbon mixture, simplest formula $C_{10}H_{16}$) bursts into flame in a jar of chlorine, liberating clouds of soot (carbon). The chlorine oxidizes the turpentine to carbon:

$$C_{10}H_{16(l)} + 8Cl_{2(g)} \rightarrow 10C_{(s)} + 16HCl_{(g)}$$

In both of these cases the presence of hydrogen chloride can be detected by the dense white fumes of ammonium chloride produced when a piece of filter paper soaked in aqueous ammonia is held near the top of the gas jar:

$$HCl_{(g)} + NH_{3(g)} \rightarrow NH_4^+ Cl^-_{(s)}$$

(ii) When hydrogen sulphide is passed into chlorine water, a creamy precipitate of sulphur is produced as the chlorine oxidizes the hydrogen sulphide to sulphur:

$$H_2S_{(g)} + Cl_{2(aq)} \rightarrow 2H^+ Cl^-_{(aq)} + S_{(s)}$$

(iii) Chlorine oxidizes ammonia to nitrogen:

$$2NH_{3(g)} + 3Cl_{2(g)} \rightarrow 6HCl_{(g)} + N_{2(g)}$$

The hydrogen chloride produced reacts with excess ammonia forming white solid ammonium chloride:

$$6HCl_{(g)} + 6NH_{3(g)} \rightarrow 6NH_4^+ Cl^-_{(s)}$$

10.7 Uses of the Halogens

Chlorine. Chlorine finds widespread use in the manufacture of plastics such as poly(chloroethene) (also known as polyvinyl chloride or PVC), disinfectants, hydrochloric acid, insecticides, refrigerants and many other chlorinated compounds. The gas itself is used for sterilizing water supplies and for making 'bleach'.

Bromine. Bromine is used for the preparation of bromides, drugs, dyes and in particular for making 1,2-dibromoethane which is added to petrol.

Iodine. Iodine is used in medicine because of its antiseptic properties, in dyes, photographic film, and in insecticides. Radioactive iodine is used as a tracer (see Unit 13).

10.8 Hydrogen Chloride, Hydrochloric Acid and the Chlorides

Hydrogen chloride is a colourless gas composed of covalent molecules. Being covalent, it dissolves in covalent solvents such as methylbenzene (toluene); the resulting solution has no acidic properties. The gas dissolves readily in water and reacts with it to form an ionic solution containing the oxonium ion $H_3O^+_{(aq)}$ and the chloride ion $Cl^-_{(aq)}$:

$$HCl_{(g)} + H_2O_{(l)} \rightarrow H_3O^+_{(aq)} + Cl^-_{(aq)}$$

This solution is *hydrochloric acid*; it has acidic properties due to the oxonium ions $H_3O^+_{(aq)}$ and it forms chlorides which contain the Cl^- ion.

(a) Preparation of Hydrogen Chloride and Hydrochloric Acid

(i) By direct synthesis of hydrogen and chlorine gases (see Section 10.6e):

$$H_{2(g)} + Cl_{2(g)} \rightarrow 2HCl_{(g)}$$

Dissolving the gas in water gives hydrochloric acid.

(ii) By the action of concentrated sulphuric acid on an ionic chloride, e.g. sodium chloride:

$$Na^+Cl^-_{(s)} + H_2SO_{4(l)} \rightarrow Na^+HSO^-_{4(s)} + HCl_{(g)}$$

This type of reaction illustrates the use of sulphuric acid in the displacement of a more volatile acid from its salt. It can be used for the preparation of acids having a lower boiling point than sulphuric acid, particularly the halogen acids and nitric acid.

Dry hydrogen chloride can be prepared and collected in the apparatus shown in Fig. 10.4. The modification shown on the right enables hydrogen chloride to dissolve slowly in water, producing hydrochloric acid. Hydrogen chloride is extremely soluble in water, and without the inverted filter funnel water would be sucked back into the apparatus.

(b) Properties of Hydrogen Chloride

(i) The gas will react with ammonia gas to produce dense white fumes of ammonium chloride:

$$NH_{3(g)} + HCl_{(g)} \rightarrow NH^+_4Cl^-_{(s)}$$

This reaction can be used to detect the presence of hydrogen chloride.

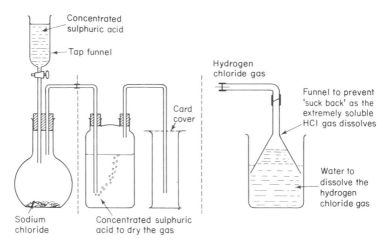

Fig. 10.4   Laboratory preparation of hydrogen chloride: (left) collection of the dry gas; (right) apparatus used for dissolving the gas in water

(ii) The gas will react with heated metals, forming chlorides and liberating hydrogen. For example, with iron (see Section 9.20):

$$Fe_{(s)} + 2HCl_{(g)} \rightarrow Fe^{2+}Cl^-_{2(s)} + H_{2(g)}$$

(iii) The gas readily dissolves in water to form hydrochloric acid:

$$HCl_{(g)} + H_2O_{(l)} \rightarrow \underbrace{H_3O^+_{(aq)} + Cl^-_{(aq)}}_{\text{(often written } H^+Cl^-_{(aq)})}$$

(c) **Properties of Hydrochloric Acid**

Hydrochloric acid is a typical mineral acid and exhibits the following properties associated with such acids.

(i) It neutralizes bases:

$$H^+Cl^-_{(aq)} + Na^+OH^-_{(aq)} \rightarrow Na^+Cl^-_{(aq)} + H_2O_{(l)}$$

(ii) It liberates carbon dioxide from carbonates and hydrogencarbonates. For example, with sodium carbonate:

$$2H^+Cl^-_{(aq)} + Na^+_2CO^{2-}_{3(s)} \rightarrow 2Na^+Cl^-_{(aq)} + CO_{2(g)} + H_2O_{(l)}$$

and with sodium hydrogencarbonate:

$$H^+Cl^-_{(aq)} + Na^+HCO^-_{3(s)} \rightarrow Na^+Cl^-_{(aq)} + CO_{2(g)} + H_2O_{(l)}$$

(iii) It reacts with metals above hydrogen in the electrochemical series, liberating hydrogen gas:

$$2H^+Cl^-_{(aq)} + Mg_{(s)} \rightarrow Mg^{2+}Cl^-_{2(aq)} + H_{2(g)}$$

10.9 Tests for Chloride, Bromide and Iodide Ions

(a) If a few drops of dilute nitric acid are added to the suspected halide solution, followed by a little dilute silver nitrate solution, the colour of any precipitate that forms has the following significance:

  (i) a *white* precipitate indicates *chloride*

$$Ag^+_{(aq)} + Cl^-_{(aq)} \rightarrow Ag^+Cl^-_{(s)}$$

 (ii) a *cream* precipitate indicates *bromide*

$$Ag^+_{(aq)} + Br^-_{(aq)} \rightarrow Ag^+Br^-_{(s)}$$

(iii) a *yellow* precipitate indicates *iodide*

$$Ag^+_{(aq)} + I^-_{(aq)} \rightarrow Ag^+I^-_{(s)}$$

(b) The presence of a bromide or an iodide in aqueous solution can be detected by adding chlorine dissolved in tetrachloromethane (carbon tetrachloride). The tetrachloromethane layer turns *red* if *bromide* ion is present in the solution under test and *purple* if *iodide* ion is present (see Experiment 10.1).

10.10 Hydrogen

Atoms of hydrogen contain one positive proton as the central nucleus surrounded by one electron in the first energy level.

Hydrogen *resembles the alkali metals* when it loses its one electron to become an ion with unit positive charge:

$$H \rightarrow H^+ + e$$

$$(\text{cf. } Na \rightarrow Na^+ + e)$$

Hydrogen *resembles the halogens* when it gains one electron to become an ion with unit negative charge having the noble-gas structure of helium:

$$H + e \rightarrow H^-$$

$$(\text{cf. } Cl + e \rightarrow Cl^-)$$

and also when it shares its one electron in covalent-bond formation:

$$H^{\cdot} + H^{\cdot} \rightarrow H:H \text{ (or } H_2)$$

These different types of behaviour make it difficult to assign hydrogen to a particular chemical family. It is often found placed above *both* the alkali metals (Group 1) and the halogens (Group 7).

10.11 Preparation of Hydrogen

(a) By the Action of Reactive Metals on Water

All the alkali metals (see Unit 9) will liberate hydrogen from *cold water*:

$$2Na_{(s)} + 2H_2O_{(l)} \rightarrow 2Na^+OH^-_{(aq)} + H_{2(g)}$$

The alkaline-earth metals (see Unit 9) will liberate hydrogen from *steam* (and, in the case of calcium, from cold water):

$$Ca_{(s)}+2H_2O_{(l)} \rightarrow Ca^{2+}(OH^-)_{2(aq)}+H_{2(g)}$$

Certain other metals, e.g. iron, react with steam when they are heated.

(*b*) **By the Action of Reactive Metals on Acids**
The acids in question are hydrochloric (either dilute or concentrated) and dilute sulphuric. The metals are all above hydrogen in the electrochemical series, and suitable ones are chosen which will react steadily (e.g. iron or zinc):

$$Zn_{(s)}+H_2^+SO_{4(aq)}^{2-} \rightarrow Zn^{2+}SO_{4(aq)}^{2-}+H_{2(g)}$$

Alkali metals react explosively with acids and should be avoided.

(*c*) **By the Action of Aluminium or Zinc on aqueous Potassium or Sodium Hydroxide**
The reaction of zinc with alkalis is described in Section 9.27*a*:

$$Zn_{(s)}+2OH_{(aq)}^-+2H_2O_{(l)} \rightarrow Zn(OH)_{4(aq)}^{2-}+H_{2(g)}$$

The reaction of aluminium with alkalis is described in Section 9.14*c*:

$$2Al_{(s)}+2OH_{(aq)}^-+6H_2O_{(l)} \rightarrow 2Al(OH)_{4(aq)}^-+3H_{2(g)}$$

10.12 Properties of Hydrogen

Hydrogen is a colourless, odourless and tasteless gas. It has a boiling point of 21 K and a melting point of 14 K. It is the least dense gas known (density $0 \cdot 0899$ kg m^{-3})

(*a*) **Reactions with Non-metals**
(i) Hydrogen combines spontaneously with chlorine, the reaction being explosive in sunlight:

$$H_{2(g)}+Cl_{2(g)} \rightarrow 2HCl_{(g)}$$

(ii) Mixtures of oxygen and hydrogen are explosive when ignited:

$$2H_{2(g)}+O_{2(g)} \rightarrow 2H_2O_{(g)}$$

(iii) Nitrogen and hydrogen will combine together to form ammonia gas in the presence of an iron catalyst, under high pressure and at a temperature of 500 °C:

$$N_{2(g)}+3H_{2(g)} \rightarrow 2NH_{3(g)}$$

This is the basis of the *Haber process* for the manufacture of ammonia (see Unit 14).

(*b*) **Reactions with Metals**
Hydrogen will accept electrons from very reactive metals (particularly Group 1).

For example, when sodium is heated in hydrogen the product is a white ionic solid called sodium hydride:

$$2Na_{(s)} + H_{2(g)} \rightarrow 2Na^+H_{(s)}^-$$

(c) **Hydrogen as a Reducing Agent**

Hydrogen reduces the heated oxides of metals low in the electrochemical series, *removing the oxygen* and leaving the metal:

$$Cu^{2+}O_{(s)}^{2-} + H_{2(g)} \rightarrow Cu_{(s)} + H_2O_{(g)}$$

10.13 Uses of Hydrogen

Hydrogen is used extensively in the manufacture of *ammonia* (Haber process) and in the manufacture of *hydrogen chloride* and *hydrochloric acid*.

Other important uses include the synthesis of *methanol* and the manufacture of *margarine*. In the latter process, hydrogen in the presence of a nickel catalyst is added to vegetable oils in order to 'harden' them, i.e. produce a solid fat.

10.14 Oxygen and Sulphur (Group 6)

The symbols, atomic numbers, electronic configurations and important physical properties of oxygen and sulphur are given in Table 10.2.

Table 10.2 Physical properties of oxygen and sulphur

	Electronic configuration	Atomic number	Melting point (K)	Boiling point (K)	Density (kg m^{-3})
Oxygen (O)	2.6	8	55	90	1·33
Sulphur (S)	2.8.6	16	386	718	2070

Both oxygen and sulphur are typical non-metals with six electrons in their outer energy level. They each accept two electrons to attain a noble-gas structure and hence form an anion with two negative charges:

$$O + 2e \rightarrow O^{2-}$$

$$S + 2e \rightarrow S^{2-}$$

Alternatively they can form two covalent bonds by sharing electrons. Thus in the water molecule each oxygen atom shares two electrons with two electrons from two separate hydrogen atoms. Sulphur shares electrons similarly in the hydrogen sulphide molecule, see Fig. 10.5.

Although oxygen and sulphur are both in Group 6, their chemical properties differ so much that it is convenient to study the elements separately.

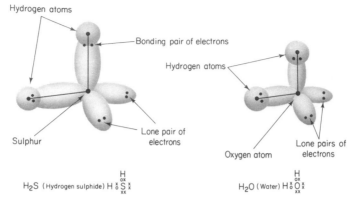

H₂S (Hydrogen sulphide) H ° S ˟

H₂O (Water) H ° O ˟

Fig. 10.5   Molecular structure of hydrogen sulphide compared with the molecular structure of water

10.15 Oxygen and Oxidation

The preparation, properties and uses of oxygen are described in Sections 6.3–6.6, and the classification of oxides is discussed in Section 8.3.

References are made throughout this book to *oxidation* and *reduction*. These terms can be defined in various ways.

(a) Oxidation and Reduction in Terms of Oxygen

When magnesium burns in oxygen, white solid magnesium oxide is formed:

$$2Mg_{(s)} + O_{2(g)} \rightarrow 2Mg^{2+}O^{2-}_{(s)}$$

The magnesium has been *oxidized* by *addition of oxygen*.

In general, whenever oxygen is added to an element or compound the process is called *oxidation*. Conversely, whenever oxygen is removed from an element or compound the process is called *reduction*.

When hydrogen is passed over heated copper(II) oxide, copper remains and water is produced:

$$Cu^{2+}O^{2-}_{(s)} + H_{2(g)} \rightarrow Cu_{(s)} + H_2O_{(g)}$$

The copper oxide has been *reduced* to copper by *loss of oxygen*, and at the same time the hydrogen has been *oxidized* to water by *oxygen gain*.

Since oxidation and reduction always occur simultaneously, the term **redox reaction** is often used to describe this type of process.

(b) Oxidation and Reduction in Terms of Hydrogen

Whenever hydrogen is added to an element or compound the process is called *reduction*, and the removal of hydrogen is *oxidation*. For example, ethene (ethylene) is *reduced* to ethane by the *addition* of hydrogen:

$$C_2H_{4(g)} + H_{2(g)} \rightarrow C_2H_{6(g)}$$

These simple early ideas concerning oxidation and reduction have been largely superseded by the following more generalized definitions involving *electron transfer* and *change in oxidation number*.

(c) Oxidation and Reduction in Terms of Electron Transfer

Let us consider the reactions which take place during the electrolysis (described in Unit 5) of molten lead(II) bromide.

At the anode	*At the cathode*
Bromine is liberated:	Lead is deposited:

$$2Br^- \rightarrow Br_2 + 2e \qquad\qquad Pb^{2+} + 2e \rightarrow Pb$$

This is *oxidation* of the bromide ion to bromine by electron loss	This is *reduction* of the lead(II) ion to lead by electron gain.

In general, *all* anode processes in electrolysis are oxidations and *all* cathode processes are reductions. Moreover, the definition of *oxidation as a loss of electrons* and *reduction as a gain of electrons* is not restricted just to electrolysis reactions. Consider for example the liberation of hydrogen from dilute sulphuric acid when zinc is added:

$$Zn_{(s)} + H_2^+ SO_{4(aq)}^{2-} \rightarrow Zn^{2+} SO_{4(aq)}^{2-} + H_{2(g)}$$

Neglecting the sulphate ions, since they are merely 'spectator ions', the reaction can be written as

$$Zn_{(s)} + 2H_{(aq)}^+ \rightarrow Zn_{(aq)}^{2+} + H_{2(g)}$$

The zinc metal is *oxidized* to zinc ions by the *loss* of two electrons:

$$Zn \rightarrow Zn^{2+} + 2e$$

and the hydrogen ions are *reduced* to hydrogen atoms by the *gain* of electrons:

$$2H^+ + 2e \rightarrow H_2$$

(d) Oxidation and Reduction in Terms of Oxidation Number

Oxidation is defined as an *increase in oxidation number*; an element which increases its oxidation number is said to be oxidized. Conversely, *reduction* is a *decrease in the oxidation number*.

For example, when chlorine gas is passed over heated iron, red-black crystals of iron(III) chloride are produced:

$$2Fe_{(s)} + 3Cl_{2(g)} \rightarrow 2Fe^{3+}Cl_{3(s)}^-$$

oxidation numbers: 0   0   +3   −1

The metallic iron *increases* its oxidation number from 0 to +3 when it is converted to iron(III) ions: thus *iron is oxidized*. The chlorine gas *decreases* its oxidation number from 0 to −1 when it is converted to chloride ions: thus *chlorine is reduced*.

10.16 Sulphur

Many *compounds* of sulphur occur naturally, the most common are metallic sulphides and sulphates. *Free* sulphur is found in the volcanic regions of southern Italy and Sicily, while vast deposits are located underground in Louisiana and Texas. Most of the world's free sulphur is obtained from this latter source by the **Frasch process.**

In the Frasch process, three concentric pipes inside a casing are sunk into the sulphur beds which are about 150 m below the surface. Superheated water (170°C) is pumped through the outermost 150 mm pipe to melt the sulphur. Hot compressed air is pumped down the central 25 mm pipe, and molten 'foamy' sulphur gushes under pressure up the 75 mm pipe between the other two (see Fig. 10.6).

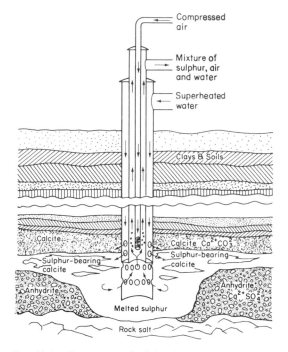

Fig. 10.6   Extraction of sulphur by the Frasch process

On reaching the surface the molten sulphur is run off into tanks where it is allowed to cool and solidify. A considerable economic advantage of the Frasch process is that the product is obtained in a high state of purity (typically 99·5% sulphur). This is because the superheated water liquefies the sulphur but not the chief contaminants such as sand and rock, so the latter remain behind in the ground.

10.17 Properties of Sulphur

(a) Allotropy

The element sulphur exists in a number of different *allotropic forms*. Each *allotrope* has its own physical properties but gives rise to identical chemical compounds.

The stable allotrope at ordinary temperatures is **rhombic** sulphur, a yellow solid which forms octahedral crystals as shown in Fig. 10.7*a*.

(a) Rhombic sulphur   (b) Monoclinic sulphur   (c) S_8 molecule

Fig. 10.7   Allotropy of sulphur: (a) and (b) the two common crystal forms; (c) a ring-shaped molecule containing eight sulphur atoms

Experiment 10.2   *Preparation of rhombic sulphur*
A little finely powdered sulphur is placed in a test tube. The tube is half filled with carbon disulphide, CS_2, and warmed in a water bath in a fume cupboard. After repeated shaking, the suspension is filtered into an evaporating basin and, still in the fume cupboard, the solution is allowed to cool. The carbon disulphide readily evaporates, leaving octahedral crystals of rhombic sulphur.

Caution: carbon disulphide is highly flammable and has a poisonous vapour; it should be handled with care.

Monoclinic sulphur is the other major allotrope of sulphur. This form is unstable at ordinary temperatures and slowly reverts to the rhombic form. Monoclinic sulphur is stable above 95·6 °C (the *transition temperature* between these two allotropes).

Experiment 10.3   *Preparation of monoclinic sulphur*
Powdered sulphur is heated in a boiling tube until it just melts. The liquid sulphur is then poured into a folded filter paper in a dry funnel on a ring-stand. When a solid crust has formed on the surface of the liquid, the filter paper is opened to expose the long needle-like crystals of monoclinic sulphur (see Fig. 10.7*b*).

Alternatively the powdered sulphur can be recrystallized from boiling methyl-benzene (toluene). This solvent boils at 111 °C, which is above the rhombic → monoclinic transition temperature. Hence monoclinic crystals are obtained.

Both rhombic and monoclinic sulphur are composed of ring-shaped S_8 molecules (see Fig. 10.7*c*), but the rings are arranged in a different way in the two allotropes. On heating, sulphur melts to a pale-yellow, mobile liquid in which the S_8 rings are able to move freely. As heating proceeds, the liquid darkens until, at 200 °C, it is nearly black and becomes very viscous. It is believed that the S_8 rings begin to open and form chains, and as these chains become entangled with each other, so the viscosity increases.

On further heating, the molecular chains break down and the liquid becomes mobile and eventually boils (444·6 °C). When boiling sulphur is poured into cold water, a dark-brown rubbery solid called *plastic sulphur* is formed. Plastic sulphur hardens on standing as it slowly changes to the stable rhombic allotrope.

(b) **Reaction of Sulphur with Metals**
Sulphur combines with almost all metals when heated. The reactions are usually vigorous and exothermic. For example, when a mixture of iron filings and powdered sulphur is heated on a small sand tray, the mixture glows even when heating is discontinued. Black solid iron(II) sulphide remains:

$$Fe_{(s)} + S_{(s)} \rightarrow Fe^{2+}S^{2-}_{(s)}$$

(c) **Reaction of Sulphur with Non-metals**
Sulphur combines directly with several non-metals, the most important of which is oxygen. For example, when powdered sulphur is heated in air or oxygen it burns with a blue flame, producing sulphur dioxide:

$$S_{(s)} + O_{2(g)} \rightarrow SO_{2(g)}$$

Sulphur dioxide is a colourless gas with a choking smell.

10.18 Uses of Sulphur

Most of the sulphur produced in the world is employed in the synthesis of sulphuric acid. It is also needed for the industrial preparation of disulphur dichloride (sulphur monochloride) S_2Cl_2, which is used in the vulcanization of rubber and in the manufacture of tetrachloromethane (carbon tetrachloride).

10.19 Hydrogen Sulphide, H_2S

(a) **Preparation**
Hydrogen sulphide gas can be prepared by treating a metal sulphide, usually iron(II) sulphide, with hydrochloric or sulphuric acids in moderate concentration:

$$Fe^{2+}S^{2-}_{(s)} + 2H^+Cl^-_{(aq)} \rightarrow Fe^{2+}Cl^-_{2(aq)} + H_2S_{(g)}$$

(b) **Properties**
(i) The most characteristic physical property of the gas is its odour, reminiscent of rotten eggs. It is colourless and *poisonous*. In water it dissolves fairly readily to give an acid solution.

(ii) Hydrogen sulphide burns in air with a blue flame to form sulphur dioxide and water:

$$2H_2S_{(g)} + 3O_{2(g)} \rightarrow 2H_2O_{(g)} + 2SO_{2(g)}$$

If the air is in limited supply, sulphur is produced:

$$2H_2S_{(g)} + O_{2(g)} \rightarrow 2S_{(s)} + 2H_2O_{(g)}$$

(iii) The sulphides of many metals are insoluble and are precipitated when hydrogen sulphide is passed through solutions containing ions of these metals. For example, lead gives a brownish-black precipitate of lead(II) sulphide, $Pb^{2+}S^{2-}$:

$$Pb^{2+}_{(aq)} + H_2S_{(g)} \rightarrow Pb^{2+}S^{2-}_{(s)} + 2H^+_{(aq)}$$

Copper gives a black precipitate of copper(II) sulphide, $Cu^{2+}S^{2-}$:

$$Cu^{2+}_{(aq)} + H_2S_{(g)} \rightarrow Cu^{2+}S^{2-}_{(s)} + 2H^+_{(aq)}$$

Zinc gives a greyish-white precipitate of zinc sulphide, $Zn^{2+}S^{2-}$:

$$Zn^{2+}_{(aq)} + H_2S_{(g)} \rightarrow Zn^{2+}S^{2-}_{(s)} + 2H^+_{(aq)}$$

The blackening of a filter paper soaked in lead(II) ethanoate (acetate) or lead(II) nitrate is used as a test for hydrogen sulphide.

The blackening of old paintings containing white lead-based paints is due to the formation of black lead(II) sulphide produced by the action of minute traces of hydrogen sulphide in the atmosphere, particularly in industrial regions. These paintings can be restored by treating them with a dilute solution of hydrogen peroxide, H_2O_2. This powerful oxidizing agent converts the black lead(II) sulphide to white lead(II) sulphate:

$$Pb^{2+}S^{2-}_{(s)} + 4H_2O_2 \rightarrow Pb^{2+}SO^{2-}_{4(s)} + 4H_2O_{(l)}$$

(iv) Hydrogen sulphide acts as a reducing agent and in each case is itself oxidized to sulphur. For example, when the gas is passed through a yellow-brown solution of iron(III) chloride the solution turns green as Fe^{3+} ions are reduced to Fe^{2+} ions:

$$\underset{\text{yellow-brown}}{2Fe^{3+}Cl^-_{3(aq)}} + H_2S_{(g)} \rightarrow \underset{\text{green}}{2Fe^{2+}Cl^-_{2(aq)}} + 2H^+Cl^-_{(aq)} + S_{(s)}$$

When a gas jar containing moist sulphur dioxide is inverted over a gas jar containing hydrogen sulphide, a precipitate of sulphur is formed. The hydrogen sulphide has reduced the sulphur dioxide to sulphur:

$$2H_2S_{(g)} + SO_{2(g)} \rightarrow 2H_2O_{(l)} + 3S_{(s)}$$

10.20 Sulphur Dioxide, SO_2

The two common oxides of sulphur are sulphur dioxide, SO_2, and sulphur trioxide, SO_3.

(a) Preparation

Although sulphur dioxide for laboratory use is normally obtained from a cylinder of the gas, the methods available for its preparation should be known.

(i) By burning sulphur in air or oxygen:

$$S_{(s)} + O_{2(g)} \rightarrow SO_{2(g)}$$

(ii) By the action of a dilute acid (except nitric acid, which is an oxidizing acid) on a sulphite or a hydrogensulphite. For example, the action of dilute hydrochloric acid on sodium sulphite:

$$Na_2^+ SO_{3(s)}^{2-} + 2H^+ Cl_{(aq)}^- \xrightarrow{\text{heat}} 2Na^+ Cl_{(aq)}^- + SO_{2(g)} + H_2O_{(l)}$$

(iii) By the action of hot concentrated sulphuric acid on copper:

$$Cu_{(s)} + 2H_2SO_{4(l)} \rightarrow Cu^{2+} SO_{4(aq)}^{2-} + 2H_2O_{(l)} + SO_{2(g)}$$

(b) Properties

(i) Sulphur dioxide is a colourless pungent gas which is easily liquefied under pressure.

(ii) The gas dissolves readily in water to form an acidic solution of the weak sulphurous acid:

$$SO_{2(g)} + H_2O_{(l)} \rightarrow H_2^+ SO_{3(aq)}^{2-}$$

Sulphur dioxide is the *anhydride* of sulphurous acid, but the free acid cannot be isolated.

(iii) Sulphur dioxide in water reduces orange *dichromate*(VI) ions to green *chromium*(III) ions:

$$Cr_2O_{7(aq)}^{2-} + 3SO_{3(aq)}^{2-} + 8H_{(aq)}^+ \rightarrow 2Cr_{(aq)}^{3+} + 3SO_{4(aq)}^{2-} + 4H_2O_{(l)}$$

Thus a filter paper soaked in potassium dichromate(VI) solution turns from orange to green when exposed to sulphur dioxide, and this reaction can be used to test for the gas.

(iv) Sulphur dioxide is absorbed by a solution of sodium hydroxide, forming sodium sulphite:

$$SO_{2(g)} + 2Na^+ OH_{(aq)}^- \rightarrow Na_2^+ SO_{3(aq)}^{2-} + H_2O_{(l)}$$

With excess sulphur dioxide the acid salt sodium hydrogensulphite is obtained:

$$SO_{2(g)} + Na^+ OH_{(aq)}^- \rightarrow Na^+ HSO_{3(aq)}^-$$

(These are both essentially *neutralization* reactions in which sulphurous acid is neutralized by the alkali sodium hydroxide.)

(c) Uses

By far the most important use of sulphur dioxide is in the manufacture of sulphuric acid (see Sections 10.22 and 14.7).

In addition it is used as a preservative in a wide variety of foods from lemon juice to sausages, and as a bleaching agent.

It is worth noting that the presence of small quantities of the gas in the atmosphere, from industrial pollution, is a health hazard to both humans and animals. Because of its acid nature, it also causes deterioration in stonework and metal structures.

10.21 Sulphur Trioxide (Sulphur(VI) Oxide), SO_3

(*a*) **Preparation**

(i) Sulphur trioxide is prepared by passing a mixture of dry sulphur dioxide and dry oxygen over a platinized-asbestos or vanadium(V) oxide (vanadium pentoxide) catalyst at 450°C:

$$2SO_{2(g)} + O_{2(g)} \rightarrow 2SO_{3(s)}$$

The gases are then passed through a cooled receiver, and the sulphur trioxide condenses as white needle-shaped crystals (see Fig. 10.8).

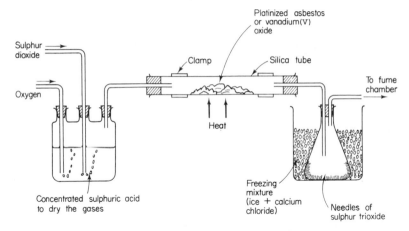

Fig. 10.8   Laboratory preparation of sulphur trioxide

(ii) Sulphur trioxide is formed when certain metallic sulphates are heated. For example, heating iron(III) sulphate:

$$Fe_2^{3+}(SO_4^{2-})_{3(s)} \rightarrow Fe_2^{3+}O_{3(s)}^{2-} + 3SO_{3(g)}$$

When iron(II) sulphate is heated, both sulphur trioxide and sulphur dioxide are produced:

$$2Fe^{2+}SO_{4(s)}^{2-} \rightarrow Fe_2^{3+}O_{3(s)}^{2-} + SO_{2(g)} + SO_{3(g)}$$

(*b*) **Properties**

Sulphur trioxide is a volatile white solid which fumes in moist air. The solid reacts explosively with water to form sulphuric acid:

$$H_2O_{(l)} + SO_{3(g)} \rightarrow H_2^+SO_{4(aq)}^{2-}$$

Sulphur trioxide is the *anhydride* of sulphuric acid.

Gaseous sulphur trioxide does not readily dissolve in water but it does dissolve in concentreated sulphuric acid forming *fuming sulphuric acid* (oleum):

$$SO_{3(g)} + H_2SO_{4(l)} \rightarrow H_2S_2O_{7(l)}$$

10.22 Sulphuric Acid, H_2SO_4

The reactions outlined in Section 10.21 form the basis of the manufacture of sulphuric acid, one of the world's most important chemicals. These reactions are discussed in detail in Unit 14.

Concentrated sulphuric acid is a dense, oily liquid which combines violently with water. The reaction is extremely exothermic, and the addition of small quantities of water to the concentrated acid can cause the production of steam with resultant 'spattering' of the acid. For this reason concentrated sulphuric acid is always diluted by *adding to water* with stirring.

The **chemical properties** of sulphuric acid can be conveniently discussed under the following headings: (*a*) its reactions *as an acid* in dilute aqueous solution, (*b*) oxidizing properties, (*c*) dehydrating properties, and (*d*) its use in the preparation of other acids.

(*a*) Sulphuric Acid as an Acid

A solution of the acid in water shows all the properties of a strong mineral acid (see Unit 8). The fact that this solution is a good conductor of electricity indicates the presence of a high concentration of ions:

$$H_2SO_{4(l)} + H_2O_{(l)} \rightarrow H_3O^+_{(aq)} + HSO^-_{4(aq)}$$
$$HSO^-_{4(aq)} + H_2O_{(l)} \rightarrow H_3O^+_{(aq)} + SO^{2-}_{4(aq)}$$

From these equations it can be seen that there is a high concentration of oxonium ions and that sulphuric acid forms two series of salts: **sulphates** and **hydrogensulphates**. The preparation of sodium sulphate and sodium hydrogensulphate has been described in Section 8.13.

Test for a Sulphate

To a solution of the suspected sulphate, dilute hydrochloric acid and barium chloride solution are added. A white precipitate of insoluble barium sulphate confirms the presence of a sulphate ion in the original solution.

$$SO^{2-}_{4(aq)} + Ba^{2+}Cl^-_{2(aq)} \rightarrow \underset{\text{(white)}}{Ba^{2+}SO^{2-}_{4(s)}} + 2Cl^-_{(aq)}$$

The hydrochloric acid is added to prevent the precipitation of barium sulphite should sulphite ions be present. (This compound is soluble in dilute hydrochloric acid.)

(*b*) Oxidizing Properties of Sulphuric Acid

Hot concentrated sulphuric acid is an oxidizing agent, and in its reactions is usually reduced to sulphur dioxide.

(i) It oxidizes copper to copper(II) sulphate:

$$Cu_{(s)}+2H_2SO_{4(l)} \rightarrow Cu^{2+}SO_{4(aq)}^{2-}+2H_2O_{(l)}+SO_{2(g)}$$

(ii) It oxidizes carbon to carbon dioxide, and sulphur to sulphur dioxide:

$$C_{(s)}+2H_2SO_{4(l)} \rightarrow CO_{2(g)}+2SO_{2(g)}+2H_2O_{(g)}$$

$$S_{(s)}+2H_2SO_{4(l)} \rightarrow 3SO_{2(g)}+2H_2O_{(g)}$$

(c) Dehydrating Properties of Sulphuric Acid

Concentrated sulphuric acid has a strong affinity for water (it is hygroscopic) and for this reason is often used to dry gases which do not react with it. Its dehydrating nature is so strong that it will remove water of crystallization from hydrated salts and the elements of water from carbohydrates, methanoic (formic) acid and ethanedioic (oxalic) acid.

(i) When crystals of copper(II) sulphate are warmed with concentrated sulphuric acid, the blue crystals turn white as water of crystallization is removed. The white anhyrous salt remains:

$$\underset{\text{(blue)}}{Cu^{2+}SO_4^{2-}.5H_2O} \rightleftharpoons \underset{\text{(white)}}{Cu^{2+}SO_{4(s)}^{2-}}+5H_2O_{(l)}$$

(ii) Sugar ($C_{12}H_{22}O_{11}$) is a *carbohydrate* and, as its name suggests, contains the elements of water. When a little sugar is treated with concentrated sulphuric acid in a breaker and warmed gently, the sugar turns yellow and then darkens. A vigorous exothermic reaction begins and a steaming black mass of carbon rises out of the beaker:

$$C_{12}H_{22}O_{11(s)} \rightarrow 12C_{(s)}+11H_2O_{(g)}$$

(iii) Methanoic (formic) and ethanedioic (oxalic) acids are both dehydrated by concentrated sulphuric acid:

$$\underset{\text{methanoic acid}}{HCOOH_{(s)}} \rightarrow CO_{(g)}+H_2O_{(l)}$$

$$\underset{\text{ethanedioic acid}}{\begin{matrix} COOH \\ | \\ COOH_{(s)} \end{matrix}} \rightarrow CO_{2(g)}+CO_{(g)}+H_2O_{(l)}$$

(d) Preparation of Other Acids using Sulphuric Acid

Concentrated sulphuric acid is used in the preparation of nitric, hydrochloric and ethanoic (acetic) acids. In each case concentrated sulphuric acid is heated with a salt of the acid to be prepared. For example, sodium nitrate yields nitric acid:

$$Na^+NO_{3(s)}^-+H_2SO_{4(l)} \rightarrow Na^+HSO_{4(s)}^-+HNO_{3(g)}$$

The volatile nitric acid vapour is removed from the reaction mixture because nitric acid has a lower boiling point than sulphuric acid.

10.23 Uses of Sulphuric Acid

Sulphuric acid is one of a large group of chemicals known in industry as 'heavy' chemicals, because they are manufactured on such a large scale.

Major uses of the acid include:

(a) manufacture of fertilizers, particularly ammonium sulphate and 'superphosphate';

(b) manufacture of explosives, dyestuffs and drugs;

(c) 'pickling' of iron and steel (i.e. the removal of surface oxide);

(d) manufacture of rayon;

(e) the lead storage accumulator (car battery);

(f) manufacture of paints and detergents.

10.24 Nitrogen and Phosphorus (Group 5)

The symbols, atomic numbers, electronic configurations and important physical properties of nitrogen and phosphorus are given in Table 10.3.

Table 10.3 Physical properties of nitrogen and phosphorus

	Electronic configuration	Atomic number	Melting point (K)	Boiling point (K)	Density (kg m^{-3})
Nitrogen (N)	2.5	7	63	77	1·165
Phosphorus (P)	2.8.5	15	317	552	2200 (red) 1800 (yellow)

Both elements are typical non-metals with five electrons in their outer energy level. Each can accept three electrons into its outer energy level to attain a noble-gas structure, forming the nitride ion:

$$N + 3e \rightarrow N^{3-}$$

or the phosphide ion:

$$P + 3e \rightarrow P^{3-}$$

However, it is much more likely that nitrogen and phosphorus should form covalent bonds by electron sharing, as illustrated by the hydrides NH_3 (ammonia, see Fig. 2.13) and PH_3 (phosphine).

10.25 Nitrogen

In the gaseous state nitrogen occurs as diatomic molecules (N_2). Because the two atoms in each molecule are joined together by extremely strong covalent bonding, molecular nitrogen is very unreactive. The chemistry of nitrogen is discussed in Section 6.9.

10.26 Ammonia, NH_3

Ammonia is the most important hydride of nitrogen. Its solution in water, aqueous ammonia, is an important common alkali.

(a) Preparation

(i) Ammonia gas is prepared in the laboratory by warming a mixture of an alkali and an ammonium salt. Calcium hydroxide (slaked lime) and ammonium chloride is a convenient mixture:

$$Ca^{2+}(OH^-)_{2(s)} + 2NH_4^+Cl_{(s)}^- \rightarrow Ca^{2+}Cl_{2(s)}^- + 2NH_{3(s)} + 2H_2O_{(g)}$$

$$(\text{In general:}\quad OH^- + NH_4^+ \rightarrow NH_3 + H_2O)$$

The gas may be dried by passing it through a drying tower containing calcium oxide (quicklime), as shown in Fig. 10.9. Common drying agents, e.g. calcium chloride and concentrated sulphuric acid, cannot be used to dry ammonia as they react with the gas.

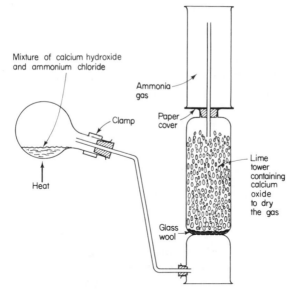

Mixture of calcium hydroxide and ammonium chloride

Ammonia gas

Clamp

Paper cover

Heat

Lime tower containing calcium oxide to dry the gas

Glass wool

Fig. 10.9   Laboratory preparation and collection of ammonia gas

(b) Properties of Ammonia Gas

(i) Ammonia is a colourless gas which is easily liquefiable. It has a characteristic pungent odour and causes irritation to the eyes and mucous membranes. The remarkable solubility of ammonia in water is demonstrated by the 'fountain experiment' shown in Fig. 10.10.

In this experiment a flask is filled with ammonia gas and closed with a cork carrying a straight glass tube. The lower end of the tube dips into a beaker of

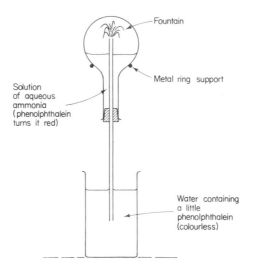

Fig. 10.10   The 'fountain' experiment demonstrating the solubility of ammonia in water

water to which a little phenolphthalein indicator has been added. As the ammonia dissolves, the pressure inside the flask is reduced and a 'fountain' sprays from the jet as more water enters. The alkaline nature of the aqueous ammonia produced causes the phenolphthalein to turn red.

(ii) *Combustion.* Ammonia will burn in oxygen with a yellowish flame, forming nitrogen and water vapour:

$$4NH_{3(g)} + 3O_{2(g)} \rightarrow 2N_{2(g)} + 6H_2O_{(g)}$$

In the presence of a platinum catalyst, ammonia is oxidized by oxygen of the air to form nitrogen monoxide (nitric oxide):

$$4NH_{3(g)} + 5O_{2(g)} \rightarrow 4NO_{(g)} + 6H_2O_{(g)}$$

This is the first stage in the manufacture of nitric acid (see Unit 14).

(iii) *Ammonia as a reducing agent.* When ammonia is passed over heated copper(II) oxide the ammonia is oxidized to nitrogen, and the copper(II) oxide reduced to copper:

$$3Cu^{2+}O^{2-}_{(s)} + 2NH_{3(g)} \rightarrow 3Cu_{(s)} + 3H_2O_{(g)} + N_{2(g)}$$

Ammonia is also oxidized to nitrogen by chlorine:

$$3Cl_{2(g)} + 2NH_{3(g)} \rightarrow N_{2(g)} + 6HCl_{(g)}$$

$$6HCl_{(g)} + 6NH_{3(g)} \rightarrow 6NH_4^+Cl_{(s)}^-$$

(c) Properties of Aqueous Ammonia

As already indicated, ammonia dissolves readily in water. Ammonia molecules form hydrogen bonds with water molecules, and some dissociate to form ammonium ions (NH_4^+) and hydroxide ions (OH^-) according to the following equilibrium reaction:

$$NH_{3(g)} + H_2O_{(l)} \rightleftharpoons NH_{4(aq)}^+ + OH_{(aq)}^-$$

(i) Because of the presence of these hydroxide ions aqueous ammonia acts as an alkali, producing ammonium salts with acids. For example:

$$2NH_{4(aq)}^+ + 2OH_{(aq)}^- + H_2^+SO_{4(aq)}^{2-} \rightarrow (NH_4^+)_2SO_{4(aq)}^{2-} + 2H_2O_{(l)}$$

(ii) It precipitates the hydroxides of most metals. For example:

$$Fe_{(aq)}^{2+} + 2OH_{(aq)}^- \rightarrow Fe^{2+}(OH^-)_{2(s)}$$

(iii) It forms soluble *ammines* with copper(II) ions and with zinc ions. (The hydroxides of these metals will therefore dissolve in excess aqueous ammonia, see Unit 9.) Copper(II) gives the deep royal-blue tetra-ammine-copper(II) ion:

$$Cu_{(aq)}^{2+} + 4NH_{3(aq)} \rightarrow Cu(NH_3)_{4(aq)}^{2+}$$

Zinc gives the colourless tetra-ammine-zinc ion:

$$Zn_{(aq)}^{2+} + 4NH_{3(aq)} \rightarrow Zn(NH_3)_{4(aq)}^{2+}$$

(d) Tests for Ammonia

Ammonia can often be detected by its characteristic odour. The gas produces dense white fumes of ammonium chloride with hydrogen chloride:

$$NH_{3(g)} + HCl_{(g)} \rightarrow NH_4^+Cl_{(s)}^-$$

Also, it is the only common *alkaline* gas.

(e) Uses of Ammonia and Ammonium Compounds

Ammonia gas is used in the manufacture of nitric acid; liquid ammonia is used as a refrigerant. Ammonium sulphate, ammonium nitrate and ammonium phosphate can be obtained from ammonia for use as 'artificial' fertilizers.

Urea is manufactured from ammonia and can be used as a fertilizer and in the manufacture of plastics. Ammonia is also used in the manufacture of nylon and as a constituent of proprietary household cleaners.

10.27 Nitric Acid, HNO_3

Nitric acid is a colourless liquid when pure; there is, however, often a yellowish coloration due to dissolved nitrogen dioxide formed when sunlight causes slight decomposition. Chemically it acts both as an acid and as an oxidizing agent.

(a) Preparation

Nitric acid is manufactured by the catalytic oxidation of ammonia gas (see

Unit 14). In the laboratory it can be prepared by warming sodium nitrate or potassium nitrate with concentrated sulphuric acid:

$$NO_3^- + H_2SO_{4(l)} \rightarrow HSO_4^- + HNO_{3(g)}$$

The nitric acid distils over and is collected in a water-cooled receiver (see Fig. 10.11).

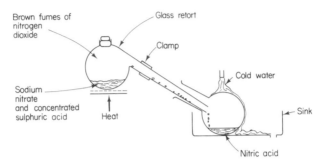

Fig. 10.11   Laboratory preparation of nitric acid

Because concentrated nitric acid attacks rubber and plastics, the preparation is carried out in an all-glass apparatus.

The flask fills with brown fumes of nitrogen dioxide produced by the decomposition of some of the nitric acid during distillation:

$$4HNO_{3(l)} \rightarrow 4NO_{2(g)} + 2H_2O_{(g)} + O_{2(g)}$$

The concentrated nitric acid obtained is yellow owing to the presence of dissolved nitrogen dioxide.

(b) Properties of Nitric Acid as an Acid

(i) A dilute aqueous solution of nitric acid behaves as a strong monobasic acid:

$$HNO_{3(l)} + H_2O_{(l)} \rightarrow H_3O^+_{(aq)} + NO^-_{3(aq)}$$

It forms only one series of salts, called *nitrates*. It will neutralize a base (e.g. sodium hydroxide) forming a salt and water only:

$$H^+NO^-_{3(aq)} + Na^+OH^-_{(aq)} \rightarrow Na^+NO^-_{3(aq)} + H_2O_{(l)}$$

(ii) It reacts with carbonates and hydrogencarbonates producing carbon dioxide. For example, with sodium hydrogencarbonate:

$$H^+NO^-_{3(aq)} + Na^+HCO^-_{3(s)} \rightarrow Na^+NO^-_{3(aq)} + CO_{2(g)} + H_2O_{(l)}$$

(iii) In general it does not react with metals to liberate hydrogen. This is because it is a powerful oxidizing agent, even in aqueous solution. (Its reaction with magnesium, however, does give hydrogen, as explained in Section 9.10.)

(c) **Properties of Nitric Acid as an Oxidizing Agent**

(i) **Reaction with metals.** Moderately concentrated nitric acid reacts with most metals to form the nitrate, *nitrogen monoxide* and water. With copper, for example:

$$3Cu_{(s)} + 8H^+NO_{3\,(aq)}^- \rightarrow 3Cu^{2+}(NO_3^-)_{2(aq)} + 2NO_{(g)} + 4H_2O_{(l)}$$

Using concentrated nitric acid, *nitrogen dioxide* is produced together with the metal nitrate and water:

$$Cu_{(s)} + 4HNO_{3(l)} \rightarrow Cu^{2+}(NO_3^-)_{2(aq)} + 2NO_{2(g)} + 2H_2O_{(l)}$$

Both *iron* and *aluminium* are rendered 'passive' by concentrated nitric acid, i.e. a thin impervious oxide layer is formed which prevents further reaction.

(ii) **Reaction with non-metals.** Warm concentrated nitric acid oxidizes sulphur to sulphuric acid and carbon to carbon dioxide, brown fumes of nitrogen dioxide being produced in each case:

$$S_{(s)} + 6HNO_{3(l)} \rightarrow H_2^+SO_{4(aq)}^{2-} + 6NO_{2(g)} + 2H_2O_{(l)}$$

$$C_{(s)} + 4HNO_{3(l)} \rightarrow CO_{2(g)} + 4NO_{2(g)} + 2H_2O_{(l)}$$

(iii) **Other oxidizing reactions.** Concentrated nitric acid oxidizes an acidified solution of an iron(II) salt to an iron(III) salt:

$$3Fe^{2+}Cl_{2(aq)}^- + 3H^+Cl_{(aq)}^- + HNO_{3(l)} \rightarrow 3Fe^{3+}Cl_{3(aq)}^- + NO_{(g)} + 2H_2O_{(l)}$$

(d) **Uses of Nitric Acid**

Nitric acid is widely used throughout the chemical industry. It is important in the manufacture of nitrogenous fertilizers such as ammonium nitrate, and explosives such as TNT and nitroglycerine (methyl-2,4,6-trinitrobenzene and propane-1,2,3,-triyl trinitrate, respectively).

10.28 Nitrates

Nitrates of metals high in the electrochemical series (e.g. sodium and potassium) decompose on heating to give the corresponding nitrite and oxygen:

$$2Na^+NO_{3(s)}^- \rightarrow 2Na^+NO_{2(s)}^- + O_{2(g)}$$

Most other nitrates give the metal oxide, nitrogen dioxide and oxygen. Notable examples are the nitrates of calcium, magnesium, zinc, lead(II) and copper(II). Thus blue-green copper(II) nitrate decomposes on heating and gives black copper(II) oxide:

$$2Cu^{2+}(NO_3^-)_{2(s)} \rightarrow 2Cu^{2+}O_{(s)}^{2-} + 4NO_{2(g)} + O_{2(g)}$$

The nitrates of metals low in the electrochemical series (e.g. mercury and silver) decompose on strong heating to give the metal, nitrogen dioxide and oxygen:

$$2Ag^+NO_{3(s)}^- \rightarrow 2Ag_{(s)}+2NO_{2(g)}+O_{2(g)}$$

Ammonium nitrate decomposes into dinitrogen oxide (nitrous oxide) and water vapour:

$$NH_4^+NO_{3(s)}^- \rightarrow N_2O_{(g)}+2H_2O_{(g)}$$

This reaction is potentially dangerous since ammonium nitrate explodes on strong heating.

10.29 Oxides of Nitrogen

(a) Nitrogen Dioxide, NO_2

(i) *Preparation*
This gas can be prepared in the laboratory by the action of heat on lead(II) nitrate:

$$2Pb^{2+}(NO_3^-)_{2(s)} \rightarrow 2Pb^{2+}O_{(s)}^{2-}+4NO_{2(g)}+O_{2(g)}$$

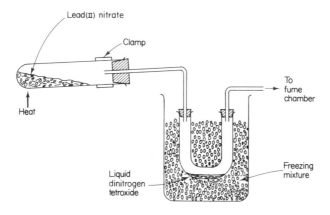

Fig. 10.12   Laboratory preparation of nitrogen dioxide (dinitrogen tetroxide)

The apparatus used in the preparation is shown in Fig. 10.12. As the white crystalline lead(II) nitrate is heated it decrepitates (crackles) and decomposes to yellow lead(II) oxide. Brown fumes of nitrogen dioxide are given off and condense in the freezing mixture to form the pale-yellow liquid dinitrogen tetroxide (N_2O_4).

(ii) *Properties*

Brown nitrogen dioxide gas exists in equilibrium with its *dimer*, dinitrogen tetroxide:

$$2NO_{2(g)} \rightleftharpoons N_2O_{4(g)}$$

The percentage of each gas in the equilibrium mixture depends on the temperature. At low temperatures the percentage of dinitrogen tetroxide is high and the mixture is pale yellow in colour. As the temperature is increased the percentage of nitrogen dioxide increases and the colour darkens until, at 150 °C, the gas is entirely nitrogen dioxide and is almost black.

The gas mixture is very soluble in water, producing an acid solution containing nitric acid and nitrous acid:

$$2NO_{2(g)} + H_2O_{(l)} \rightarrow H^+NO_{3(aq)}^- + H^+NO_{2(aq)}^-$$

Nitrogen dioxide supports the combustion of strongly burning materials such as phosphorus and magnesium:

$$4Mg_{(s)} + 2NO_{2(g)} \rightarrow 4Mg^{2+}O_{(s)}^{2-} + N_{2(g)}$$

(b) Nitrogen Monoxide, NO

(i) *Preparation*

Nitrogen monoxide (nitric oxide) can be prepared by the action of moderately concentrated nitric acid on copper:

$$3Cu_{(s)} + 8H^+NO_{3(aq)}^- \rightarrow 3Cu^{2+}(NO_3^-)_{2(aq)} + 2NO_{(g)} + 4H_2O_{(l)}$$

The gas can be collected over water so that nitrogen dioxide impurities will be dissolved.

(ii) *Properties*

Nitrogen monoxide is a colourless gas which rapidly turns brown on exposure to air or oxygen, owing to the formation of nitrogen dioxide:

$$\underset{\text{colourless}}{2NO_{(g)}} + O_{2(g)} \rightarrow \underset{\text{brown}}{2NO_{2(g)}}$$

A cold solution of iron(II) sulphate absorbs nitrogen monoxide, becoming dark brown in colour. This reaction is reversible, and pure nitrogen monoxide is liberated on warming the solution. The compound responsible for the dark-brown colour can be represented by the formula $Fe^{2+}SO_4^{2-}$. NO.

10.30 Phosphorus and Some of its Compounds

Phosphorus exhibits allotropy, i.e. it exists in several physically different but chemically identical forms. The two common forms are **white phosphorus**, which is in fact a yellowish waxy solid, and **red phosphorus**, a reddish powder. White phosphorus ignites spontaneously in air and is therefore stored under water.

Great care must be taken not to handle it with the fingers. Red phosphorus is much less reactive and can be safely stored in contact with the air.

(*a*) **Phosphorus(v) Oxide (phosphorus pentoxide)**
This compound is formed as a white solid when phosphorus burns in an excess of air or oxygen:

$$P_{4(s)} + 5O_{2(g)} \rightarrow P_4O_{10(s)}$$

White phosphorus exists as P_4 molecules with a tetrahedral structure, and the phosphorus(v) oxide exists as P_4O_{10} molecules.

Phosphorus(v) oxide reacts vigorously with water, forming phosphoric acid:

$$P_4O_{10(s)} + 6H_2O_{(l)} \rightarrow 4H_3^+PO_{4(aq)}^{3-}$$

Because of this affinity for water it is often used for drying gases which do not react with it.

(*b*) **Chlorides of Phosphorus**
Phosphorus trichloride (PCl_3) is a typical covalent chloride of a non-metal. It is prepared by passing chlorine over white phosphorus in the absence of air and moisture:

$$P_{4(s)} + 6Cl_{2(g)} \rightarrow 4PCl_{3(l)}$$

The colourless liquid fumes in moist air and reacts readily with water to form phosphonic acid (phosphorous acid):

$$PCl_{3(l)} + 3H_2O_{(l)} \rightarrow H_2^+PHO_{3(aq)}^{2-} + 3H^+Cl_{(aq)}^-$$

Phosphorus pentachloride (PCl_5) is obtained by the direct addition of chlorine to the ice-cold trichloride:

$$PCl_{3(l)} + Cl_{2(g)} \rightleftharpoons PCl_{5(s)}$$

Like the trichloride, phosphorus pentachloride is attacked by water. If the water is in excess, phosphoric acid results:

$$PCl_{5(s)} + 4H_2O_{(l)} \rightarrow H_3^+PO_{4(aq)}^{3-} + 5H^+Cl_{(aq)}^-$$

(*c*) **Uses of Phosphorus**
Much of the phosphorus produced is burned to phosphorus(v) oxide (phosphorus pentoxide) in order to prepare pure phosphoric acid. The acid can be converted into *phosphate fertilizers*; however, as outlined in Unit 14, most of these are prepared directly from phosphate rock (calcium phosphate).

Phosphates are used commercially in water softening, washing powders, baking powder and in dietary supplements.

10.31 Carbon and Silicon (Group 4)

The symbols, atomic numbers, electronic configurations and important physical properties of these two elements are given in Table 10.4.

Table 10.4 Physical properties of carbon and silicon

	Electronic configuration	Atomic number	Melting point (K)	Boiling point (K)	Density (kg m^{-3})
Carbon (C)	2.4	6	–	5100	2300
Silicon (Si)	2.8.4	14	1680	2628	2300

Carbon and silicon are both non-metals with four electrons in their outer energy level. Both are characterized by their ability to form four covalent bonds, but their chemical properties are completely different. Carbon atoms readily form multiple covalent bonds *with each other*, resulting in long chains and rings (see Unit 11). The chemistry of silicon, however, is dominated by the silicon–oxygen covalent bond, as illustrated by the giant molecules of silicon(IV) oxide (silica) and the silicones.

10.32 Carbon Compounds

A study of the compounds of carbon is called 'organic chemistry' and is the subject of Units 11 and 12. In this Unit we shall deal only with the oxides and related compounds.

The chemistry of carbon dioxide (CO_2) is discussed in Unit 6 as one of the component gases of the air.

10.33 Carbon Monoxide

Carbon monoxide (CO) is not a normal component of the air, and its presence in atmospheric pollution is entirely due to human activity. It is present in the exhaust gases from internal combustion engines, and escapes from coal and coke fires when incomplete combustion occurs. It is a *very* poisonous gas, and particularly dangerous because it has no odour and causes drowsiness. (For this reason a car engine should never be run in a closed garage.) Carbon monoxide is poisonous because it reacts with the haemoglobin in the blood, forming 'carboxyhaemoglobin'. The haemoglobin is no longer available to carry oxygen throughout the body, and death results from oxygen starvation. Carbon monoxide poisoning can be recognized by the cherry-red colour it imparts to the blood.

(a) Laboratory Preparation
(i) *By the reduction of carbon dioxide with carbon*
The complete combustion of carbon gives carbon dioxide:

$$C_{(s)} + O_{2(g)} \rightarrow CO_{2(g)}$$

This carbon dioxide can react with more carbon at red heat to form carbon monoxide:

$$CO_{2(g)} + C_{(s)} \rightarrow 2CO_{(g)}$$

These reactions take place in a coke fire or when carbon dioxide is passed over red-hot carbon in a silica tube.

(ii) *By the dehydration of methanoic acid or ethanedioic acid with concentrated sulphuric acid*
Concentrated sulphuric acid is a powerful dehydrating agent (see Section 10.22), and its action on methanoic acid (formic acid) removes the elements of water, releasing carbon monoxide:

$$HCOOH_{(l)} - H_2O_{(l)} \rightarrow CO_{(g)}$$

In the laboratory preparation of carbon monoxide, solid sodium methanoate (sodium formate) is preferable to liquid methanoic acid which has a pungent smell and will blister the skin on contact. Concentrated sulphuric acid first liberates the more volatile methanoic acid from its salt and then dehydrates it:

$$HCOO^-Na^+_{(s)} + H_2SO_{4(l)} \rightarrow HCOOH_{(l)} + Na^+HSO^-_{4(s)}$$

$$HCOOH_{(l)} - H_2O_{(l)} \rightarrow CO_{(g)}$$

If ethanedioic acid (oxalic acid) is used in place of methanoic acid, the product is a mixture of carbon monoxide and carbon dioxide:

$$\begin{matrix} COOH \\ | \\ COOH_{(s)} \end{matrix} - H_2O_{(l)} \rightarrow CO_{(g)} + CO_{2(g)}$$

The carbon dioxide is removed by passing the gases through sodium hydroxide solution: the carbon dioxide dissolves but the carbon monoxide is unaffected.

(b) **Properties**
(i) Carbon monoxide burns with a blue flame, forming carbon dioxide (the reaction is exothermic):

$$2CO_{(g)} + O_{2(g)} \rightarrow 2CO_{2(g)}$$

(ii) It is a powerful reducing agent, being used in the reduction of certain heated metal oxides to the metal. Thus it plays an important role in the extraction of iron:

$$Fe_2O_{3(s)} + 3CO_{(g)} \rightarrow 3CO_{2(g)} + 2Fe_{(l)}$$
(in iron ore)

10.34 Gaseous Fuels

Fuels are substances that provide heat energy. They may be solid (e.g. coal or coke), liquid (e.g. petrol and oils) or gases (e.g. methane, hydrogen, carbon monoxide). Gaseous fuels are often preferred in industry because they are easily controlled during burning and can be piped to any locality. Some important fuel gases are discussed below.

(a) Water Gas and Producer Gas

Carbon monoxide is an important ingredient of both *water gas* and *producer gas*, two gaseous fuels used in industry. When a blast of steam is forced through a white-hot bed of coke, hydrogen and carbon monoxide are produced in equal volumes:

$$C_{(s)} + H_2O_{(g)} \rightarrow CO_{(g)} + H_{2(g)}$$

Both gases will burn in air, producing heat. This mixture of gases is called *water gas* and is formed only when the coke is at white heat.

The reaction is endothermic and the coke cools to about 1000 °C after a few minutes. A blast of air is used to reheat the coke, forming *producer gas* by the exothermic reaction:

$$2C_{(s)} + \underbrace{O_{2(g)} + 4N_{2(g)}}_{air} \rightarrow 2CO_{(g)} + 4N_{2(g)}$$

Producer gas is a much less efficient fuel than water gas because it contains only one-third carbon monoxide (by volume) and is diluted by nitrogen of the air.

(b) Coal Gas

Coal gas is a mixture of gases produced when coal is destructively distilled in the absence of air. Its composition depends on the temperature of distillation and on the grade of coal; a typical analysis is given in Table 10.5.

Table 10.5 Composition of coal gas

Gas	Percentage by volume
Hydrogen	50
Methane	35
Carbon monoxide	7
Nitrogen	5
Other hydrocarbon gases	3

The destructive distillation of coal can be illustrated using the apparatus shown in Fig. 10.13.

After heating, coke remains in tube *A*. A brownish-yellow liquid with a black oil (tar) is seen in the cooled tube *B*. The brownish-yellow liquid is aqueous ammonia and the black oil (tar) contains a variety of important organic compounds. Coal gas is collected in the gas jar *C*.

Before coal gas is used as a fuel, the tar, ammonia and sulphur compounds (particularly hydrogen sulphide) must be removed.

(c) Natural Gas

Until the discovery of large natural-gas fields in the Sahara and beneath the North Sea, the most important domestic fuel gas in Great Britain was *town gas* (coal gas with various additions). Now the most important fuel gas is natural gas,

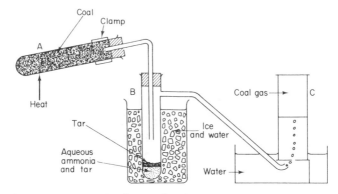

Fig. 10.13   Destructive distillation of coal in the laboratory: this experiment shows the principles underlying the manufacture of coal gas

composed of almost pure methane (CH_4). In the United States and Canada natural gas has been used as a major industrial and domestic fuel for many years.

Natural gas burns with a hot blue flame and is the cleanest and most efficient of all fuel gases. Its heat value is considerably higher than that of water gas or town gas.

10.35 Silicon and Silicon(IV) Oxide

Silicon occurs abundantly in the earth's crust as the oxide SiO_2 (silica) and as silicates in rocks and clays. The chemistry of silicon is characterized by its ability to form covalent silicon–oxygen bonds. In the giant covalent silicon(IV) oxide molecule, each silicon atom is covalently bonded to four oxygen atoms (see Fig. 10.14).

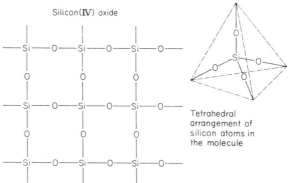

Fig. 10.14   Molecular structure of silicon(IV) oxide (silica)

This results in an extremely hard and stable compound existing in a number of different crystalline forms, the most common of which is *quartz*. Sand is mainly impure silicon(IV) oxide.

10.36 Silicones

Since silicon and carbon occur in the same group in the periodic table, it might be expected that silicon would form complex structures corresponding to the 'polymeric' carbon compounds described in Unit 11. This is partly true, but unlike carbon, which can form multiple carbon–carbon bonds, silicon tends to form stable silicon–oxygen multiple bonds as shown in the silicon(IV) oxide molecule (Fig. 10.14). Also unlike carbon, silicon compounds can exist with two hydroxyl groups attached to the same silicon atom:

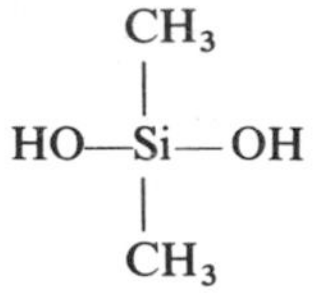
CH3
HO—Si—OH
CH3

Molecules of this type are able to *polymerize* (join together to form larger molecules) by eliminating the elements of water (see Unit 11):

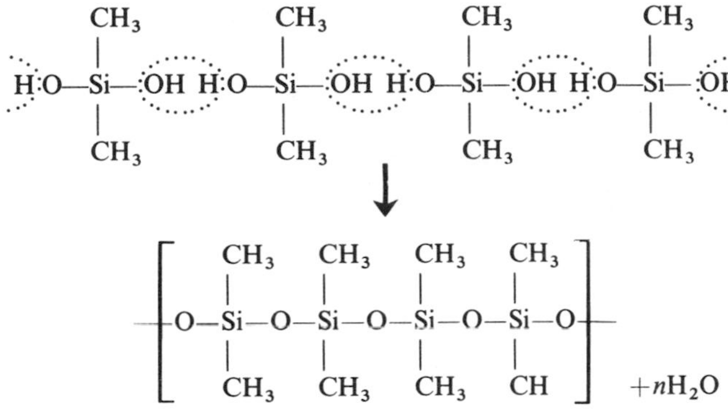
CH3
H:O—Si—:OH H:O—Si—:OH H:O—Si—:OH H:O—Si—:OH
CH3
O—Si—O—Si—O—Si—O—Si—O
CH
+nH2O

Polymeric silicon molecules like that shown above, together with those which can be produced with cross-linking like that shown below, are called **silicones**.

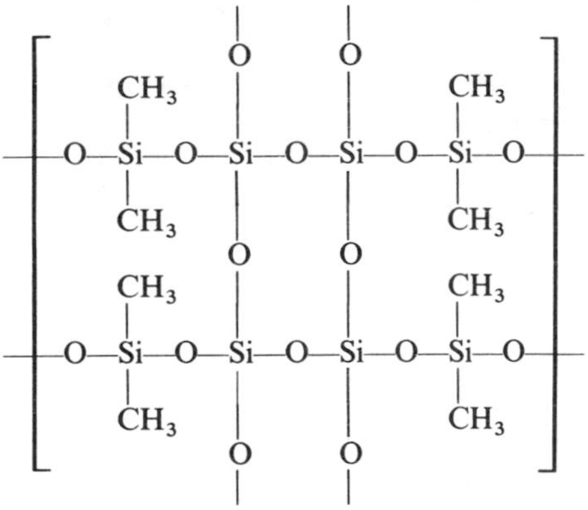
O
CH3
O—Si—O—Si—O—Si—O—Si—O
CH3

(Part of a silicone with cross-linking)

Silicones can be oils or rubbery solids, according to the length of the polymer chain and the degree of cross-linking. Because of their ability to withstand heat and retain their lubricating properties even under very cold conditions, they are used as high- and low-temperature lubricants. In addition, they find widespread applications in electrical insulation, water-repellent finishes and in heat-resisting resins and lacquers.

10.37 Glass

'Soda glass' or ordinary 'soft' glass is prepared from sand (silicon(IV) oxide), limestone and sodium carbonate. When these ingredients are heated strongly they fuse (melt), forming a transparent liquid mixture of sodium silicate, calcium silicate and silicon(IV) oxide (silica). The transparent mixture known as *soft glass* does not crystallize on cooling but hardens into a transparent solid. Molten 'glass' can be blown or shaped into any desired form.

The addition of appropriate metal oxides during manufacture gives colour to the glass. Cobalt(II) oxide, for example produces a deep-blue colour.

'Pyrex' glass is made by fusing silicon(IV) oxide (sand), disodium tetra-borate-10-water (borax) and aluminium oxide. The glass produced can withstand wide temperature variations because of its small coefficient of expansion, and is therefore widely used in ovenware and laboratory glassware.

Summary of Unit 10

1. Most **non-metals** are gases consisting of separate covalently bonded diatomic molecules with only weak attractive forces between them. Such a structure accounts for their low electrical and thermal conductivity.
2. Non-metals readily *gain electrons* to form *anions*. They also form acidic oxides.
3. Members of the **halogen** family each have seven electrons in their outermost energy level, and react by accepting an electron to become the anion X^-.
4. *Fluorine* is a stronger electron acceptor than any other element. The other halogens are strong electron acceptors, the ease of acceptance decreasing as the atomic number increases.
5. The halogens are *powerful oxidizing agents*; the order of oxidizing ability is fluorine > chlorine > bromine > iodine.
6. Halogens are prepared by the removal of electrons from halide ions. In the case of fluorine such removal of electrons requires an electrolytic process, whereas the other halogens are liberated by chemical methods.
7. Chemical reactivity of the halogens is usually associated with the formation of the halide ion.
8. The *halogen hydrides* are gases (hydrogen fluoride boils at 19 °C) and their aqueous solutions are acidic.
9. **Hydrogen** can be obtained from *water, acids* or *alkalis.*
10. Pure hydrogen burns quietly with a blue flame, but forms explosive mixtures

with oxygen or air. It is a *reducing* agent. Reaction with *non-metals* produces the appropriate hydride.

11. **Oxygen** and **sulphur** are elements in Group 6, having six electrons in their outer energy level. They form both *ionic* and *covalent* bonds.

12. **Oxidation** can be defined in terms of *oxygen gain, hydrogen loss,* a *loss of electrons* or an *increase in oxidation number.* **Reduction** is the converse of oxidation.

13. *Sulphur* is an *allotropic* element.

14. *Hydrogen sulphide* has a characteristic odour and is a *reducing agent.*

15. The two common oxides of sulphur are *sulphur dioxide* (SO_2) and *sulphur trioxide* (SO_3).

16. *Sulphur dioxide* is the *anhydride* of *sulphurous acid.* Both the gas and the acid have *reducing properties.*

17. *Sulphur trioxide* (sulphur(VI) oxide) is the *anhydride* of *sulphuric acid.*

18. **Sulphuric acid** has the following properties:
    (*a*) it is an *acid* in dilute solution
    (*b*) it is an *oxidizing agent* when hot and concentrated
    (*c*) it is a *dehydrating agent* when concentrated.

19. Sulphuric acid is used in the preparation of the more volatile acids.

20. **Nitrogen** and **phosphorus** are elements in Group 5, having five electrons in their outer energy level.

21. **Ammonia** (NH_3) is the most important hydride of nitrogen. It is a *reducing agent.*

22. Ammonia dissolves readily in water; its solution, *aqueous ammonia,* is an alkali.

23. **Nitric acid** acts both as an *acid* and as an *oxidizing agent.*

24. Common oxides of nitrogen include *nitrogen dioxide* (NO_2), *nitrogen monoxide* (nitric oxide, NO) and *dinitrogen oxide* (nitrous oxide, N_2O).

25. *Phosphorus* is an *allotropic* element.

26. Phosphorus forms an extremely reactive acidic oxide (*phosphorus*(V) *oxide,* phosphorus pentoxide) and two *chlorides* (phosphorus pentachloride and phosphorus trichloride).

27. **Carbon** and **silicon** are elements in Group 4, having four electrons in their outer energy level. Both form strong *covalent* bonds.

28. *Diamond* and *graphite* are allotropic forms of carbon.

29. **Carbon dioxide** is the product of complete combustion of carbon, whereas **carbon monoxide** is produced by the incomplete combustion of carbon.

30. Carbon monoxide is a powerful *reducing agent.* It is a main constituent of *water gas* and *producer gas,* two important *gaseous fuels.*

31. *Coal gas* and *natural gas* are two other important gaseous fuels.

32. **Silicon** forms covalent bonds with oxygen in the giant covalent *silicon*(IV) *oxide* (silica) molecule and also in *silicones.*

33. *Glass* is produced from silicon(IV) oxide.

Test Yourself on Unit 10

1. You are provided with the following chemicals: copper, manganese(IV) oxide, sulphuric acid, sodium hydroxide, ammonium chloride.
   Which of these would you mix together (heating if necessary) to prepare:
   (a) hydrogen chloride
   (b) chlorine
   (c) ammonia
   (d) sulphur dioxide?
Write an equation for each preparation.

2. This question refers to the halogens (fluorine, chlorine, bromine and iodine).
   (a) Bromine has an atomic number of 35. Give its electronic configuration.
   (b) Which halogen has the lowest boiling point?
   (c) Which halogen is the most powerful oxidizing agent?
   (d) Name the compound formed when chlorine reacts with iron.
   (e) Write an electronic equation showing the formation of a fluoride ion from a fluorine atom. Is this an oxidation or a reduction process?
   (f) Which halogen exists as a liquid at room temperature and atmospheric pressure?
   (g) Which of these halogens has the largest number of electrons in the neutral atom?

3. Sulphuric acid can act as (i) an acid, (ii) an oxidizing agent, and (iii) a dehydrating agent. In which of these ways is it reacting when:
   (a) Sugar blackens and swells, becoming hot and liberating steam.
   (b) Sulphur is converted to sulphur dioxide.
   (c) Blue copper(II) sulphate crystals turn white.
   (d) Carbon dioxide is liberated from a carbonate.
   (e) Warmed with copper, sulphur dioxide is liberated.

4. Name:
   (a) The nitrogen-containing compound formed when calcium hydroxide is warmed with ammonium chloride.
   (b) The white solid formed when sulphur dioxide and oxygen are passed over a heated catalyst and the product is cooled.
   (c) The white fumes produced when phosphorus burns in oxygen.
   (d) The colourless gas obtained when concentrated sulphuric acid is added to sodium methanoate (sodium formate).

5. Compounds A, B and C are all white crystalline sodium salts.
   (a) A solution of A reacts with silver nitrate solution to give a yellow precipitate (D), which is insoluble in dilute nitric acid. An aqueous solution of A reacts with chlorine dissolved in tetrachloromethane (carbon tetrachloride) to give a purple coloration.
   (b) B reacts with warm dilute hydrochloric acid to liberate a pungent-smelling colourless gas (E), which turns acidified dichromate(VI) solution from orange to green.

(*c*) C reacts with dilute hydrochloric acid to liberate an unpleasant-smelling gas (F), which blackens a filter paper soaked in lead(II) ethanoate (lead acetate) solution.

Identify the sodium salts A, B and C, and the products D, E and F, whose reactions are described above.

6. State whether each of the following statements is true or false:

(*a*) Sulphur is an allotropic element.

(*b*) Silicon forms multiple silicon–oxygen bonds in both silicon(IV) oxide and silicones.

(*c*) All non-metals are gases.

(*d*) All non-metals are held together by weak attractive van der Waals' forces.

(*e*) Oxidation is defined as an increase in oxidation number.

(*f*) Ammonia forms a deep-blue ammine with Cu^{2+} ions.

(*g*) All metal nitrates yield the metal oxide, nitrogen dioxide and oxygen on heating.

(*h*) Water gas and producer gas are gaseous fuels containing carbon monoxide.

(*i*) Chlorine is absorbed by sodium hydroxide solution.

Mark this test out of 40 with the answers given on page 379.

Classification of Matter IV: Organic Chemistry—The Chemistry of Carbon

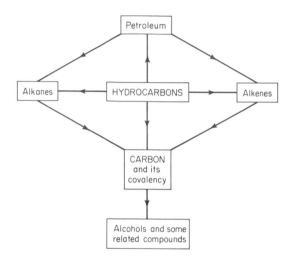

The term 'organic' chemistry was originally confined to a study of those compounds of carbon which occurred in living organisms. However, in 1828 a German chemist, Friedrich Wöhler, synthesized the organic compound *urea* from the non-living 'inorganic' substance ammonium cyanate, and so the meaning of *organic chemistry* has had to be extended to include the compounds of carbon whether or not they occur in living organisms. A few carbon compounds such as carbon dioxide and the carbonates have been described in previous Units and are not usually studied as part of organic chemistry.

Because of the unique ability of carbon atoms to bond covalently in chains and rings, there is an enormous number of different organic compounds. Indeed the compounds containing carbon outnumber those that do not.

This Unit looks at *petroleum*, a major source of the *hydrocarbons* (compounds containing carbon and hydrogen only). It begins to classify these hydrocarbons according to their structure and properties and finally describes some other simple but important organic compounds.

Attempts are made throughout Units 11 and 12 to illustrate the shape of organic molecules. For a more complete picture, three-dimensional molecular models would greatly assist an appreciation of the spatial arrangement of the atoms within the molecule. These may be constructed with polystyrene balls and pipe-cleaners or cocktail-sticks.

11.1 Covalency of Carbon

A carbon atom contains four electrons in its outer energy level and can form four covalent bonds by sharing with four electrons from other atoms. The carbon atom thus has a share in eight electrons and attains the stable configuration of the noble gas neon. Each covalent bond is formed by the sharing of a pair of electrons and is directed in space (see Fig. 11.1).

(a) Carbon–Hydrogen Covalent Bonds

As described in Unit 2, *methane* (CH_4) contains four carbon–hydrogen covalent bonds directed towards the corners of a regular tetrahedron. In this molecule each of the four carbon electrons forms an electron-pair covalent bond with the four electrons of four hydrogen atoms:

$$\overset{\circ}{\underset{\circ}{\overset{\circ}{C}}}\circ + 4H^x \rightarrow H\,{}^{\circ}_{x}\overset{ox}{\underset{ox}{C}}\,{}^{\circ}_{x}H \qquad \text{or} \qquad H-\overset{\displaystyle H}{\underset{\displaystyle H}{\overset{|}{\underset{|}{C}}}}-H$$

Each atom thus attains a noble-gas electronic configuration, and the resulting molecule is very stable.

The above 'dot-and-cross' representation for methane does not give any indication of the three-dimensional shape of the molecule. Fig. 11.1 shows three ways in which the spatial arrangement of the atoms in methane can be represented.

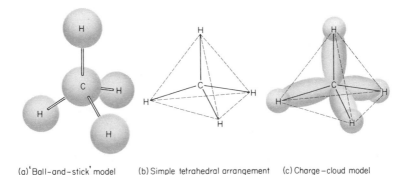

(a) 'Ball-and-stick' model   (b) Simple tetrahedral arrangement   (c) Charge–cloud model

Fig. 11.1   Three methods of representing the molecular structure of methane CH_4

(b) Carbon–Carbon Covalent Single Bonds

Carbon can also form strong covalent bonds with other carbon atoms. This is the unique property of carbon which is fundamental to its chemistry. The *ethane* molecule (C_2H_6) contains one carbon–carbon single covalent bond and six carbon–hydrogen covalent bonds. Fig. 11.2 shows a 'dot-and-cross' representation and a spatial arrangement of the ethane molecule.

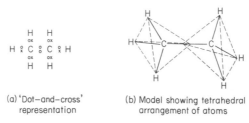

(a) 'Dot–and–cross'
representation

(b) Model showing tetrahedral
arrangement of atoms

Fig. 11.2   Molecular structure of ethane C_2H_6

There is seemingly no limit to the number of carbon atoms which can link up to form carbon chains. For example, Fig. 11.3 shows three carbon atoms linked together in a molecule of the hydrocarbon *propane* (C_3H_8) and nine in the hydrocarbon *nonane* (C_9H_{20}).

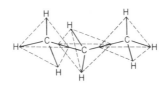

(a) Propane showing tetrahedral arrangement of atoms

(b) Propane

(c) Nonane

Fig. 11.3   The hydrocarbons propane C_3H_8 and nonane C_9H_{20}

In Figs. 11.3b and 11.3c the tetrahedral outline of the carbon atoms has been omitted for ease of drawing, but the three-dimensional configuration of the molecule and the fact that the carbon chain is *not* straight should still be apparent. For even greater convenience it is often sufficient to represent carbon chains in only two dimensions: thus the formula of propane may be written as

$$
\begin{array}{ccc}
\text{H} & \text{H} & \text{H} \\
| & | & | \\
\text{H}-\text{C}-\text{C}-\text{C}-\text{H} \\
| & | & | \\
\text{H} & \text{H} & \text{H}
\end{array}
$$

It must be borne in mind, however, that this is much further from a true picture than the structures shown in Fig. 11.3.

The structures of hydrocarbon compounds are not restricted to 'straight' chains of carbon atoms. Rings and branched chains often occur, as shown in Fig. 11.4.

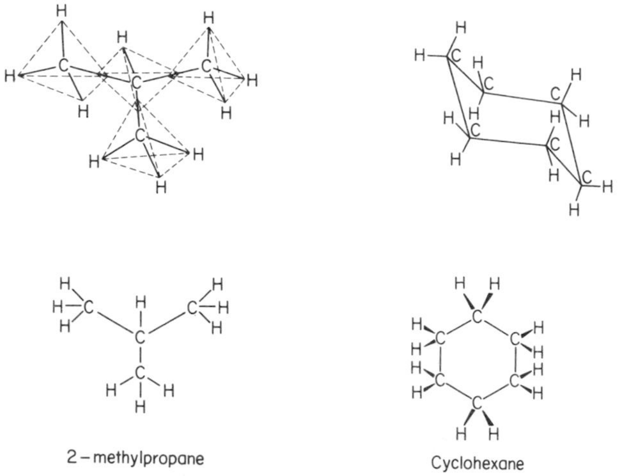

2 – methylpropane   Cyclohexane

Fig. 11.4   Alternative representations of (left) a branched-chain hydrocarbon and (right) a ring hydrocarbon

(c) Carbon–Carbon Covalent Double Bonds

The carbon–carbon bonds described so far are formed by the sharing of two electrons. In the *ethene* (ethylene) molecule, C_2H_4, there are four electrons shared between the two carbon atoms, two electrons being provided by each atom. This can be represented as

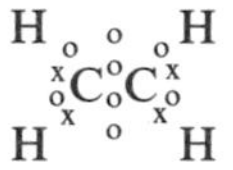

or simply as

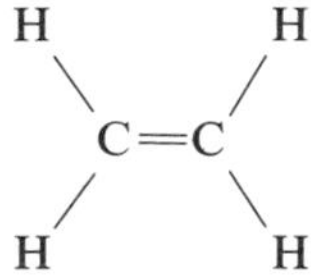

All the atoms in the molecule attain a noble-gas structure. In this case there are two pairs of electrons and hence two covalent bonds between the two carbon atoms. This is a carbon–carbon *double* bond. The presence of such a double bond results in a planar (flat) structure (Fig. 11.5).

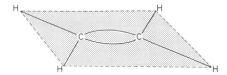

Fig. 11.5   Molecular structure of ethene C_2H_4: all six atoms lie in the same plane

(d) Carbon–Carbon Covalent Triple Bonds

The sharing of three pairs of electrons between two carbon atoms results in the formation of a carbon–carbon *triple* bond, as illustrated in the *ethyne* (acetylene) molecule C_2H_2:

$$H_x^oC_o^oC_x^oH \qquad \text{or} \qquad H-C\equiv C-H.$$

This molecule is linear, with all four atoms in a straight line.

(e) Carbon–Oxygen Covalent Bonds

Oxygen, having six electrons in its outer energy level, must share with two electrons of other atoms (or atom) to gain a stable noble-gas structure. Thus oxygen will form two electron-pair covalent bonds, either two single (as in methanol) or one double (as in propanone).

Methanol (methyl alcohol) CH_3OH has a structure which can be represented as

$$H_x^oC_o^oO: \qquad \text{or} \qquad H-C-O$$

Propanone (acetone) CH_3COCH_3 has a structure containing a double covalent carbon–oxygen bond

$$H_x^oC_o^oC_o^oC_x^oH \qquad \text{or} \qquad H-C-C-C-H$$

(f) Other Elements which Bond Covalently with Carbon

Nitrogen has five electrons in its outer energy level, and the halogens have seven. Nitrogen requires three more electrons to form a noble-gas configuration, and the halogens require one. Thus nitrogen will form three covalent bonds and halogens a single covalent bond. For example:

(i) *Methylamine* CH_3NH_2 contains a carbon–nitrogen single bond

$$\begin{array}{ccc}
\text{H} & & \text{H} \\
\text{OX} \quad \cdot\cdot & & | \\
\text{H}^{o}_{X}\text{C}^{o}_{X}\text{N}^{x}_{X}\text{H} & \text{or} & \text{H}-\text{C}-\text{N}-\text{H} \\
\text{OX} \quad \cdot\text{X} & & | \quad | \\
\text{H} \; \text{H} & & \text{H} \; \text{H}
\end{array}$$

(ii) *Trichloromethane* (chloroform) $CHCl_3$ contains three carbon–chlorine single covalent bonds

$$\begin{array}{ccc}
& & \text{Cl} \\
:\ddot{\text{C}}\text{l}: & & | \\
\text{H}^{o}_{X}\text{C}^{o}_{o}\ddot{\text{C}}\text{l}: & & \text{H}-\text{C}-\text{Cl} \\
:\ddot{\text{C}}\text{l}: & & | \\
& & \text{Cl}
\end{array}$$

11.2 The Alkanes

Organic compounds containing only carbon and hydrogen are called *hydrocarbons*. These compounds can be further classified according to their varying structures. The *alkanes* (also known as paraffins) are those hydrocarbons which contain only carbon–carbon covalent single bonds. Table 11.1 lists the names and molecular formulas for the first ten alkanes.

Table 11.1 The alkanes

Methane	CH_4
Ethane	C_2H_6
Propane	C_3H_8
Butane	C_4H_{10}
Pentane	C_5H_{12}
Hexane	C_6H_{14}
Heptane	C_7H_{16}
Octane	C_8H_{18}
Nonane	C_9H_{20}
Decane	$C_{10}H_{22}$

Alkanes are not normally prepared in the laboratory because they are easily obtained on a large scale from petroleum or, in the case of methane, from natural gas.

(a) General Formula

The alkanes have a general formula C_nH_{2n+2} where n is the number of carbon atoms in the molecule. For example, when n is 3, $(2n+2) = 8$ and the alkane has a formula C_3H_8 (propane).

(b) Structure

The structure of methane and ethane has already been discussed and is illustrated in Figs. 11.1 and 11.2. In all the alkanes the distribution of bonds around each carbon atom is tetrahedral.

(c) Isomerism

It is found that there are two different alkanes having the molecular formula C_4H_{10}: one boiling at $-0.5\,°C$, the other at $-11.7\,°C$. This is because there are two, and only two, ways in which the carbon atoms can combine with one another: they must join up either in a straight chain or in a branched chain (see Fig. 11.6).

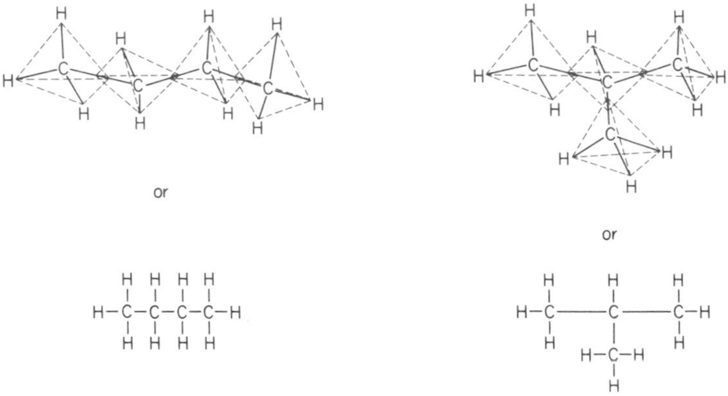

Fig. 11.6   Structures of the two alkanes having the same molecular formula C_4H_{10}

These two compounds having the *same molecular formula* are described as **isomers**. All the alkanes containing four or more carbon atoms exist in *isomeric forms*. The more carbon atoms in the molecule, the greater the number of isomers possible. It thus becomes clearer why there is such an enormous number of organic compounds.

(d) Nomenclature

The names of all the alkanes end with the suffix -ane, e.g. methane, ethane, propane, butane. With the exception of the first four members of the series (i.e. the four listed above) the names of the alkanes begin with a Greek prefix indicating the number of carbon atoms in the main chain, e.g. pentane (five carbon atoms), hexane (six), octane (eight).

Hydrocarbons containing simple substituents, e.g. a halogen or a hydroxyl group, and branched-chain hydrocarbons are named according to the following system.

First, select the longest continuous chain of carbon atoms in the molecule and use this to deduce the parent name of the compound.

Secondly, number the carbon atoms of this chain so that the substituents are attached to the carbon atoms having the lowest number.

Finally, name the substituent and write down the name of the compound as one word.

Example (i). We have already seen that the molecular formula C_4H_{10} can apply to two different structures, as illustrated in Fig. 11.6. The first of these is

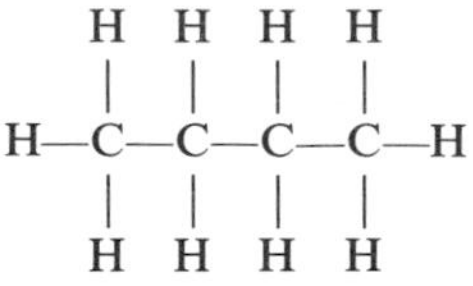

$$
\begin{array}{ccccc}
 & H & H & H & H \\
 & | & | & | & | \\
H- & C- & C- & C- & C-H \\
 & | & | & | & | \\
 & H & H & H & H \\
\end{array}
$$

The longest carbon chain has *four* atoms, so the parent name is *butane*. As there are no substituents it is unnecessary to number the carbon atoms, hence the name of the compound is simply *butane*.

The other isomer of C_4H_{10} is

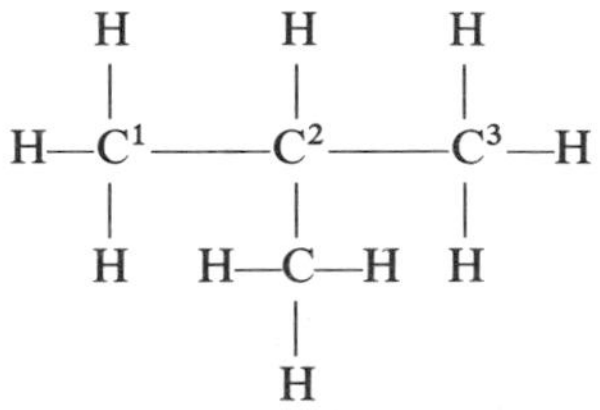

$$
\begin{array}{ccc}
H & H & H \\
| & | & | \\
H-C^1\!-\!-\!-\!-C^2\!-\!-\!-\!-C^3-H \\
| & | & | \\
H & H-C-H & H \\
 & | \\
 & H \\
\end{array}
$$

The longest carbon chain here has only *three* atoms, so the parent name is *propane*. On the central carbon atom in the longest chain there is a *methyl* (CH_3) substituent. Whichever way we number the chain, the carbon atom to which the substituent is attached is number 2. Hence the name of the compound is *2-methylpropane*.

Example (ii). The molecular formula $C_2H_4Br_2$ can apply to two different structures. One is

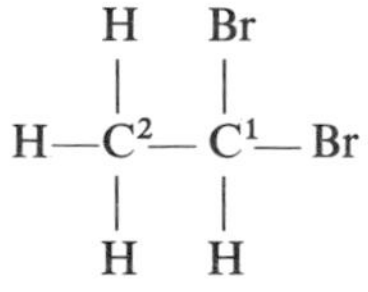

$$
\begin{array}{cc}
H & Br \\
| & | \\
H-C^2\!-\!C^1\!-\!Br \\
| & | \\
H & H \\
\end{array}
$$

The longest carbon chain has *two* atoms, so the parent name is *ethane*. Two *bromine* substituents are located on one of the carbon atoms: this carbon is given the lowest number possible, i.e. 1. So the name of the compound is *1,1-dibromoethane* (1,1- indicates that both bromine atoms are attached to the same carbon atom).

The other isomer of $C_2H_4Br_2$ is

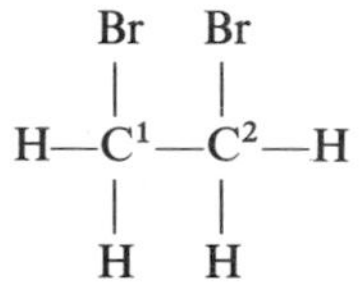

$$
\begin{array}{cc}
Br & Br \\
| & | \\
H-C^1-C^2-H \\
| & | \\
H & H \\
\end{array}
$$

Again the longest carbon chain has two atoms, so the parent name is ethane. Whichever way they are numbered, there is a bromine substituent on carbon atom number 1 and on carbon atom number 2. Hence the name of the compound is 1,2-*dibromoethane*.

(e) Homologous Series

The alkanes listed in Table 11.1 differ from each other by —CH_2—. Thus methane CH_4 differs from ethane C_2H_6 by —CH_2—, and ethane in turn differs from propane C_3H_8 by —CH_2—. Such a series of compounds is called a *homologous series*.

The members of a homologous series have similar chemical properties and show a distinct gradation in physical properties. Thus the melting and boiling points of the alkanes increase as the number of carbon atoms increases. Because they have similar chemical properties, a knowledge of the properties of one member of a homologous series enables the properties of another member to be predicted.

(f) Functional Groups

A functional group is a part of a compound which has a characteristic set of properties. Thus when a bromine atom replaces a hydrogen atom in an alkane, it imparts to the compound new chemical and physical properties. Six important functional groups are listed in Table 11.2.

Table 11.2 Functional groups

Name	Structural formula	Abbreviated formula
Chloro-	—Cl	—Cl
Bromo-	—Br	—Br
Iodo-	—I	—I
Hydroxy-	—O$\diagdown$H	—OH
Carboxy-	$-C\diagup^{\parallel O}_{\diagdown O-H}$	—COOH
Amino-	$-N\diagup^{H}_{\diagdown H}$	—NH$_2$

(g) Alkyl Groups

A group formed by the removal of a hydrogen atom from a hydrocarbon is called an *alkyl* group. These groups do not exist on their own but are always attached to another atom or group. Alkyl groups are named by removing the ending -*ane* from the parent alkane and replacing it with -*yl*. Thus the alkyl group from methane (CH_4) is methyl (CH_3—), and that from propane (C_3H_8) is propyl (C_3H_7—).

11.3 Properties of Alkanes

Compounds which contain only covalent single bonds are said to be *saturated*. The alkanes contain only carbon–carbon and carbon–hydrogen single bonds and are therefore *saturated hydrocarbons*. As they have no functional groups they are not very reactive.

The simplest members of the alkanes are colourless gases; the higher members are colourless liquids or waxy solids (e.g. paraffin wax is a mixture of alkanes).

Alkanes burn readily in air, combining with oxygen to produce carbon dioxide, water vapour and large quantities of heat.

$$C_3H_{8(g)} + 5O_{2(g)} \rightarrow 3CO_{2(g)} + 4H_2O_{(g)} \; (\Delta H = -2058 \text{ kJ mol}^{-1})$$

For this reason they are widely used as fuels.

Substitution reactions. For a chemical reaction to occur, bonds must be broken. In alkanes the carbon–hydrogen bonds are weaker than the carbon–carbon bonds, and for this reason replacement of a hydrogen atom or atoms by some other functional group is to be expected. This type of reaction is called a *substitution*.

Experiment 11.1   Substitution of bromine in hexane
A sample of hexane (C_6H_{14}) is placed in a boiling tube and to it is added some bromine dissolved in tetrachloromethane (carbon tetrachloride). The tube is stoppered, and the mixture shaken and allowed to stand. At first there is no change and the red coloration due to bromine is clearly visible. After a while the red colour begins to fade and the pungent smell of hydrogen bromide gas can be detected when the stopper is removed. The presence of this acid gas is further demonstrated if moist Universal indicator paper is held in the vapour inside the tube.

The bromine is substituted for hydrogen atoms in hexane, producing bromo-hexane and hydrogen bromide:

$$C_6H_{14(l)} + Br_2 \underset{\text{(in CCl}_4)}{\rightarrow} C_6H_{13}Br_{(l)} + HBr_{(g)}$$

Further substitution may take place as more hydrogen atoms are replaced by bromine.

The rate of this type of reaction is increased when energy, in the form of heat or sunlight, is supplied. This can be demonstrated by exposing the reaction mixture of bromine and hexane to the radiation from a tungsten or ultra-violet lamp: the colour of the bromine is then found to disappear much more rapidly.

11.4 Petroleum

Crude oil or *petroleum* is a complex mixture of hydrocarbons. It may contain sulphur and appreciable amounts of other substances. This oil was formed millions of years ago by bacterial action on microscopic organisms living in seas. It accumulated first as droplets widely dispersed in porous rock layers, usually together with water and gas. Gradually the oil and gas rose through the porous rock and much of it escaped. However, some was caught in faults and anticlines where layers of porous rock were covered by layers of impervious (non-porous) rock, as shown in Fig. 11.7.

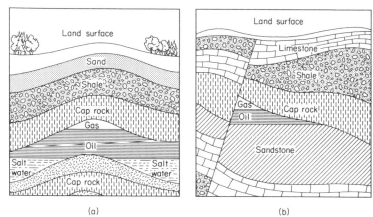

(a)   (b)

Fig. 11.7   Oil traps: (a) an anticline, (b) a fault

The gas and petroleum can be extracted by drilling a hole through the cap rock into the deposit. Natural gas confined by the impervious rock may collect under great pressure, and this compressed gas is often sufficient to force the oil up to the surface. As the pressure diminishes the oil is recovered by pumping.

Rich oil deposits are found in the Middle East, the United States, Canada, Mexico, Venezuela and under the North Sea.

11.5 Refining Petroleum

(a) Fractionation
The oil industry takes crude petroleum and separates it into 'fractions' having different boiling points. This is accomplished by the process of *fractional distillation* (see Section 1.5).

In the laboratory this fractionation of crude oil may be illustrated by using the apparatus shown in Fig. 1.7. On an industrial scale the separation is carried out in vertical columns containing series of perforated horizontal trays (see Fig. 11.8).

The crude oil is first heated so that all the fractions to be removed are vaporized. This mixture passes into a fractionating column which is divided

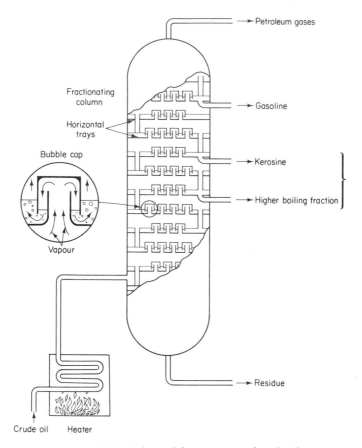

Fig. 11.8   Industrial fractionation of crude oil

horizontally by plates or trays. Each tray contains a number of 'bubble caps'. These force the rising vapour through the liquid condensed on the tray causing a continuous process of condensation and re-evaporation. Consequently each tray contains a lower boiling-point fraction than the one below it. The different fractions are taken from the column as indicated in Fig. 11.8.

No attempt is made in fractionation to separate the petroleum into individual hydrocarbons. The main fractions include:

(i) *petroleum gases*, containing simple chain alkanes;

(ii) *gasoline*, a very volatile, low-boiling-point liquid containing hexanes, heptanes and octanes;

(iii) *naphtha*, *kerosine* and other higher-boiling-point fractions, containing a mixture of complex hydrocarbons.

In the modern world the demand for these different fractions varies. At the present time, with ever-increasing numbers of motor cars, the demand for gasoline exceeds that which can be obtained solely by distillation. To increase the proportion of gasoline obtained from crude oil a process called **cracking** is employed to break large molecules of high-boiling-point hydrocarbons into smaller molecules of the more volatile gasoline fraction.

(b) Cracking

Cracking is the breaking of carbon–carbon or carbon–hydrogen bonds in large hydrocarbon molecules to produce smaller molecules of more simple hydrocarbons, together with hydrogen. In *thermal cracking*, crude oil is decomposed by heating to a high temperature under high pressure; in the more efficient *catalytic cracking*, a catalyst (e.g. a mixture of aluminium and silicon oxides) enables the decomposition to be carried out at lower temperatures and pressures. The product of cracking is a complex mixture of alkanes (e.g. hexane, C_6H_{14}) and alkenes (e.g. ethene and propene).

Experiment 11.2   *Demonstration of the 'cracking' process*
The apparatus shown in Fig. 11.9 can be used to demonstrate the cracking of a number of different substances.

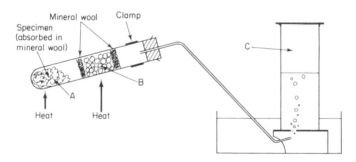

Fig. 11.9   Apparatus used in the laboratory to demonstrate the 'cracking' of oil

(i) When A is *crude oil* it is cracked by heating and passing the vapour over heated pumice stone located at B. The product collected at C is a mixture of alkanes and alkenes. These gases burn with a yellow flame. If *bromine water* is added to a further sample of these gases and the mixture shaken, the red-brown colour of the bromine water rapidly disappears. This immediate decolorization indicates the presence of an *alkene*, e.g. ethene, C_2H_4.

(ii) A similar experiment is performed using *hexane* as A in Fig. 11.9 and a silicon(IV) oxide and aluminium oxide catalyst at B. Cracking the hexane vapour produces a mixture of hydrocarbons, of which the major component is *ethene*.

(iii) When A is *ethanol*, C_2H_5OH, B can be porous pot or pumice stone. Cracking of the ethanol vapour produces *ethene*. (This method can be used to prepare a sample of ethene in the laboratory.)

Petrochemicals. An oil refinery takes crude oil and from it produces gasoline, refinery gases, and other high-boiling-point fractions. It converts the less useful fractions into gasoline, ethene, propene and other basic materials used in the production of plastics, detergents, solvents, rubbers and an enormous variety of equally important compounds. In fact most areas in our technical society use petrochemicals in one form or another.

11.6 Alkenes

We have seen that two of the most important products of cracking are *ethene* (ethylene) C_2H_4, and *propene* (propylene) C_3H_6. Ethene and propene are the first two members of a homologous series of hydrocarbons called **alkenes**. The general formula for the alkenes is C_nH_{2n}, although the compound with $n = 1$ (CH_2) does not exist.

Each of these compounds contains two pairs of electrons shared between two carbon atoms, i.e. a carbon–carbon double bond, as shown in Fig. 11.10. They are therefore described as *unsaturated*. (This is in contrast to the alkanes, which contain only single covalent bonds and are described as *saturated*.)

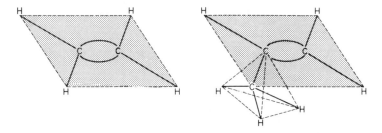

Fig. 11.10   Molecular structure of ethene C_2H_4 (left) compared with that of propene C_3H_6 (right)

Nomenclature of the alkenes. The presence of a double covalent bond in the alkenes is indicated by the suffix -*ene* in the name of each compound, while the position of the double bond is indicated by using the lowest number of the two carbon atoms joined by the double bond. Let us consider for example a molecule having the structure

$$\begin{array}{ccccc} H & H & H & H & H \\ | & | & | & | & | \\ H-C^1-C^2 & = & C^3-C^4-C^5-H \\ | & & | & | \\ H & & H & H \end{array}$$

(i) The carbon chain contains a double bond; therefore the compound is an alk*ene*.

(ii) The longest carbon chain containing the double bond comprises five atoms; therefore the compound is a *pentene*.

(iii) The double bond is situated between carbon atoms 2 and 3; therefore the compound is *pent-2-ene*.

There is no necessity to number the position of the double bond in either ethene

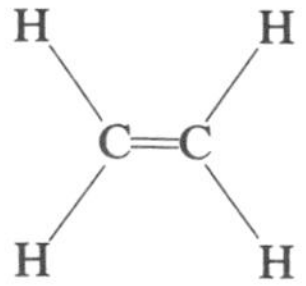

$$\underset{H}{\overset{H}{\diagdown}}C=C\underset{H}{\overset{H}{\diagup}}$$

or propene

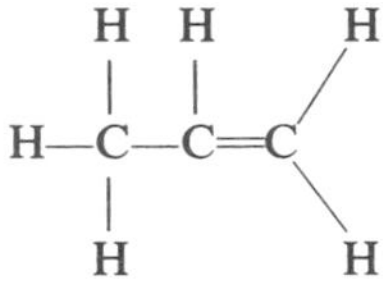

$$H-\underset{H}{\overset{H}{\underset{|}{C}}}-\underset{H}{\overset{H}{\underset{|}{C}}}=C\underset{H}{\overset{H}{\diagup}}$$

The structure

$$\underset{H}{\overset{H}{\diagdown}}C=\underset{H}{\overset{H}{\underset{|}{C}}}-\underset{H}{\overset{H}{\underset{|}{C}}}-H$$

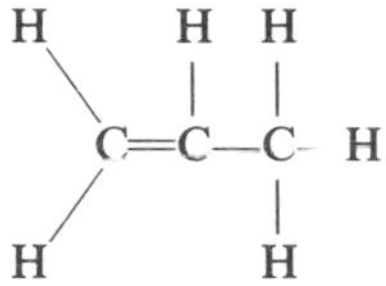

is identical with the propene molecule shown above: it is merely drawn the other way round.

11.7 Properties of Alkenes

(*a*) Alkenes burn readily in air (as do the alkanes) producing carbon dioxide, water vapour and large quantities of heat. With ethene, for example:

$$C_2H_{4(g)} + 3O_{2(g)} \rightarrow 2CO_{2(g)} + 2H_2O_{(g)}\ (\Delta H = -1329\ \text{kJ mol}^{-1})$$

(*Caution*: mixtures of air and ethene can be explosive and must be handled very carefully.)

(*b*) In other reactions the alkenes differ from the alkanes because of their double bond. Unlike the alkanes, alkenes, such as ethene, undergo a *rapid* reaction with bromine dissolved in tetrachloromethane (carbon tetrachloride), and the red-brown colour of the bromine quickly disappears:

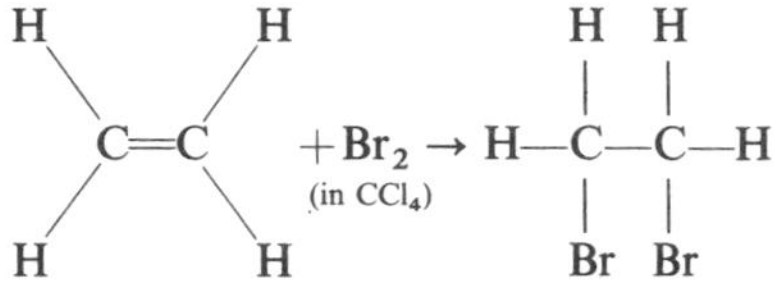

$$\underset{H}{\overset{H}{\diagdown}}C=C\underset{H}{\overset{H}{\diagup}}\ \underset{\text{(in CCl}_4)}{+Br_2} \rightarrow H-\underset{\underset{Br}{|}}{\overset{\overset{H}{|}}{C}}-\underset{\underset{Br}{|}}{\overset{\overset{H}{|}}{C}}-H$$

ethene (g)   1,2-dibromoethane(l)

This *addition* of bromine producing a compound containing two more atoms than were present in the original alkene is a typical reaction of *unsaturated hydrocarbons*. Note that it is in complete contrast to the alkanes, which can only react by *substitution*.

Many other reagents will undergo similar *addition reactions* with alkenes. Thus hydrogen in the presence of a finely divided nickel catalyst *adds across the double bond* of ethene to produce the saturated alkane, ethane:

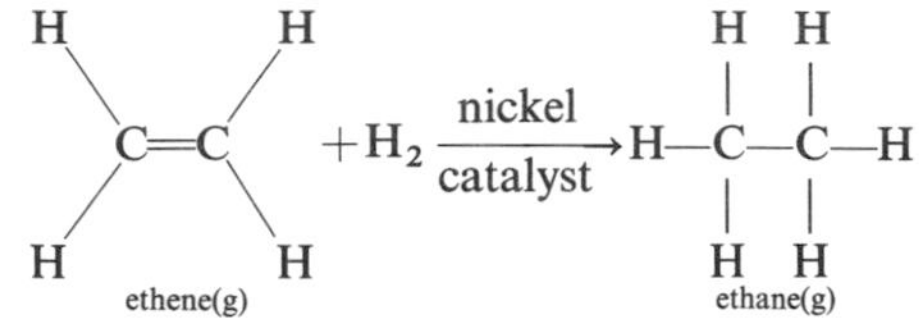

ethene(g) ethane(g)

A Test For Unsaturation

A compound containing a carbon–carbon covalent double bond rapidly decolorizes a red-brown solution of bromine in tetrachloromethane (carbon tetrachloride). An **alkyne**, e.g. *ethyne* (acetylene) C_2H_2, containing a carbon–carbon *triple* bond reacts similarly.

Alkanes react extremely slowly, decolorizing the bromine and liberating pungent hydrogen bromide gas (see Experiment 11.1). There is no liberation of hydrogen bromide with an alkene or an alkyne.

11.8 Alcohols

An alcohol may be considered as an alkane in which a hydrogen has been replaced by a hydroxyl (—OH) group. Table 11.3 gives the names and structural formulas of the four simplest alcohols.

(*a*) Nomenclature

An alcohol is named by replacing the final *-e* of the parent hydrocarbon with *-ol*. The position of each hydroxyl group is indicated by the number of the chain carbon atom to which the hydroxyl group is attached. This number is placed before the suffix *-ol* (see Table 11.3). No number is necessary for methanol or ethanol.

(*b*) Isomerism

Compounds which have the same *molecular* formula but different *structural* formulas are said to be **isomeric**. Thus propan-1-ol and propan-2-ol are two isomers, each having the molecular formula C_3H_8O. Methoxyethane (methyl ethyl ether) is also an isomer having this molecular formula, but it is not an alcohol.

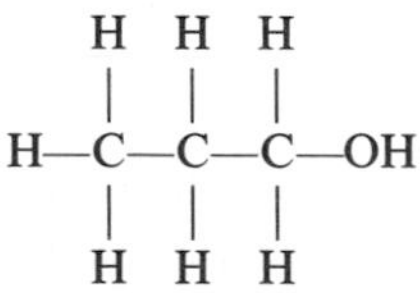

propan-1-ol

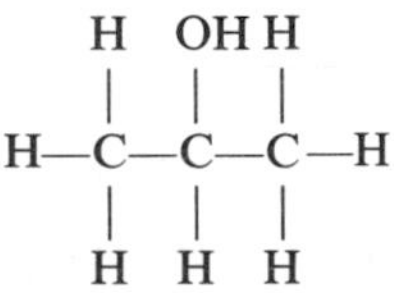

propan-2-ol

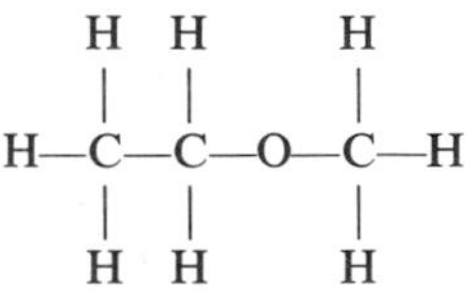

methoxyethane

Table 11.3 The alcohols

Name	Structural formula
Methanol	
Ethanol	
Propan-1-ol	
Propan-2-ol	

11.9 Ethanol

Of all the alcohols the most important is *ethanol*, C_2H_5OH. Its presence in intoxicating beverages is well known: beers and wines contain up to 10% of ethanol, whereas spirits such as whisky contain about 40% by volume. Pure ethanol (*absolute ethanol*) is described as '200° proof', so a drink labelled '80° proof' contains 40% ethanol by volume. However, most of the ethanol manufactured is used industrially as a starting material for other chemicals such as ethanoic acid (acetic acid).

The two most important methods used for the manufacture of ethanol are (*a*) the *fermentation of sugar or starch*, and (*b*) the *hydration of ethene* from petroleum.

(a) Fermentation

Fermentation is a chemical action brought about by bacteria or yeasts. Living yeast produces biological catalysts called *enzymes*. When yeast is added to a dilute solution of ordinary table sugar a reaction occurs which proceeds most readily at about $38\,^{\circ}C$. The enzyme *sucrase*, produced by the yeast, catalytically breaks down ordinary table sugar (sucrose, $C_{12}H_{22}O_{11}$) into the simpler sugars *glucose* and *fructose*. These are isomers having the molecular formula $C_6H_{12}O_6$. *Zymase*, a second enzyme produced by the yeast, then converts the glucose and fructose into ethanol and carbon dioxide.

$$C_{12}H_{22}O_{11(aq)}+H_2O_{(l)} \xrightarrow{\text{sucrase}} C_6H_{12}O_{6(aq)}+C_6H_{12}O_{6(aq)}$$
$$\text{(sucrose)} \qquad\qquad\qquad\qquad \text{(glucose)} \qquad \text{(fructose)}$$

$$C_6H_{12}O_{6(aq)} \xrightarrow{\text{zymase}} 2C_2H_5OH_{(aq)}+2CO_{2(g)}$$
$$\text{(glucose or fructose)} \qquad\qquad \text{(ethanol)}$$

When the reaction mixture contains about 12% by volume of ethanol the activity of the yeast ceases. The ethanol can then be concentrated by fractional distillation if required.

(b) Ethanol from Petroleum

Large quantities of synthetic ethanol are manufactured from ethene, a gas produced by the cracking of petroleum. The ethene is *hydrated* (a molecule of water is added) to produce ethanol:

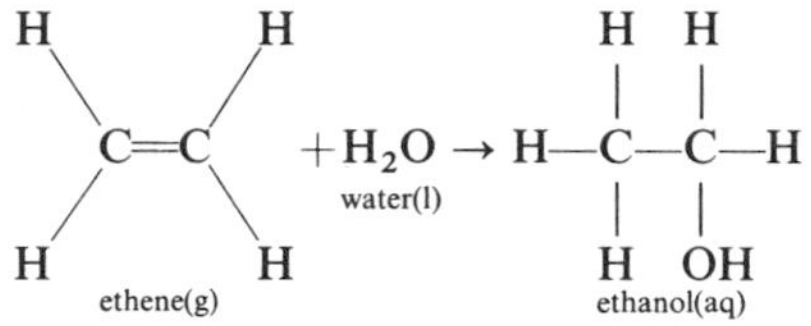

Sulphuric acid assists in the hydration process.

11.10 Properties of Ethanol

The principal reactions of ethanol are those of its functional group, the hydroxyl group (—OH). Thus in many of its reactions it has similar properties to water (H—OH). The following reactions of ethanol can easily be carried out as small-scale test-tube experiments.

(a) Reaction with Sodium

A small piece of freshly cut sodium sinks when dropped into a test tube containing ethanol. It reacts immediately, liberating a steady stream of bubbles of hydrogen, and eventually dissolves leaving a clear solution of sodium ethoxide:

$$2C_2H_5OH_{(l)}+2Na_{(s)} \rightarrow 2C_2H_5O^-Na^+ +H_{2(g)}$$

This reaction can be used for the safe disposal of unwanted sodium.

(b) Reaction with Phosphorus Pentachloride

When a little solid phosphorus pentachloride is added to ethanol, a vigorous reaction occurs in which hydrogen chloride is liberated together with volatile ethyl chloride:

$$C_2H_5OH_{(l)}+PCl_{5(s)} \rightarrow C_2H_5Cl_{(g)}+POCl_{3(l)}+HCl_{(g)}$$

The product remaining in solution is phosphorus trichloride oxide (phosphorus oxychloride).

(*Caution*: phosphorus pentachloride should be handled with care and not allowed to come into contact with the skin.)

Its reactions with sodium and with phosphorus pentachloride indicate the presence of a hydroxyl group in ethanol.

(*c*) Oxidation of Ethanol

When a little acidified potassium manganate(VII) (potassium permanganate) is added to a little ethanol in a test tube, and the mixture is warmed in a water bath, the purple colour of the manganate(VII) (permanganate) ion disappears. The characteristic pungent smell of ethanoic acid (acetic acid) soon becomes noticeable.

$$5C_2H_5OH_{(l)} + 4K^+MnO_{4(aq)}^- + 6H_2^+SO_{4(aq)}^{2-} \rightarrow 4Mn^{2+}SO_{4(aq)}^{2-} + 2K_2^+SO_{4(aq)}^{2-} +$$
$$5CH_3COOH_{(aq)} + 11H_2O_{(l)}$$

(*d*) Esterification

A little ethanol is mixed with an equal quantity of *glacial* (concentrated) ethanoic (acetic) acid in a clean dry test tube, together with one or two drops of concentrated sulphuric acid to act as a catalyst. On warming and pouring the mixture into a large volume of water, the fruity smell of ethyl ethanoate (ethyl acetate) can be detected:

$$C_2H_5OH_{(l)} + CH_3COOH_{(l)} \rightleftharpoons \underset{\text{ethyl ethanoate}}{CH_3COOC_2H_{5(l)}} + H_2O_{(l)}$$

Ethyl ethanoate (ethyl acetate) is one member of a class of organic compounds called **esters**, all of which have pleasant fruity smells.

In general the reaction between an alcohol and a carboxylic (—COOH) acid produces an ester and water, the process being termed *esterification*. Thus if acetic acid is warmed with a variety of alcohols, different fruity smells can be detected characteristic of the particular ester produced.

Summary of Unit 11

1. Carbon forms *strong covalent bonds* with *itself* and with many other *non-metals* including hydrogen, oxygen, halogens and nitrogen. Carbon–carbon covalent bonds can be single (with two shared electrons), double (with four shared electrons) or triple (with six shared electrons). The spatial arrangement of individual molecules varies according to their structure.

2. **Alkanes** are *saturated hydrocarbons* with general formula C_nH_{2n+2}. They belong to a *homologous series*, the successive members of which differ by $-CH_2-$, have similar chemical properties and a gradation in physical properties. The distribution of bonds around each carbon atom is tetrahedral.

3. Alkanes react in two important ways:
   (*a*) they burn, producing heat; hence they are often used as fuels,
   (*b*) the hydrogen atoms can be *substituted* for other *functional groups* such as —OH or —Cl. A functional group is a part of a compound which has a characteristic set of properties.

4. **Petroleum** is a complex naturally occurring mixture of hydrocarbons. It is refined by *fractional distillation*, and it can be converted to useful **petrochemicals** by *cracking*.
5. **Alkenes** are *unsaturated hydrocarbons* containing a *double* carbon–carbon covalent bond. They belong to a homologous series with general formula C_nH_{2n}.
6. Alkenes react similarly to alkanes, with the one important difference that they can undergo *addition reactions* across the carbon–carbon double bond.
7. An **alcohol** is derived from an alkane in which a hydrogen atom has been replaced by a hydroxyl group.
8. **Ethanol** is the most important alcohol and is manufactured either by *fermentation* or by the *hydration* of ethene.
9. The reactions of alcohols are dominated by the functional group —OH. Thus they react with sodium to form the *alkoxide*, with phosphorus pentachloride to form the *chloride*, with oxidizing agents to form the *carboxylic acid*, and with carboxylic acids to form the *ester*.
10. Many organic compounds exhibit **isomerism**. Isomers are compounds which have the *same molecular formula* but *different structural formulas*.

Test Yourself on Unit 11

1. Name the following compounds:

(a)

$$
\begin{array}{ccccc}
H & H & H & H & H \\
| & | & | & | & | \\
H—C—C—C—C—C—H \\
| & | & | & | & | \\
H & H & H & H & H
\end{array}
$$

(b)

$$
\begin{array}{c}
H \quad\quad\quad H \quad H \\
\backslash \quad\quad\quad\; | \quad\; | \\
C{=}C—C—H \\
/ \quad\quad\quad\quad\; | \\
H \quad\quad\quad\quad H
\end{array}
$$

(c)

$$
\begin{array}{ccc}
H & Br & H \\
| & | & | \\
H—C—C—C—Br \\
| & | & | \\
H & H & H
\end{array}
$$

(d)

$$
\begin{array}{ccc}
H & H & H \\
| & | & | \\
H—C—C—C—Cl \\
| & | & | \\
H & H & Cl
\end{array}
$$

(e)

2. Which of the following could *not* be an alkane:
   (a) C_5H_{12} (b) C_4H_{10} (c) C_8H_{18} (d) C_7H_{14} (e) C_9H_{20}

3. How many *different* compounds can be produced when one chlorine atom is substituted into a molecule of butane (C_4H_{10})?
   (a) 1   (b) 2   (c) 3   (d) 10   (e) 4.

4. Write down the names and formulas of the isomers having the molecular formula C_4H_{10}.

5. The following tests were carried out on some unknown organic compounds:
   (a) Compound A was shaken with bromine in tetrachloromethane (carbon tetrachloride). There was an immediate decolorization of the red bromine solution.
   (b) Compound B was shaken with bromine in tetrachloromethane (carbon tetrachloride). There was only a very slow decolorization of the red bromine solution, and after a while pungent acid fumes could be detected.
   (c) A small piece of freshly cut sodium was added to the non-acidic compound C. A gas was liberated which formed an explosive mixture with air.
   (d) On warming compound D with ethanol and a few drops of concentrated sulphuric acid, a pleasant fruity smell could be detected when the mixture was poured into water.
   Assign the following formulas to compounds A, B, C and D:
   (i) $CH_3CH_2CH_2OH$,   (ii) $CH_3CH_2CH_2CH_2CH_2CH_3$,
   (iii) $CH_3CH_2CH_2COOH$,   (iv) $CH_2{=}CHCH_2CH_3$.

6. A sample of crude oil was heated and its vapour passed over red-hot pumice stone. A mixture of gases was evolved, which decolorized bromine in tetrachloromethane (carbon tetrachloride) and burned in air with a yellow flame.
   (a) Is the process taking place when the vapour from the crude oil passes over the heated pumice stone: (i) polymerization, (ii) distillation, (iii) cracking, or (iv) refining?
   (b) Is the type of compound causing decolorization of the bromine solution most probably: (i) an alkene, (ii) an alkane, (iii) an alcohol, or (iv) an acid?
   (c) Name two compounds which could be formed when the gas mixture burns in air.

7. Write down the names of two alcohols having the molecular formula C_3H_8O.

Mark this test out of 20 with the answers given on page 380.

Classification of Matter V: Large Molecules Containing Carbon

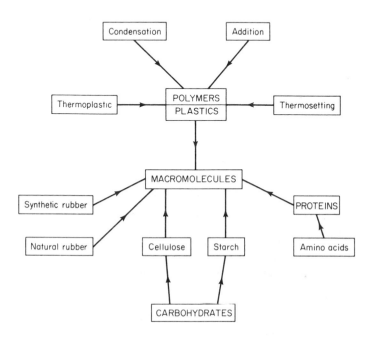

Unit 11 introduced the chemistry of the compounds of carbon. The majority of compounds studied contained a relatively small number of carbon atoms, but carbon's ability to form multiple bonds was pointed out. This Unit looks at the various types of compound containing many carbon–carbon bonds, some man-made and others naturally occurring.

One of the two major areas of study in this Unit deals with synthetic *macromolecules*, often referred to as *plastics*. The term 'plastics' is in common use to describe a variety of materials, including polythene, nylon and Perspex, which at some stage during their processing become mobile or plastic. The other area briefly covers naturally occurring macromolecules such as starch and proteins. Many, but not all, of these compounds are *polymers:*

12.1 What are Polymers?

A polymer is a large molecule built up from many hundreds or thousands of *monomer* units joined together. Thus the well-known plastic poly(ethene) or polythene is composed of large molecules formed by the repeated combination of ethene molecules:

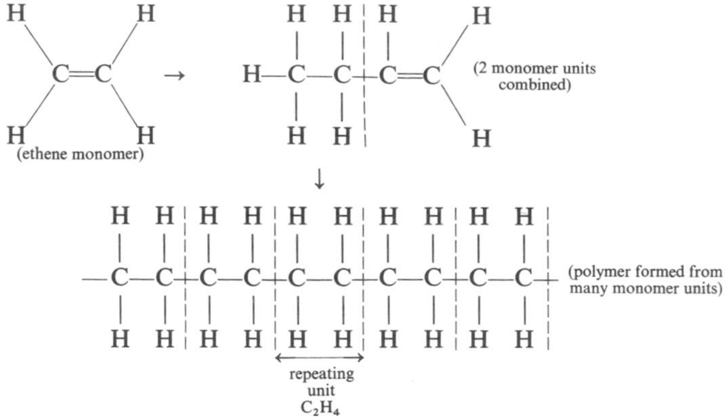

Polymers are classified as either *addition polymers* or *condensation polymers*, depending on their method of formation.

12.2 Addition Polymers

Poly(ethene), described above, is an example of an 'addition' polymer: the formula for the repeating unit (in this case C_2H_4) is the same as that of the starting monomer (in this case *ethene*). This is true for all addition polymers, the only difference being the nature of the repeating unit. Thus the repeating unit in poly(chloroethene) or polyvinyl chloride (PVC) is C_2H_3Cl.

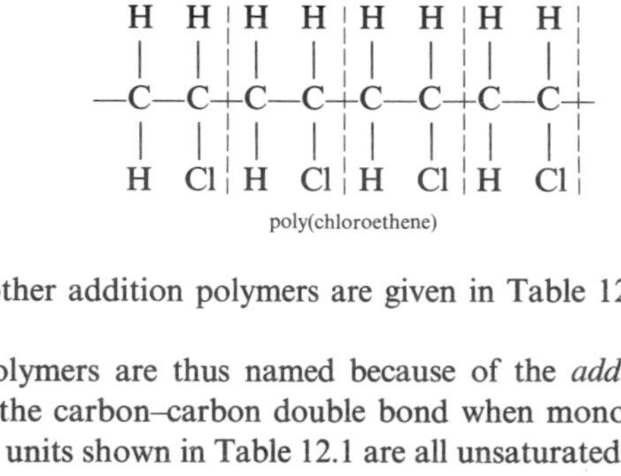

Examples of other addition polymers are given in Table 12.1.

Addition polymers are thus named because of the *addition reaction* which occurs across the carbon–carbon double bond when monomer units combine. The monomer units shown in Table 12.1 are all unsaturated compounds derived from ethene. Experiments 12.1 and 12.2 describe the preparation of two of these polymers.

Table 12.1 Nomenclature of polymers

Common name	International name	Formula of monomer unit
Polytetrafluoro-ethylene(PTFE)	poly(tetrafluoro-ethene)	F and F on top; $C=C$; F and F on bottom
Polystyrene	poly(phenylethene)	H and H on top; $C=C$; C_6H_5 and H on bottom (Note: C_6H_5- is the phenyl group derived from benzene, C_6H_6)
Perspex (polymethyl methacrylate)	poly(methyl 2-methylpropen-oate)	H and CH_3 on top; $C=C$; H and $COOCH_3$ on bottom
Polypropylene	poly(propene)	CH_3 and H on top; $C=C$; H and H on bottom

Experiment 12.1 Preparation of poly(methyl 2-methylpropenoate), Perspex
A little methyl 2-methylpropenoate (methyl methacrylate) is poured into a test tube and maintained at approximately 60 °C in a water bath. About 1% dodecanoyl peroxide (lauroyl peroxide) is added to catalyse the polymerization, and the mixture is shaken to dissolve the catalyst. After approximately one hour the liquid monomer polymerizes into solid transparent Perspex:

$$
\underset{\text{methyl 2-methylpropenoate}}{\begin{array}{c} H \\ | \\ C=C \\ | \\ H \end{array} \begin{array}{c} CH_3 \\ \\ \\ COOCH_3 \end{array}} \rightarrow \underset{\text{poly(methyl 2-methylpropenoate), Perspex}}{\left[\begin{array}{cccc} H & CH_3 & H & CH_3 \\ | & | & | & | \\ -C-&C-&C-&C- \\ | & | & | & | \\ H & COOCH_3 & H & COOCH_3 \end{array} \right]_n}
$$

Perspex can be *depolymerized* by heating, see Fig. 12.1. At about 300 °C the polymer softens and undergoes rapid depolymerization to the monomer. The clear liquid collected in the water-cooled tube contains the monomer.

Caution: the monomer vapour is harmful and the experiment should be performed in a fume chamber.

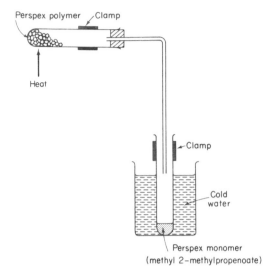

*Fig. 12.1   Depolymerization of Perspex: the solid polymer is broken down
into its liquid constituent monomer*

Experiment 12.12   *Preparation of poly(phenylethene), polystyrene*
This polymer is prepared by adding about 1% of dodecanoyl peroxide (lauroyl
peroxide) to a little phenylethene (styrene) monomer in a test tube and warming
to 100 °C in a boiling-water bath. Polymerization of the clear liquid phenylethene
monomer into colourless solid poly(phenylethene), commonly known as polystyrene, occurs within one hour:

12.3 Condensation Polymers

In the chemistry of carbon there are many reactions where the combination of two
or more substances is accompanied by the elimination of a small, simple molecule
such as water, hydrogen chloride, ammonia or methanol. Such a reaction is often
called a *condensation reaction*. We have already seen a typical example in Unit 11
when discussing esterification, the reaction between an organic carboxylic acid
such as ethanoic (acetic) acid and an alcohol such as ethanol:

The product here is ethyl ethanoate (ethyl acetate) and the simple substance that is eliminated is water. In general this type of reaction can be written as

$$\cdots-\overset{\displaystyle O}{\underset{\underset{\cdots\cdots}{OH\ \ H}}{C}}\diagdown O-\cdots\rightarrow\ \cdots-\overset{\displaystyle \overset{O}{\parallel}}{C}-O-\cdots+H_2O$$

where ... represents some form of carbon-chain backbone.

Another example of a condensation reaction is

$$\cdots-\overset{\overset{\displaystyle H}{\diagup}\ \ \overset{\displaystyle O}{\diagdown}}{\underset{\underset{\cdots\cdots}{H\ \ \ Cl}}{N}}\ \ C-\cdots\rightarrow\cdots-\overset{\overset{\displaystyle H}{|}\ \ \overset{\displaystyle O}{\parallel}}{N-C}-\cdots+HCl$$

where the eliminated small molecule is hydrogen chloride.

Each of the carbon-chain backbones in the above examples is attached to only one functional group (—COOH, —OH, —NH$_2$ etc.). But if each carbon-chain backbone has *two* functional groups attached to it, a condensation reaction can occur involving *polymerization*. For example:

Whenever condensation occurs together with polymerization, the polymer produced is called a *condensation polymer*.

Nylon is a condensation polymer produced by the reaction between a diamine

$$\overset{H}{\underset{H}{\diagup}}N-\cdots-N\overset{\diagdown H}{\underset{\diagup H}{}}$$

and a dibasic organic carboxylic acid (i.e. having two COOH groups per molecule)

$$\overset{O}{\underset{HO}{\diagdown}}C-\cdots-C\overset{\diagup O}{\underset{\diagdown OH}{}}$$

For example:

If $\overset{H}{\underset{H}{\diagdown}}N\text{---}\cdots\text{---}N\overset{H}{\underset{H}{\diagup}}$ is $H_2N\text{---}(CH_2)_6\text{---}NH_2$, 1,6-diaminohexane, and

$\overset{O}{\underset{HO}{\diagdown}}\overset{\diagup}{C}\text{---}\cdots\text{---}C\overset{O}{\underset{OH}{\diagup}}$ is $\overset{O}{\underset{HO}{\diagdown}}\overset{\diagup}{C}\text{---}(CH_2)_4\text{---}C\overset{O}{\underset{OH}{\diagup}}$, hexanedioic acid

(adipic acid), the condensation polymer is *nylon 6.6* (six carbon atoms in each monomer).

Terylene (or Dacron) is a condensation polymer formed from a diester

$\overset{O}{\underset{H_3CO}{\diagdown}}\overset{\diagup}{C}\text{---}\cdots\text{---}C\overset{O}{\underset{OCH_3}{\diagup}}$ and a diol HO----$\cdots$----OH. Thus

$$\overset{O}{\underset{H_3CO}{\diagdown}}C\text{---}\cdots\text{---}C\overset{O}{\underset{OCH_3}{\diagup}} + HO\text{---}\cdots\text{---}OH + \overset{O}{\underset{H_3CO}{\diagdown}}C\text{---}\cdots\text{---}C\overset{O}{\underset{OCH_3}{\diagup}} + HO\text{---}\cdots\text{---}OH +$$

$$\downarrow$$

$$\left[\overset{O}{\underset{O}{\diagdown}}C\text{---}\cdots\text{---}C\overset{O}{\underset{O\text{---}\cdots\text{---}O}{\diagup}}\overset{O}{\underset{O}{\diagdown}}C\text{---}\cdots\text{---}C\overset{O}{\underset{O\text{---}\cdots\text{---}O}{\diagup}}\right] + nCH_3OH$$

For Terylene the diester is $\overset{O}{\underset{H_3CO}{\diagdown}}C\text{---}C_6H_4\text{---}C\overset{O}{\underset{OCH_3}{\diagup}}$ (dimethyl ester

of benzene-1,4-dicarboxylic acid, or dimethyl terephthalate) and the diol is $HO\text{---}CH_2CH_2\text{---}OH$ (ethane-1,2-diol, or ethylene glycol).

Experiment 12.3   *Preparation of nylon 6.10 ('the nylon rope trick')*
50 cm³ of a 2% (by volume) solution of decanedioyl dichloride (sebacoyl chloride) in tetrachloromethane (carbon tetrachloride) is measured into a 100 cm³ tall-form beaker. 25 cm³ of an aqueous solution containing 2·2 g of 1,6-diaminohexane is added carefully to the beaker so that the aqueous solution floats on top of the tetrachloromethane solution, without mixing. A thread of nylon is drawn from the interface between the two liquids, using a pair of forceps, and wound around a thick glass rod (see Fig. 12.2).

$$\overset{H}{\underset{H}{\diagdown}}N\text{---}(CH_2)_6\text{---}N\overset{H}{\underset{H}{\diagup}} + \overset{O}{\underset{Cl}{\diagdown}}C\text{---}(CH_2)_8\text{---}C\overset{O}{\underset{Cl}{\diagup}}$$

(1,6-diaminohexane containing   (decanedioyl dichloride containing
6 carbon atoms)   10 carbon atoms)

$$\downarrow$$

$$\left[\overset{H}{\underset{N}{|}}\text{---}(CH_2)_6\text{---}\overset{H}{\underset{N}{|}}\text{---}\overset{O}{\underset{C}{\|}}\text{---}(CH_2)_8\text{---}\overset{O}{\underset{C}{\|}}\right] + nHCl$$

(nylon 6.10)

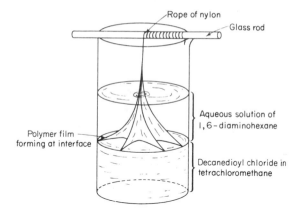

Fig. 12.2   A simple method of preparing nylon 6.10

If the thread is not easily withdrawn from the interface the beaker may be warmed gently in a water bath, taking care that the solutions do not mix.

12.4 Characteristics of Synthetic Polymers

(a) Thermoplastic Polymers
A thermoplastic polymer is one which softens on heating and becomes rigid again on cooling. This is because there are only weak attractive forces between the long polymer molecules and these are readily disrupted on heating. Most addition polymers and some condensation polymers are thermoplastic. Examples include nylon, polythene and polystyrene.

(b) Thermosetting Polymers
A thermosetting polymer is one which becomes hard on heating. It cannot be softened by heat. Polymers of this type are often prepared in two stages. The first stage is the production of long-chain molecules which are capable of further reaction with each other. These intermediate polymers usually flow and can be placed into moulds. Colouring is often added at this stage. The second stage is the application of heat which causes a reaction to occur between the chains, thus producing a complex-network polymer. Fig. 12.3 illustrates these two stages in the production of a thermosetting plastic polymer.

12.5 Uses of Synthetic Polymers

(a) Polythene, poly(ethene),
was first made in 1933 by Fawcett and Gibson of Imperial Chemical Industries. Low-density polythene is used in packaging, housewares such as buckets and bottles, carpet backing, cable insulation and many other applications. High-density polythene has greater rigidity and is therefore used in the manufacture of piping, dustbins, crates, etc., where mechanical strength is essential.

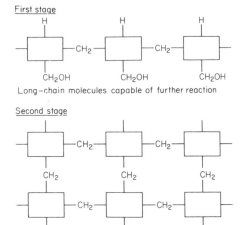

Fig. 12.3   Complex network structure of a thermosetting polymer

(*b*) **PVC or polyvinyl chloride**, poly(chloroethene), is widely used in imitation leathers, floor coverings, corrugated roofing material, gramophone records, etc.

(*c*) **Polystyrene**, poly(phenylethene), is used in moulded objects such as disposable drinking cups, radio and television cabinets, brush handles and switches.

(*d*) **PTFE or Teflon**, poly(tetrafluoroethene), is familiar because of its use as an anti-adhesive ('non-stick') coating, particularly for cooking utensils. Because of its low chemical reactivity, allied with its excellent toughness, electrical and heat resistance, it is used as insulation for electrical items and in the manufacture of gaskets and valves.

(*e*) **Perspex**, poly(methyl 2-methylpropenoate), is a transparent glass-like plastic which, because of its good optical characteristics, is used in lenses. Other uses include dentures and car rear-light mouldings.

(*f*) **Nylon** is well known as a synthetic fibre in carpets, fabrics, rope, stockings and other clothing. Because of its mechanical strength, nylon is also used in moulded machine parts such as gears and bearings.

12.6 Natural and Synthetic Rubbers

(*a*) Natural Rubber

With the expansion of the motor-car industry the demand for rubber has increased enormously. Chemists found that natural rubber is a hydrocarbon polymer built up from the monomer unit methylbuta-1,3-diene (isoprene):

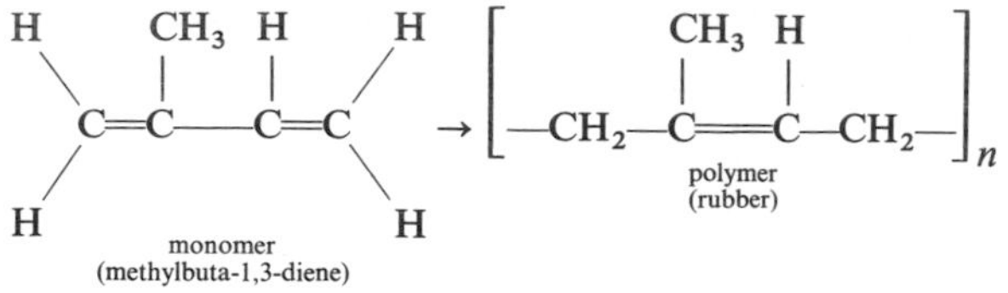

'Raw' rubber obtained from *latex* tapped from the rubber tree, particularly *Hevea brasiliensis*, does not possess the characteristics of the rubber with which we are familiar. In order to give it strength and elasticity it has to be *vulcanized*. In the vulcanization process, raw rubber is mixed with small amounts of sulphur and heated. The sulphur reacts with the polymer molecules forming a cross-linked network:

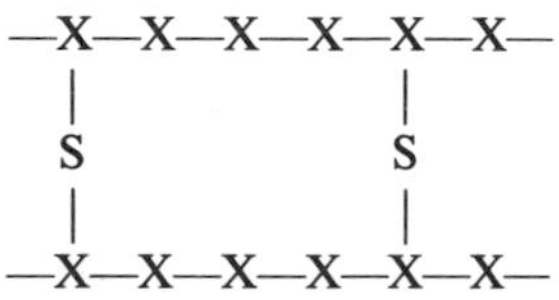

where X represents a monomer unit. This cross linking gives mechanical strength to the rubber. In addition, 'fillers' such as carbon black and zinc oxide are usually added to the crude rubber before vulcanization in order to improve its wearing characteristics.

(b) Synthetic Rubbers

(i) *Neoprene*, poly(2-chlorobuta-1,3-diene), was one of the first synthetic rubbers manufactured on a large scale. The monomer unit, 2-chlorobuta-1,3-diene (chloroprene), is made from ethyne which itself is produced easily from coal and limestone.

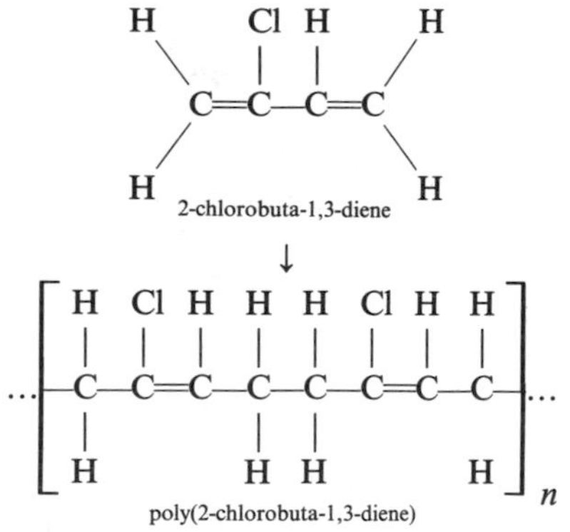

This polymer is particularly resistant to chemical action and is therefore used in making hoses for petrol and oil and containers for corrosive chemicals.

(ii) *Styrene–butadiene rubber* and *butyl rubber* are manufactured from the C_4 and C_2 hydrocarbons from petroleum.

Styrene–butadiene rubber is made by copolymerizing the two monomer units phenylethene (styrene) and buta-1,3-diene:

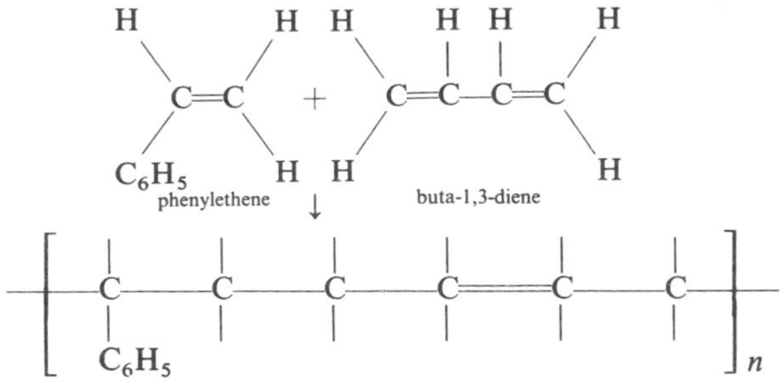

copolymer structure (hydrogen atoms omitted)

Butyl rubber is made from 2-methylpropene (isobutylene) monomer polymerized with a little methylbuta-1,3-diene (isoprene):

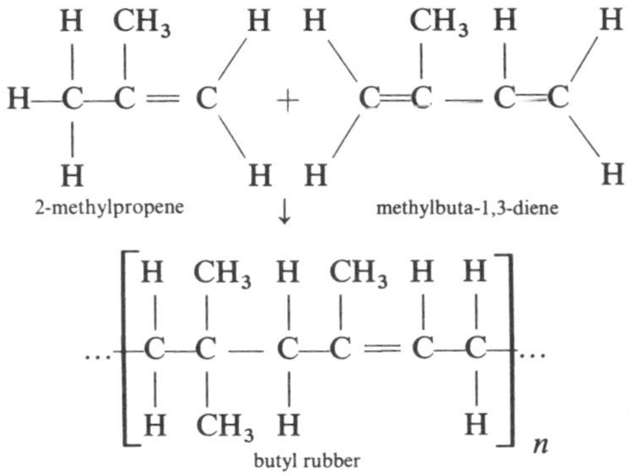

12.7 Naturally Occurring Large Molecules

The basic unit of all living matter is the *cell*. Each cell is itself built up from a variety of materials, many of which are large polymeric molecules. Included among these naturally occurring large molecules are *carbohydrates, proteins* and *nucleic acids*.

The chief function of carbohydrates is that of a fuel. An organism provides itself with energy when the carbohydrate is broken down into carbon dioxide and water. Simple carbohydrates can also be used as a starting material for the biological synthesis within the body of more complex carbohydrates such as *glycogen*, the storage form of carbohydrate in animals.

Proteins are used mainly for the construction of tissue; they occur in muscle, skin, nerves, brain and blood. Nucleic acids are responsible for synthesizing protein and for transmitting hereditary characteristics. Also present in living cells are substances called *nucleoproteins*: compounds whose molecules are part protein and part nucleic acid.

12.8 Carbohydrates

A carbohydrate is a compound of carbon, hydrogen and oxygen, where, as the name suggests, the hydrogen and oxygen are present in the same ratio (2:1) as in water. Glucose ($C_6H_{12}O_6$), sucrose ($C_{12}H_{22}O_{11}$) and other sugars are common examples of carbohydrates.

Just as glycogen is the storage form of carbohydrates in animals, *starch* is the storage form of carbohydrates in plants. Under the influence of the sun's energy and certain biological catalysts, carbon dioxide taken in through a plant's leaves combines with water to form simple carbohydrates. The process is called *photosynthesis*. Further biochemical processes combine these simple carbohydrates into starch, which is then stored by the plant in its roots, tubers, seeds and fruits.

Starch is a complex polymer and its molecule is composed of thousands of glucose monomer units. If glucose $C_6H_{12}O_6$ is written $HO-(C_6H_{10}O_4)-OH$, its polymerization to starch can be represented by the following condensation reaction:

$$n \; HO-(C_6H_{10}O_4)-OH$$
$$\downarrow$$
$$HO-(C_6H_{10}O_4)-OH \quad HO-(C_6H_{10}O_4)-OH \quad HO-(C_6H_{10}O_4)-OH$$
$$\downarrow$$
$$-(C_6H_{10}O_4)-O-(C_6H_{10}O_4)-O-(C_6H_{10}O_4)-O-+n \; H_2O$$

In the complex biochemical process known as **digestion** starch is broken down to glucose in a *hydrolysis* reaction. This is effectively a depolymerization reaction in which the addition of water breaks the polymer into monomer units. This reaction may be regarded as the reverse of the condensation reaction shown above. The following experiment illustrates two methods of breaking the polymeric molecule starch into the monomer glucose.

Experiment 12.4   Breakdown of starch
In this experiment starch is hydrolysed in two different ways and the products of hydrolysis are identified using paper chromatography.

(i) *Hydrolysis of starch using saliva*
About 1 g of starch is made into a paste and this mixture is stirred into a beaker containing about $100 \; cm^3$ of hot water. The solution is allowed to cool to about $38 \, °C$ and $10 \; cm^3$ of it is measured into a small beaker together with a little saliva. (The enzymes in saliva are most effective at a body temperature of approximately $38 \, °C$.)

(ii) *Hydrolysis of starch using 2 M hydrochloric acid*
A second $10 \; cm^3$ of the starch solution is measured into a small beaker and a few drops of 2 M hydrochloric acid are added. The solution is boiled.

The solutions from parts (i) and (ii) of the experiment are tested at regular intervals to ensure that hydrolysis is complete. This is carried out by removing a drop from each solution with a clean glass rod and transferring it to a white spotting tile. A drop of a solution of iodine dissolved in potassium iodide is added and the colour noted. Since starch produces a deep-blue colour with iodine in potassium iodide, the absence of this deep-blue colour indicates that the reaction is complete and no starch remains. At this stage the solution from experiment (ii) is neutralized with a few drops of sodium hydroxide.

(iii) *Identification of the products of hydrolysis using paper chromatography*
Paper chromatography as a separating technique has been described in Section 1.7. In this experiment the separation is carried out on a strip of filter paper supported in a gas jar (see Fig. 12.4).

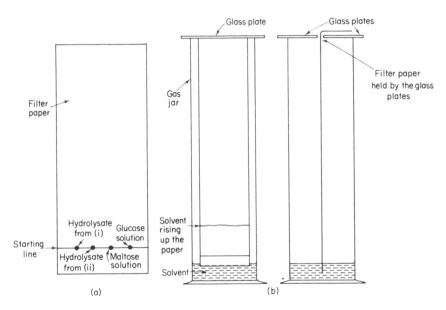

Fig. 12.4   Separation of starch hydrolysates by paper chromatography: (a) initial preparation of the paper; (b) solvent rising up the paper and carrying the samples with it

The solvent, a mixture of 1 volume water, 1 volume glacial ethanoic acid (acetic acid), and 3 volumes of propan-2-ol, is poured into a large gas jar and the glass plates placed over the top. This enables the vapour from the solvent to saturate the atmosphere in the gas jar.

A rectangle of filter paper is marked out as shown in Fig. 12.4a with spots of hydrolysates from parts (i) and (ii) of the experiment, together with spots of glucose and maltose solutions.

In order to concentrate the sugar in the hydrolysed solutions (i) and (ii), successive spots are applied and the paper is dried between each application. The paper is then supported in the gas jar so that it just dips into the solvent, as shown in Fig. 12.4b. After leaving the chromatogram to run overnight the paper is removed and dried.

At this stage no spots are visible on the paper. To locate the position of the sugar spots the paper is sprayed with a mixture containing 2% phenylamine (aniline) in propanone (acetone), 2% diphenylamine in propanone (acetone) and 85% phosphoric acid in the ratio by volume of 5:5:1. When the paper is warmed in front of an electric fire, coloured spots develop and these indicate the location of the sugars. See Fig. 12.4c.

Conclusion. From the chromatogram it can be seen that solution (i) contains maltose, and solution (ii) contains glucose. Thus saliva hydrolyses starch to maltose, and hydrochloric acid hydrolyses starch to glucose.

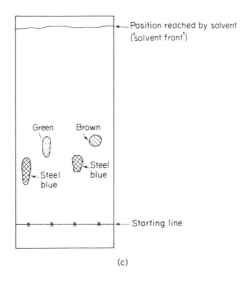

(c)

Fig. 12.4c   The chromatogram after development

This is because saliva contains the enzyme *amylase* which is capable of hydrolysing starch to maltose, a sugar containing two glucose units. Hydrochloric acid, however, breaks down the starch completely into the simple monomer units, glucose.

Whereas starch is the form in which plants store carbohydrates, *cellulose* (another polymeric carbohydrate) is the structural material of plants and is found in cell walls. Human beings can digest starch but do not have the necessary biological catalysts (enzymes) to break down cellulose. Ruminants (such as cows), however, are able to utilize cellulose since they produce the enzymes necessary to break it down by hydrolysis into simpler carbohydrates, including glucose.

12.9 Amino-acids

All simple amino-acids contain a primary amino group ($-NH_2$) and a carboxylic acid group ($-COOH$) attached to a carbon skeleton:

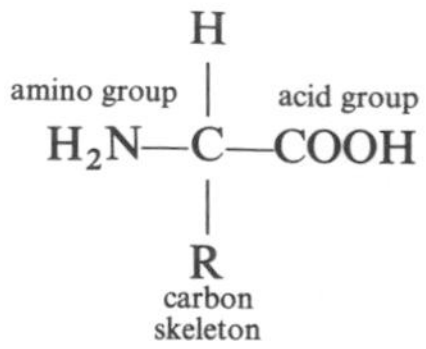

$$\underset{\substack{\text{carbon}\\\text{skeleton}}}{R}$$

amino group $\quad$ acid group

$$H_2N\!-\!\underset{\underset{R}{|}}{\overset{\overset{H}{|}}{C}}\!-\!COOH$$

Some examples of these are shown in Fig. 12.5.

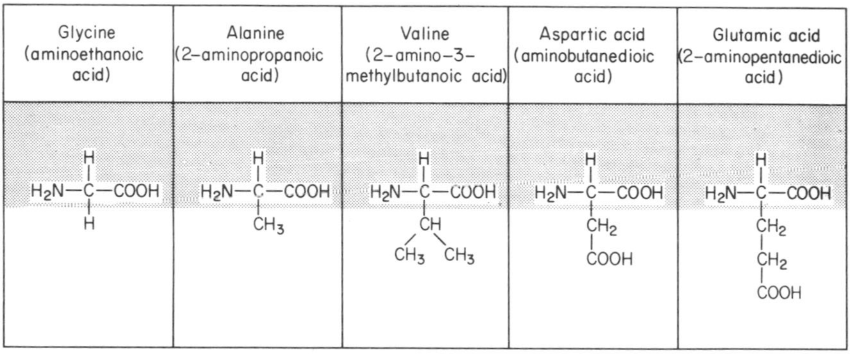

Glycine (aminoethanoic acid)	Alanine (2-aminopropanoic acid)	Valine (2-amino-3-methylbutanoic acid)	Aspartic acid (aminobutanedioic acid)	Glutamic acid (2-aminopentanedioic acid)
H | H₂N—C—COOH | H	H | H₂N—C—COOH | CH₃	H | H₂N—C—COOH | CH / \ CH₃ CH₃	H | H₂N—C—COOH | CH₂ | COOH	H | H₂N—C—COOH | CH₂ | CH₂ | COOH

Fig. 12.5   Structures of simple amino-acids

One of the most important reactions of this class of substances is the condensation of two amino-acid molecules with the elimination of water:

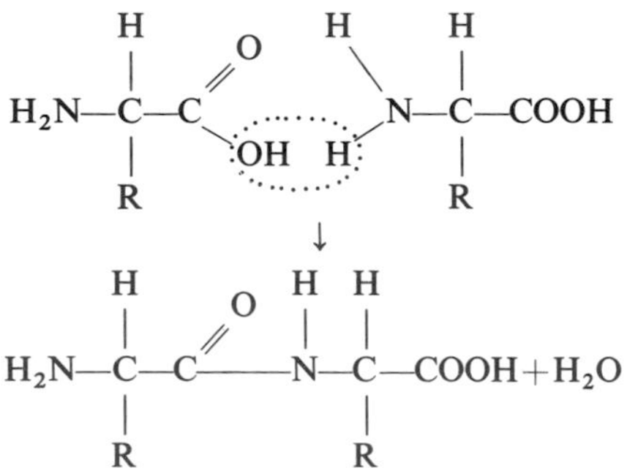

The resulting $-\overset{\overset{\displaystyle O}{\|}}{C}-\overset{\overset{\displaystyle H}{|}}{N}-$ linkage is called a *peptide linkage* and the compound formed from *two* amino-acids is a *dipeptide*. Since the dipeptide still contains reactive carboxylic and amino groups, it can react with further amino-acids to form a polymer with many peptide linkages, called a *polypeptide*. Such polymers form the basis of the vital naturally occurring macromolecules which we know as proteins.

12.10 Proteins

Proteins are polymers consisting largely (or entirely) of chains of amino-acids united by peptide linkages. The constituent amino-acids can be obtained from the protein by hydrolysis.

Individual amino-acids may be considered analogous to the monomer unit in carbohydrates (e.g. the monomer unit glucose in the polymer starch) with one important difference. In carbohydrates the monomer unit is continually repeated, but in proteins there may be 20 or more individual amino-acids present in characteristic proportions and linked in a specific sequence. Thus part of the amino-acid sequence in beef insulin has been characterized as:

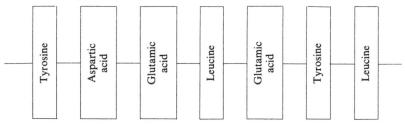

All these amino-acids are joined by peptide linkages.

Because there are so many possible permutations of the ways in which the various amino-acids may combine, the number of possible proteins is enormous. In fact there are tens of thousands, perhaps as many as 100 000, different kinds of proteins in the human body. Hydrolysis of a protein by acid or alkali or enzymes results in the breakdown of the amino-acid chain and the liberation of the free amino-acids. These amino-acids may be identified by paper chromatography, but the determination of the *sequence* of amino-acids in the chain is extremely complex. Chemical research on proteins is at present seeking to solve this problem and also the way in which the polypeptide chain of amino-acids is coiled and twisted in space.

Proteins and Food

Proteins are probably the most important compounds in plants and animals. Plants can synthesize proteins from carbon dioxide, water and inorganic nitrogen compounds, but animals are unable to do this and must rely on eating plants or other animals to obtain their protein (see Unit 6: the nitrogen cycle). Structural proteins are found in body tissue (skin, muscle, etc.) and cellular membranes. Enzymes and some *hormones* (regulators of body reactions) contain a large proportion of protein. Haemoglobin, a protein which is found in red blood cells, has already been referred to in its function as an oxygen carrier and in carbon monoxide poisoning.

Proteins are indispensable to life. People or animals will sicken and die unless their food contains protein. This may be obtained from protein-rich foods such as meat, eggs and milk, but in overcrowded areas of the world these sources of protein are not readily available. Research is going on to find new sources of protein, such as soya beans and coconuts, and the use of yeasts to convert carbohydrates into proteins is being explored. In addition, improvements in livestock and plant production are being developed to help meet world protein shortages. It is becoming clear that indiscriminate consumption of protein is wasteful, and one of the greatest needs is education in nutrition science.

Summary of Unit 12

1. A **polymer** is a large molecule (a macromolecule) built up from many hundreds or thousands of *monomer* units joined together.
2. **Addition polymers** are those in which the molecular formula of the recurring unit is the same as that of the monomer, e.g. poly(ethene) commonly called polythene.
3. **Condensation polymers** are formed from monomers with the elimination of a small molecule such as water, e.g. nylon and Terylene.
4. A **thermoplastic polymer** is one which softens on heating and becomes rigid again on cooling.
5. A **thermosetting polymer** is one which becomes hard on heating.
6. **Natural rubber** is a hydrocarbon polymer obtained from the monomer methylbuta-1,3-diene (isoprene).
7. *Vulcanization* is the name given to the strengthening process which occurs when natural rubber is heated with sulphur.
8. **Synthetic rubbers** are obtained by the polymerization of dienes, with or without the addition of a copolymer.
9. Naturally occurring large molecules include *proteins* and certain *carbohydrates*.
10. A **carbohydrate** is a compound containing carbon, hydrogen and oxygen only, with the hydrogen and oxygen present in the same ratio as in water.
11. *Cellulose*, the main structural material in plants, is a polymeric carbohydrate.
12. *Starch*, the principal energy store in plants, is another polymeric carbohydrate.
13. Starch can be hydrolysed by (*a*) *saliva*, to *maltose*, a carbohydrate consisting of two monomer units, and (*b*) *acid*, such as hydrochloric acid, to the monomer *glucose*.
14. Simple **amino-acids** contain a *primary amino group* ($-NH_2$) and a *carboxylic acid group* ($-COOH$) attached to a carbon skeleton.
15. A *peptide linkage* is formed during a condensation reaction between two amino-acids.
16. A *dipeptide* is the name given to compounds formed from two amino-acids during the condensation reaction to produce a single peptide linkage.

17. A *polypeptide* contains many peptide linkages.
18. **Proteins** are vital naturally occurring macromolecules containing hundreds or thousands of amino-acids joined together by peptide linkages.

Test Yourself on Unit 12

1. From the following list: butane, buta-1,3-diene, ethanol, maltose, poly(ethene), starch, nylon, glucose, alanine (2-aminopropanoic acid), select the names of
   (*a*) a natural polymer   (*b*) a synthetic polymer
   (*c*) a condensation polymer   (*d*) an addition polymer
   (*e*) a monomer which polymerizes to give a rubber
   (*f*) the product obtained when starch is hydrolysed by acid
   (*g*) a carbohydrate containing two monomer units combined
   (*h*) an amino-acid.

2. Poly(tetrafluoroethene) can be made by polymerizing tetrafluoroethene (tetrafluoroethylene).
   (*a*) Is this an addition or condensation polymerization?
   (*b*) Which will have the highest relative molecular mass, tetrafluoroethene or poly(tetrafluoroethene)?
   (*c*) Draw a short section showing three repeating monomer units in the polymer.
   (*d*) Name one other polymer which has a similar structure to poly(tetrafluoroethene).
   (*e*) Give one use of poly(tetrafluoroethene).

3. Saliva will hydrolyse starch.
   (*a*) What is meant by the term hydrolysis?
   (*b*) What is the major product in this hydrolysis of starch?
   (*c*) What substance present in saliva is responsible for this hydrolysis?

4. The following table is incorrect:

Polymer	Column A	Column B
Terylene	hydrocarbon	natural
Cellulose	protein	natural
Poly(phenylethene) (polystyrene)	carbohydrate	synthetic
Insulin	polyester	synthetic

(*a*) Rearrange the terms in Column A and in Column B so that the appropriate descriptions fit the given polymers.

(*b*) For each polymer state whether it is formed by a *condensation* or *addition* process.

(*c*) Which of the above polymers contains peptide linkages?

(*d*) Which of the above polymers has more than two different monomer units?

Mark this test out of 30 with the answers provided on page 381.

Classification of Matter VI: Radioactive Elements

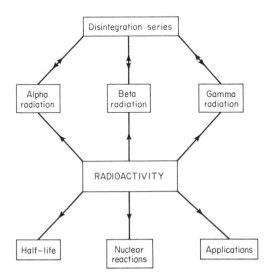

In the last five Units we have been particularly concerned with the chemical reactivity of elements and compounds. This reactivity is related to the *electron structure* of the elements, and a knowledge of the electronic structure enables the chemical properties to be predicted with a reasonable degree of accuracy.

However, certain elements undergo spontaneous changes in their *nuclei*, resulting in the formation of new elements and the emission of high-energy radiation. This is the phenomenon, known as **radioactivity**, which concerns us here in Unit 13.

13.1 Radioactivity is Discovered

The discovery of radioactivity in 1896 by the French physicist Henri Becquerel proved to be one of the most important steps in the elucidation of the structure of matter. Becquerel found that certain uranium salts affected a wrapped photographic film causing it to darken directly beneath the uranium salt sample. He deduced that uranium salts spontaneously emitted 'rays' of some kind without any external help. These rays caused the 'fogging' of the photographic film, and elements which liberated such rays were said to be *radioactive*.

Becquerel repeated the experiment with pitchblende, an ore containing oxides of uranium, and found that the photographic film was even more affected than with pure uranium salts. Marie Curie and her husband Pierre attempted to isolate from pitchblende the element causing this intense radiation. In 1898 the Curies isolated polonium, a radioactive element named after Marie Curie's native Poland. Four years later they also isolated the intensely radioactive element radium.

13.2 The Nature of Radioactivity

By 1899 Rutherford had demonstrated that the radiation which caused the 'fogging' of a photographic film consisted of two types: *alpha* (α) rays and *beta* (β) rays. Soon afterwards Villard discovered a third type, which was called *gamma* (γ) radiation. All three types of radiation are the result of the disintegration of the nuclei of radioactive atoms.

Alpha radiation consists of a flow of positively charged particles, each of which is identical with the nucleus of a helium atom, i.e. two protons and two neutrons but no electrons. If it can gain two orbital electrons an alpha particle becomes a normal atom of helium. Because of their relatively large mass (four times as great as a proton), alpha particles are easily absorbed by matter: typical penetrations are 4 cm in air and 0·002 cm in aluminium.

Beta radiation consists of a flow of negatively charged particles which are identical with electrons. Electrons do not normally exist in the nucleus of an atom, and these beta particles are produced in (and immediately expelled from) the nucleus when a neutron changes into a proton:

$$\begin{array}{llll}
\text{neutron} & \rightarrow & \text{proton} & + & \beta \text{ particle (electron)} \\
\text{(mass 1,} & & \text{(mass 1,} & & \text{(mass negligible,} \\
\text{charge 0)} & & \text{charge}+1) & & \text{charge}-1)
\end{array}$$

Beta particles have greater penetrating power than alpha particles. Their penetration varies, however, because they are emitted with a whole spectrum of energies.

Gamma radiation is unlike alpha and beta radiation in that it is *not* a flow of particles. In fact it belongs to the same family of electromagnetic radiations as visible light, X-rays and radio waves. It travels with the speed of light and only differs from the other electromagnetic radiations in its very short wavelength and very high frequency.

Note that alpha particles, relatively large and positively charged, are stopped by thin aluminium foil whereas beta particles, smaller and negatively charged, need aluminium sheet to absorb them. Gamma rays are far more penetrating than the other types of radiation, and substances which emit gamma rays need to be protected by several centimetres of lead. Both alpha and beta particles are deflected by electrostatic or magnetic fields (see Fig. 13.1). On the other hand gamma radiation, because it has no charge, is unaffected by either electrostatic or magnetic fields.

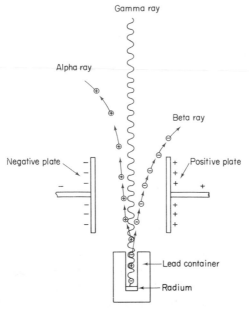

Fig. 13.1 Emission of alpha, beta and gamma radiation from radium: a strong electrostatic field deflects the alpha and beta rays, but not the gamma rays; a strong magnetic field also causes deflection of alpha and beta rays

13.3 Detection of Radioactivity

The radiation from radioactive elements and compounds produces certain effects on matter which can be used for their detection. Thus alpha and beta radiation in particular causes ionization of gases and this is used in both the *gold-leaf electroscope* and the *Geiger–Müller* tube.

(a) The Gold-leaf Electroscope
A gold-leaf electroscope consists of a metal rod to which is attached a delicate strip of gold leaf. The metal rod is insulated from the case (see Fig. 13.2).

When the electroscope is charged, by placing a charged object (e.g. an ebonite rod that has been rubbed with fur) either near to or in contact with its metal cap, the leaf rises away from the metal rod, as shown in Fig. 13.2, and remains in this position until the charge is removed.

If a radioactive material is placed near the electroscope, the radiation ionizes the air around the gold leaf so that it is no longer a good insulator. The charge on the gold leaf leaks away through the ionized air, and the leaf collapses. To detect alpha rays the source must be introduced into the electroscope chamber; beta rays can be detected if the electroscope is provided with a thin window through which the beta rays can enter. Gamma rays cause little ionization.

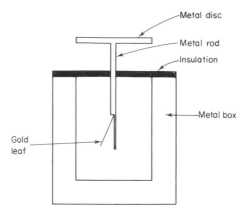

*Fig. 13.2   Section through a gold-leaf electroscope: an electrical charge
placed on the metal disc causes the gold leaf to rise*

(b) The Geiger–Müller Tube

This detector consists of a fine tungsten wire (the anode) centrally aligned inside
a metal tube (the cathode). The tube is filled with argon gas at reduced pressure.
A high voltage (about 1000 V) is maintained between the central wire and the
tube. As this voltage is just insufficient to ionize the argon gas, no current flows
(see Fig. 13.3).

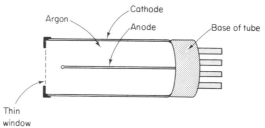

*Fig 13.3   Section through a Geiger–Müller tube: charged particles entering
the thin window result in an electrical pulse passing between anode and
cathode; pulses are counted by a circuit connected to the base of the tube*

If an alpha or beta particle enters the tube, ionization of the argon gas occurs
and a small pulse of current flows between anode and cathode. This pulse can be
amplified and recorded. As each particle produces an avalanche of ions and hence
a pulse of current, the number of particles entering the tube can be determined.
This type of detector is used for alpha and beta particles but is of little use for
gamma radiation.

(c) Scintillation Counters

When radiation reacts with certain luminescent substances, a flash of visible
light or *scintillation* is produced. These scintillations can be detected on an

extremely sensitive photocell called a *photomultiplier*. For the detection of gamma radiation a single large crystal of sodium iodide containing a trace of thallium(I) iodide is used as the luminescent material, whereas for alpha particles zinc sulphide is used. The crystals give off visible light when acted upon by the radiation.

13.4 The Effect of Particle Emission on a Radioactive Element

(*a*) Alpha Emission

An alpha particle consists, as we have seen, of two protons combined with two neutrons. When a radioactive atom emits an alpha particle its nucleus plainly loses two protons and two neutrons. As a result, the atomic number (i.e. the number of protons in the nucleus) decreases by 2 and the mass number (i.e. the sum total of protons and neutrons) decreases by 4. Thus, for example, when uranium (atomic number 92, mass number 238) emits an alpha particle its atomic number drops to 90 and its mass number drops to 234. The element whose atomic number is 90 is in fact thorium, so the emission of an alpha particle is accompanied by a transmutation from uranium to thorium.

In order to simplify the writing down of nuclear reactions the mass number and atomic number of each reacting atom are shown above and below the chemical symbol as follows:

$$\frac{\text{mass number}}{\text{atomic number}}\ \text{CHEMICAL SYMBOL}$$

and an alpha particle is written as ^4_2He since it is the same as a helium nucleus. Using this convention we can write the transmutation of uranium into thorium (see above) as

$$^{238}_{92}\text{U} \rightarrow \ ^{234}_{90}\text{Th} + ^4_2\text{He}$$

Note that the sum of the mass numbers is the same on each side of the arrow, as is the sum of the atomic numbers.

(*b*) Beta Emission

Thorium $^{234}_{90}\text{Th}$ emits a beta particle to form a new element of mass number 234 and atomic number 91. This is protactinium $^{234}_{91}\text{Pa}$.

$$^{234}_{90}\text{Th} \rightarrow \ ^{234}_{91}\text{Pa} + ^{\ 0}_{-1}\text{e}$$
$$\text{thorium} \qquad\qquad\quad \text{beta particle}$$
$$\text{(electron)}$$

The loss of a beta particle results from the conversion of a neutron into a proton and an electron. Thus the mass number remains unchanged while the atomic number (number of protons) increases by 1 unit.

(*c*) Gamma Emission

When a nucleus has ejected either an alpha particle or a beta particle it has an excess energy which it loses in part by emitting gamma radiation. There is no consequent change in the mass number or the atomic number.

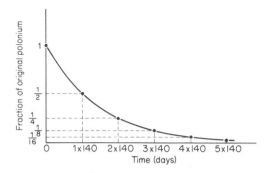

Fig. 13.4   *Radioactive decay of polonium-210: in 140 days half the atoms in the sample disintegrate; in a further 140 days half the remaining atoms disintegrate. The rate of decay is conveniently expressed by the 'half-life' period, which in this case is 140 days*

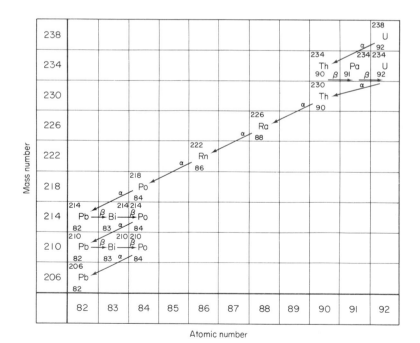

Fig. 13.5   *The uranium disintegration series: the final product is a stable isotope of lead*

Thus all atoms of radioactive elements spontaneously emit alpha or beta particles, and these particles are often accompanied by gamma rays. Whenever an alpha or beta particle is emitted, a new element is formed with quite different physical and chemical properties from the original element. These radioactive changes continue spontaneously at a steady unvarying rate. They cannot be stopped and are unaffected by those changes (e.g. changes in temperature or pressure) which normally affect the rate of a chemical reaction.

This steady unvarying rate of radioactive change is best expressed in terms of the time in which the number of nuclei is reduced to half its original value. For a particular element the time taken for half the activity to disappear is known as its **half-life** (see Fig. 13.4). The curve in Fig. 13.4 shows that the element is most active during its first half-life and after this the activity falls very quickly.

13.5 Natural Radioactive Disintegration

The example given to illustrate the loss of an alpha particle from a uranium atom ($^{238}_{92}U$) is the beginning of one of the *uranium disintegration series*. This series continues by loss of alpha or beta particles through a number of radioactive elements until it ends with a non-radioactive isotope of lead, $^{206}_{82}Pb$. Fig. 13.5 shows a representation of the uranium disintegration series.

In addition to the uranium disintegration series shown in Fig. 13.5 there are two other naturally occurring series: (*a*) a second uranium series beginning with the isotope $^{235}_{92}U$ and (*b*) a thorium series beginning with $^{232}_{90}Th$. All three series end with stable isotopes of lead. Disintegration series which begin with 'artificial' elements such as neptunium $^{237}_{93}Np$ are also known.

13.6 Nuclear Reactions

So far we have been concerned with *spontaneous* nuclear reactions occurring in radioactive elements. *Induced* nuclear reactions, however, can occur when nuclei are struck by high-velocity particles. Such reactions may be initiated either by neutral particles (neutrons) or by charged particles (e.g. protons). 'Atom-smashing' machines, such as the cyclotron and the linear accelerator, are used to give high velocities to charged particles, but they cannot accelerate uncharged neutrons. High-velocity neutrons are obtained from a nuclear reactor.

(*a*) Fission
Fission means 'splitting' and nuclear fission is a process in which a heavy nucleus is split into two fragments of approximately equal size; at the same time neutrons, gamma radiation and a considerable amount of energy in the form of heat are liberated. Hahn and Strassmann in 1939 announced that they had split a uranium-235 nucleus using slow neutrons. In this process the heavy nucleus is struck by and absorbs a neutron; it becomes unstable, pulsates and finally splits into two radioactive nuclei of roughly similar size. At the same time two or three

neutrons are released, each capable of splitting another uranium-235 atom. One way in which this may occur is

$$\underset{\substack{\text{uranium}\\\text{isotope}}}{^{235}_{92}\text{U}} + \underset{\substack{\text{neutron}}}{^{1}_{0}\text{n}} \rightarrow \underset{\substack{\text{unstable}\\\text{nucleus}}}{[^{236}_{92}\text{U}]} \rightarrow \underset{\substack{\text{krypton}\\\text{isotope}}}{^{90}_{36}\text{Kr}} + \underset{\substack{\text{barium}\\\text{isotope}}}{^{144}_{56}\text{Ba}} + \underset{\substack{\text{two}\\\text{neutrons}}}{2^{1}_{0}\text{n}}$$

The two product neutrons may collide with two other uranium nuclei to liberate four more neutrons (see Fig. 13.6). As a result the reaction becomes self propagating. Such a reaction is called a *chain reaction*.

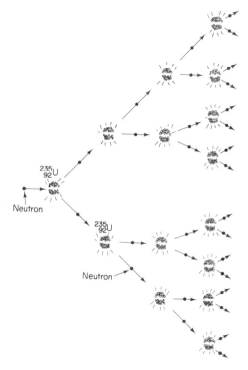

Fig. 13.6   *Example of a chain reaction: the fission of uranium-235 by slow neutrons*

Mass–Energy Relationship

When a fuel burns energy is released (i.e. the reaction is exothermic) because of the breakdown and formation of chemical bonds between atoms. However, in a *nuclear reaction* energy is released because the forces binding the nuclear particles are broken. For example, when lithium atoms are bombarded with high-speed protons, alpha particles are produced together with an immense quantity of energy which represents the *binding energy* of the lithium nucleus:

$$^{7}_{3}\text{Li} + ^{1}_{1}\text{H} \rightarrow 2\,^{4}_{2}\text{He} + \text{large amount of energy}$$

This reaction results in the loss of a minute quantity of matter and the release of a large amount of energy. It had been suggested as early as 1905 by Albert Einstein that matter and energy were different forms of the same thing and that theoretically matter could be changed into energy. The release of such large quantities of energy during nuclear fission substantiated Einstein's suggestion that mass and energy are related by the equation

$$E = mc^2$$

where E is energy (in joules), m is mass (in kilograms) and c is the velocity of light (in metres per second).

The Atomic Bomb

Natural uranium is a mixture of mainly two isotopes, $^{235}_{92}U$ and $^{238}_{92}U$, of which the former constitutes less than 1%. We have seen that the $^{235}_{92}U$ isotope undergoes fission with neutrons. However, $^{238}_{92}U$ does not undergo fission with slow neutrons and its presence prevents the occurrence of the self-sustaining chain reaction previously described. In fact, in order that a chain reaction should occur, a minimum mass of $^{235}_{92}U$ (the *critical mass*) must be present. The problem of separating the two uranium isotopes to obtain pure $^{235}_{92}U$ proved to be the major stumbling block in the production of the first atomic bomb. (The method actually employed was based on the fact that the gaseous hexafluorides of the two isotopes diffuse at very slightly different rates through a porous barrier.)

An amount of $^{235}_{92}U$ less than the critical mass does not undergo a chain reaction because sufficient neutrons escape to the surroundings to prevent the possibility of chain reaction. However, when two masses slightly less than the critical mass are brought together, sufficient neutrons are absorbed by nuclei to produce fission resulting in a wildly explosive chain reaction. This was the principle used in the first atomic bomb.

The Nuclear Reactor

In the early 1940s Enrico Fermi built the first nuclear reactor (or pile). It was constructed to use natural uranium rather than pure $^{235}_{92}U$, which was extremely difficult to obtain. Fermi argued that if the neutrons from $^{235}_{92}U$ were *slowed down* they would be more easily captured by the small amount of $^{235}_{92}U$ in natural uranium—and a chain reaction could be sustained. A *neutron moderator* of pure graphite was therefore used to slow down the neutrons without absorbing them.

To regulate the rate at which fission occurs in a nuclear reactor, control rods of neutron-absorbing boron steel are inserted whenever it is necessary to reduce the number of free neutrons. Fast neutrons released by the fission of $^{235}_{92}U$ are absorbed by $^{238}_{92}U$ which is eventually converted into plutonium, a product which is itself fissionable.

A nuclear reactor has thus two main functions:

(i) Fission of $^{235}_{92}U$ by slow neutrons to produce energy and radioactive isotopes.

(ii) The conversion of $^{238}_{92}U$ into plutonium, which can split up into further radioactive isotopes.

An immense amount of heat energy is liberated during fission and this energy can be usefully employed in driving electrical generators.

(b) Fusion of Light Nuclei

In contrast to fission, where energy is released when the nuclei of heavy atoms are *split* into smaller fragments, energy can also be liberated when very small nuclei *combine*. Such combinations of nuclei are called *fusion*.

It is believed that the source of the sun's energy is a fusion reaction in which hydrogen nuclei fuse to form a helium nucleus with a loss of mass and corresponding liberation of an immense quantity of energy. The fusion of nuclei will only take place at very high temperatures. The staggering quantity of energy produced in the hydrogen bomb is the result of a fusion reaction between a hydrogen isotope and a lithium isotope.

13.7 Applications of Radioactive Isotopes

We have seen that fission produces a variety of radioactive isotopes. These, together with naturally occurring radioactive elements, especially radium, have found numerous uses in both medicine and industry. Today radioisotopes are usually prepared from other elements by neutron bombardment.

(a) Radioactive Isotopes in Medicine

Radioactive isotopes are used either to *diagnose* disorders in the body or to *treat* disorders by radiotherapy.

(i) **Diagnosis**. Such applications include the use of $^{131}_{53}I$ (radioactive iodine-131) in the diagnosis and treatment of thyroid disorder. The thyroid gland, located in the neck, is responsible for the regulation of many body processes and has the ability to concentrate iodine. Thus radioactive iodine can be used to diagnose an overactive or underactive thyroid gland by measuring the concentration of isotope in the gland. Cancer of the thyroid can be successfully treated with the gamma radiation from $^{131}_{53}I$ which is allowed to accumulate in the gland.

(ii) **Radiotherapy**. Radioisotopes such as $^{60}_{27}Co$ (cobalt-60) have been used in the treatment of some forms of cancer. In this treatment, cancerous tissue is subjected to strong gamma radiation from a cobalt-60 source. Gamma rays destroy the cancerous cells, as well as healthy cells, but the rapidly dividing cancer cells are destroyed preferentially. It is thus important to localize the area which is subjected to radiation.

(b) Industrial Applications of Radioisotopes

The uses of radioactive isotopes in industry are many and varied. A few examples are given in this section to illustrate their diverse application.

(i) **Mixing processes**. Radioactive tracers have been used to measure the mixing efficiency in the preparation of a variety of products, including cattle food and chocolate. In foodstuffs it is important to choose a harmless radioisotope with a short half-life. A radioisotope of manganese is sometimes used to ensure the complete mixing of mineral additives in cattle foods. After mixing, a sample is tested for radioactivity before the radioactive content of the manganese tracer has disappeared. If mixing is complete all samples should show the same level of radioactivity.

(ii) **Thickness and density measurements.** We have seen already that beta particles are capable of penetrating aluminium sheet. As the number of particles absorbed by the sheet is proportional to its thickness and density, measuring the number of particles which succeed in getting through gives an indication of how much material they have penetrated. This fact is used in industry for testing the thickness and density of materials. For example, beta-particle thickness gauges are used for checking the packing of tobacco in cigarettes, and in rolling mills for maintaining a constant thickness of metal sheet (Fig. 13.7).

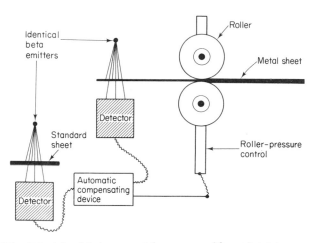

Fig. 13.7   Principle of the beta-particle gauge used for maintaining a constant thickness of metal in a rolling mill

The same principle can be used in checking the level of liquids in tanks or cylinders (Fig. 13.8) and also in ensuring that packages contain the correct amount of material.

In recent years radioactive isotopes have come to be used in more and more industrial and medical applications. The ionizing radiations emitted present a hazard to health either when they are sufficiently intense or when exposure to them is prolonged, so it has to be borne in mind that any such material can be dangerous if mishandled. Isotopes are kept in aluminium or lead containers, depending on the type of radiation emitted, and must not be touched with the bare hands. For most purposes the amount of radioactive isotope employed is very small indeed; it is 'diluted' with a non-active material for ease of handling.

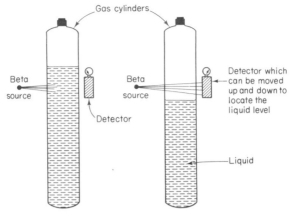

Fig. 13.8   Principle of the beta-particle gauge used for detecting the level of liquid in a closed container

Summary of Unit 13

1. Henri Becquerel discovered radioactivity in 1896 through the fogging of photographic film. This led to the isolation of the radioactive elements *polonium* and *radium* by Marie and Pierre Curie.

2. Radioactive elements may emit three types of radiation: (*a*) **alpha radiation** consisting of *helium nuclei*, (*b*) slightly more penetrating **beta radiation** consisting of *electrons*, (*c*) the intensely penetrating **gamma radiation** which is electromagnetic in nature.

3. Alpha and beta radiation can be *detected* using a *gold-leaf electroscope* or a *Geiger–Müller* tube, whereas gamma radiation is detected using *scintillation counters*.

4. The time taken for the activity of a radioactive isotope to decay to half its original value is called its **half-life**.

5. The loss of an alpha particle from an element causes the atomic number to decrease by two units and the mass number to decrease by four units. Loss of a beta particle increases the atomic number by one unit but leaves the mass number unchanged. Gamma emission causes no change in the atomic number or in the mass number.

6. **Fission** is a process in which a heavy nucleus is split into two large fragments, usually accompanied by an energy release.

7. A **chain reaction** is a self-sustaining nuclear reaction.

8. **Fusion** takes place when light atomic nuclei combine together to form a heavier nucleus. In the process a minute loss of mass is converted into a huge release of energy.

9. A nuclear reactor is a structure in which a controlled nuclear-fission reaction produces energy and radioactive isotopes.

10. **Radioactive isotopes** find widespread application in both medicine and industry.

Test Yourself on Unit 13

1. What is:

   (a) an α (alpha) particle?

   (b) a β (beta) particle?

   (c) a γ (gamma) ray?

2. In the following radioactive decay series:

$$^{228}_{88}\text{Ra} \xrightarrow{(i)} {}^{228}_{89}\text{Ac} \xrightarrow{(ii)} {}^{228}_{90}\text{Th} \xrightarrow{(iii)} {}^{224}_{88}\text{Ra}$$

   (a) What information about the nucleus of the radium atom does the symbol $^{228}_{88}\text{Ra}$ provide?

   (b) State whether α (alpha) or β (beta) emission occurs in (i), (ii) and (iii).

   (c) What name is given to the first and last members of the above series?

3. Part of the decay series of $^{232}_{90}\text{Th}$ includes:

$$^{232}_{90}\text{Th} \xrightarrow{\alpha} (\text{X}) \xrightarrow{\beta} (\text{Y}) \xrightarrow{\beta} (\text{Z}) \xrightarrow{\alpha} {}^{224}_{88}\text{Ra}$$

   Give the mass number and atomic number of the elements X, Y and Z.

4. Are the following statements true or false?

   (a) A radioactive element has a half-life of 10 minutes. One-eighth of the radioactive material remains after 80 minutes.

   (b) A Geiger tube detects α particles but not β particles.

   (c) Slow neutrons produce fission of $^{235}_{92}\text{U}$ nuclei.

   (d) The source of the sun's energy is a fusion reaction in which hydrogen nuclei fuse to form a helium nucleus.

   (e) All isotopes are radioactive.

Mark this test out of 20 with the answers provided on page 382.

Matter in Bulk—The Heavy Chemicals Industry

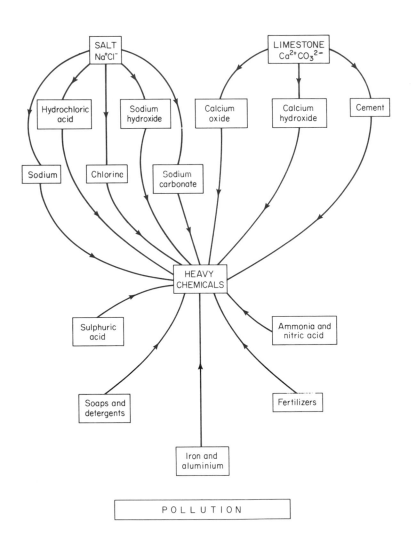

Working in competition with other manufacturers, the large-scale producer of industrial chemicals must take into account the economics of the various processes as well as their chemistry. Among the factors to be considered are:

(*a*) cost and transport of raw materials
(*b*) available labour to run the plant
(*c*) energy requirements and their most efficient use
(*d*) the most economic use of by-products
(*e*) disposal of industrial waste
(*f*) location and capital cost of the most efficient chemical plant.

This Unit looks at the manufacture of several 'heavy' chemicals. It considers not only the chemistry of the reactions, but also some of the economic and social factors affecting manufacture.

14.1 Chemicals from 'Salt'

Of the many salts dissolved in sea water the most abundant is sodium chloride (common salt), comprising some 2–3% by weight of the major seas of the world. In hot dry areas salt is extracted from sea water by solar evaporation. Common salt occurs also in vast underground deposits throughout the world, and it is from these that most of the salt for commercial use is obtained.

In Great Britain sodium chloride deposits are found in Cheshire, Lancashire and Durham. After purification the salt is used on a large scale for the manufacture of *sodium, sodium hydroxide* (caustic soda), *chlorine* and *sodium carbonate*. The first three of these are produced from salt by electrical processes. Siting of chemical plant for their manufacture is therefore influenced by the availability of salt and electricity.

14.2 Sodium from 'Salt'

Sodium is produced by the electrolysis of molten sodium chloride fed continuously into the **Downs cell**. Calcium chloride is added to lower the melting point to around 600 °C (pure sodium chloride melts at 801 °C).

The Downs cell consists of a steel shell lined with refractory brick (Fig. 14.1) enclosing a steel cathode and a cylindrical graphite anode.

Molten *sodium* is liberated at the cathode in preference to calcium, the liberation of which requires more energy. Being less dense than the electrolyte, the molten sodium rises as it is produced and is caught in a specially designed container.

Gaseous *chlorine* is liberated at the anode in the centre of the cell. A steel gauze between the two electrodes keeps the chlorine from coming into contact with the sodium and thus prevents any interaction between them. The reaction at the cathode is

$$2Na^+ + 2e \rightarrow 2Na$$

and the anode reaction is

$$2Cl^- \rightarrow Cl_2 + 2e$$

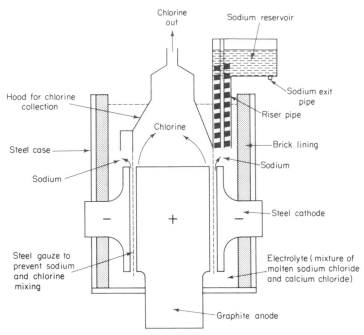

Fig. 14.1   Production of sodium and chlorine in a Downs cell

There is virtually no waste material in this reaction and therefore the problem of waste disposal does not arise. Chlorine, the only by-product, is of great commercial importance and finds a ready market: in fact chlorine is more in demand than sodium and this source provides only a fraction of the world's requirements.

14.3 Sodium Hydroxide (Caustic Soda) from 'Salt'

Sodium hydroxide (caustic soda) is manufactured by the electrolysis of sodium chloride *solution*. The electrolysis is carried out in a **mercury cell**, which consists of two compartments: one for the electrolytic decomposition of the brine, the other for the chemical decomposition of sodium amalgam (Fig. 14.2).

In the electrolytic compartment the cathode consists of a moving stream of mercury and the anode consists of a number of graphite blocks. Sodium is liberated at the cathode (in preference to hydrogen) and dissolves in the mercury forming an *amalgam*. This sodium amalgam flows out of the electrolytic chamber into the decomposition chamber where it reacts with water forming sodium hydroxide solution, hydrogen and free mercury:

$$2Na_{(amalgam)} + 2H_2O_{(l)} \rightarrow 2Na^+OH^-_{(aq)} + H_{2(g)}$$

The mercury is then recycled through the electrolytic chamber.

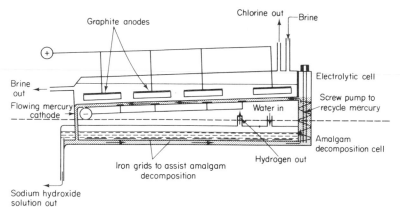

Fig. 14.2   Production of sodium hydroxide and chlorine in a mercury cell

Chlorine is liberated at the graphite anodes (in preference to oxygen) and is piped from the cell.

Cathode reaction: Na^+ and H^+ ions migrate to the cathode; sodium is discharged

$$2Na^+ + 2e \rightarrow 2Na \text{ (amalgamates with mercury)}$$

Anode reaction: Cl^- and OH^- migrate to the anode; chlorine is discharged

$$2Cl^- \rightarrow Cl_2 + 2e$$

The chlorine manufactured by this process is no longer a by-product but is itself one of the major heavy chemicals. Hydrogen is collected from the decomposition chamber in a high state of purity and finds a ready market. There are no waste products in this reaction, but care must be taken to prevent the pollution which would result from the accidental escape of either chlorine or mercury.

14.4 Chlorine and Hydrochloric Acid

Chlorine is manufactured during the electrolysis of sodium chloride in the Downs cell (Fig. 14.1) and the mercury cell (Fig. 14.2), the latter being by far the more important. The demand for chlorine, particularly in the manufacture of organic chlorine derivatives, has made it one of the most important heavy chemicals.

Hydrochloric acid is manufactured by the direct combination of hydrogen and chlorine, although quite large quantities are obtained as a by-product in the chlorination of alkanes and other organic compounds.

14.5 Sodium Carbonate and Sodium Hydrogencarbonate from 'Salt'

Sodium carbonate and sodium hydrogencarbonate are manufactured in the **Solvay** (ammonia–soda) process. The raw materials are *sodium chloride, calcium carbonate* (limestone), *coal* and *coke*.

The first stage in the process is the production of ammoniacal brine (ammonia dissolved in sodium chloride solution) by passing ammonia gas up a tower in which it meets a downward flow of brine (see Fig. 14.3*b*).

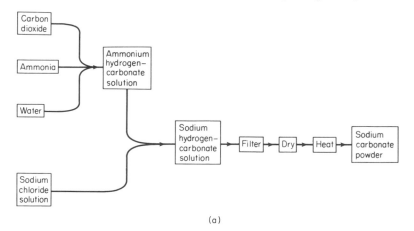

(a)

Fig. 14.3a   Flow diagram of the Solvay process for the production of sodium carbonate

The saturated ammoniacal brine is then pumped to the top of a second tower (known as the *carbonator*) and allowed to trickle down while carbon dioxide is pumped in under pressure at the base. Sodium hydrogencarbonate is precipitated in the lower, cooled, part of the carbonator. The reactions taking place in the carbonator can be summarized as follows:

$$NH_{3(g)} + CO_{2(g)} + H_2O_{(l)} \rightarrow NH_4^+ HCO_{3(aq)}^-$$

$$NH_4^+ HCO_{3(aq)}^- + Na^+ Cl_{(aq)}^- \rightarrow Na^+ HCO_{3(s)}^- + NH_4^+ Cl_{(aq)}^-$$

The suspension of sodium hydrogencarbonate is filtered, dried and heated to form the carbonate:

$$2Na^+ HCO_{3(s)}^- \rightarrow Na_2^+ CO_{3(s)}^{2-} + H_2O_{(g)} + CO_{2(g)}$$

The carbon dioxide produced in this reaction is recycled, but the main supply for the carbonator is obtained by heating limestone with coke in kilns:

$$Ca^{2+} CO_{3(s)}^{2-} \rightarrow Ca^{2+} O_{(s)}^{2-} + CO_{2(g)}$$

An alternative source of carbon dioxide is the burning of coke:

$$C_{(s)} + O_{2(g)} \rightarrow CO_{2(g)}$$

The calcium oxide produced in the lime kiln is 'slaked' with water and mixed with the filtrate from the carbonator (ammonium chloride solution). Ammonia is liberated on heating this mixture and is used to saturate more brine.

$$Ca^{2+} O_{(s)}^{2-} + H_2O_{(l)} \rightarrow Ca^{2+}(OH^-)_{2(s)}$$

$$Ca^{2+}(OH^-)_{2(s)} + 2NH_4^+ Cl_{(aq)}^- \rightarrow Ca^{2+} Cl_{2(s)}^- + 2NH_{3(g)} + 2H_2O_{(g)}$$

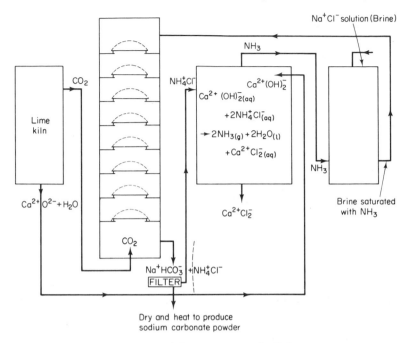

Fig. 14.3b   Location of the reactions in the Solvay process

The Solvay process is one of the best examples of an efficient chemical process. The only by-product is calcium chloride, and the raw materials (salt, limestone and coal) are readily available. It is advantageous to site the plant close to a river or lake since large quantities of cooling water are needed for removing the heat produced in the carbonator.

14.6 Chemicals from Limestone

Limestone rock is mostly calcium carbonate. It occurs in vast deposits throughout the world and is used for the production of *calcium oxide* (quicklime), *calcium hydroxide* (slaked lime) and *cement*.

(a) Calcium Oxide (Quicklime)

Quicklime is manufactured from limestone by heating it strongly in a *lime kiln* —a steel or brick tower lined with firebrick (see Fig. 14.4). Coal is used as the fuel for heating the kiln.

The limestone, introduced continuously at the top, decomposes and is removed as quicklime from the bottom of the kiln:

$$Ca^{2+}CO_{3(s)}^{2-} \rightarrow Ca^{2+}O_{(s)}^{2-} + CO_{2(g)}$$

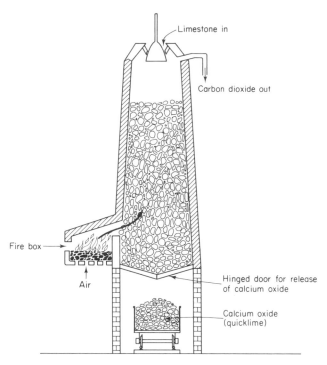

Fig. 14.4   Lime kiln used for the conversion of calcium carbonate (limestone) into calcium oxide (quicklime)

(b) Calcium Hydroxide (Slaked Lime)

When water is added to calcium oxide (quicklime) a highly exothermic reaction occurs as the oxide is 'slaked' to calcium hydroxide:

$$Ca^{2+}O^{2-}_{(s)} + H_2O_{(l)} \rightarrow Ca^{2+}(OH^-)_{2(s)}$$

Vast quantities of calcium hydroxide (slaked lime) are used in agriculture to neutralize acid soils, in the manufacture of sodium carbonate and in making mortar.

Mortar is a mixture of calcium hydroxide (slaked lime), sand and water. It hardens when the calcium hydroxide absorbs and reacts with carbon dioxide from the air, forming calcium carbonate:

$$Ca^{2+}(OH^-)_{2(s)} + CO_{2(g)} \rightarrow Ca^{2+}CO^{2-}_{3(s)} + H_2O_{(l)}$$

(c) Cement

Millions of tons of limestone are used annually in the production of cement. Limestone and clay are mixed in the necessary proportions and ground to a fine powder. This powder is then strongly heated in a huge sloping rotary kiln. The

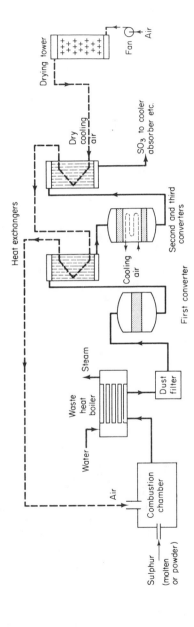

*Fig. 14.5   Manufacture of sulphuric acid: production of sulphur trioxide
SO₃ by the contact process*

incandescent mass undergoes chemical change as it passes down the kiln, finally emerging as an aggregate which is cooled and then crushed to an extremely fine powder. This is Portland cement.

Cement, a complex silicate, is used to make *concrete*, the major non-metal structural material of our technological society. Concrete is formed when cement, sand and gravel are mixed to a paste with water. The cement reacts with the water to form products which cling to the sand and gravel and in so doing harden to a synthetic rock.

Lime kilns and cement works tend to be sited as near as possible to a source of limestone, and preferably not too far from a coalfield.

14.7 Sulphuric Acid

The most important method for the manufacture of sulphuric acid is the **contact process**. The principal raw material for this process is native sulphur or, to a lesser extent, some sulphur-containing compound such as pyrites (an iron sulphide).

Sulphur is burned in excess dry air to form sulphur dioxide. The gas mixture (sulphur dioxide and air) is filtered to free it from dust and then passed through a series of converters (see Fig. 14.5) containing trays of vanadium(v) oxide (vanadium pentoxide) which acts as a catalyst. Sulphur dioxide and oxygen from the air combine on contact with the catalyst in an exothermic reaction to produce sulphur trioxide:

$$2SO_{2(g)} + O_{2(g)} \rightarrow 2SO_{3(g)}$$

Cooling air is used to maintain the temperature in the converters at approximately 450°C.

The sulphur trioxide is absorbed in 98–99% sulphuric acid to form **oleum** (fuming sulphuric acid):

$$H_2SO_{4(l)} + SO_{3(g)} \rightarrow H_2S_2O_{7(l)}$$

Suitable dilution of the oleum gives sulphuric acid of any desired concentration.

In the contact process the catalyst is *surface active*. This means that gas reactions take place on the surface of the catalyst, which must therefore provide a large surface area. Dust will reduce the effective surface area and may even react with the catalyst, 'poisoning' it and further limiting its efficiency. However, vanadium(v) oxide is a reasonably efficient catalyst for the oxidation of sulphur dioxide and is not readily poisoned.

In a sulphuric acid plant one of the most important factors is the cost of producing pure, *dust-free* sulphur dioxide. Although it is cheaper to produce sulphur dioxide from sulphur ores rather than from native sulphur, the saving may easily be offset by the extra expense of purifying and removing dust from the gas. Such problems of economics are common in industrial chemistry and the manufacturer must weigh them carefully. It may well prove cheaper in the long run to use the more expensive native sulphur.

In the United Kingdom almost 200 000 tonnes of sulphuric acid are produced annually, most of which is used in the manufacture of fertilizers.

14.8 Ammonia

The synthesis of ammonia from nitrogen and hydrogen is carried out in the **Haber process**:

$$N_{2(g)} + 3H_{2(g)} \rightleftharpoons 2NH_{3(g)}$$

This is an exothermic reaction and under normal conditions of temperature and pressure is extremely slow. To increase the rate of the reaction finely divided iron is used as a catalyst. (Note that the introduction of a catalyst also increases the rate of the *reverse* reaction; however, it enables the equilibrium position to be reached more rapidly.) The optimum conditions from the point of view of cost and efficiency are a moderately high temperature of 500 °C and a pressure of 150–300 atmospheres. Under these conditions the equilibrium mixture contains approximately 15% of ammonia. The ammonia is removed from the unchanged nitrogen and hydrogen by liquefaction, and unchanged gases are recycled over the catalyst.

In the United States almost all of the hydrogen for the ammonia synthesis is obtained from natural gas (methane) by reaction with steam, using a nickel catalyst:

$$CH_{4(g)} + H_2O_{(g)} \rightarrow CO_{(g)} + 3H_{2(g)}$$

An increasingly large proportion of hydrogen is now being produced in the United Kingdom from natural gas and petroleum products (naphtha). Some hydrogen is still produced from water gas (a mixture of hydrogen and carbon monoxide) by mixing with steam and passing it over an iron(III) oxide catalyst at 400 °C:

$$\underbrace{H_{2(g)} + CO_{(g)}}_{\text{water gas}} + H_2O_{(g)} \rightarrow 2H_{2(g)} + CO_{2(g)}$$

The carbon dioxide is removed by washing the gas mixture with water under pressure.

Nitrogen for the Haber process is obtained from the air.

At low pressures the yield of ammonia is extremely small. Thus a manufacturer wishing to increase his ammonia yield must invest in costly high-pressure plant. Efficient producers are now operating plant at 1000 atmospheres pressure.

14.9 Nitric Acid

Although much of the ammonia produced in the Haber process is used in the manufacture of fertilizers, large quantities are converted to nitric acid in the

Ostwald process. In this process ammonia is first oxidized to *nitrogen monoxide* (nitric oxide):

$$4NH_{3(g)} + 5O_{2(g)} \rightarrow 4NO_{(g)} + 6H_2O_{(g)}$$

An ammonia–air mixture containing approximately 10% ammonia is passed over a platinum–rhodium gauze catalyst. The reaction is exothermic and, once started, maintains the catalyst at approximately 900 °C.

On cooling, the nitrogen oxide reacts with air to produce *nitrogen dioxide*:

$$2NO_{(g)} + O_{2(g)} \rightarrow 2NO_{2(g)}$$

Absorption of the nitrogen dioxide in water produces 60–65% nitric acid:

$$3NO_{2(g)} + H_2O_{(l)} \rightarrow 2H^+NO_{3(aq)}^- + NO_{(g)}$$

Precautions are taken to ensure that gases leaving the absorption tower do not pollute the atmosphere with oxides of nitrogen.

14.10 Fertilizers

Fertilizers are added to the soil to provide elements which are essential to plant life. The most important of these elements are **nitrogen, phosphorus** and **potassium**.

Nitrogen is absorbed by plant roots as nitrate (NO_3^-) or as ammonium (NH_4^+). Nitrogen from the air is converted to ammonia in the Haber process and almost all of this is used in the manufacture of *nitrogenous fertilizers* such as ammonium nitrate, ammonium sulphate and ammonium phosphate.

Phosphate fertilizers are obtained from phosphate rock which is treated with sulphuric acid in order to convert it to **superphosphate**. This superphosphate is soluble (unlike the phosphate rock) and consists of a mixture of calcium dihydrogenphosphate and calcium sulphate.

The most important *potassium fertilizer* is the chloride, most of which is imported into the United Kingdom from eastern Europe. However, there are huge deposits of *sylvinite* (a mixture of potassium chloride and sodium chloride) in north Yorkshire, and it is hoped that this source will provide sufficient potassium for the United Kingdom's needs.

A whole range of *compound fertilizers* containing nitrogen, phosphorus and potassium in varying proportions is now manufactured. These are mixtures of simple fertilizers such as ammonium nitrate, potassium chloride and superphosphate. Care is taken to produce a granular, free-flowing, easily handled fertilizer to aid the farmer. Fig. 14.6 shows some of the interrelations between the heavy-chemicals industry and fertilizer manufacture.

14.11 Iron and Steel

Iron ore is one of the few metal ores found in large quantities in the United Kingdom. But the iron content of this ore, although adequate to meet the demands of the Industrial Revolution which it made possible, is relatively low and Britain now has to buy high-grade ores from abroad.

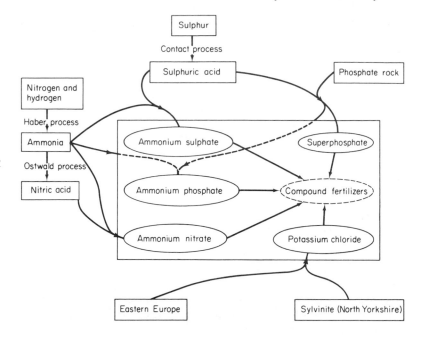

Fig. 14.6   Fertilizer manufacture

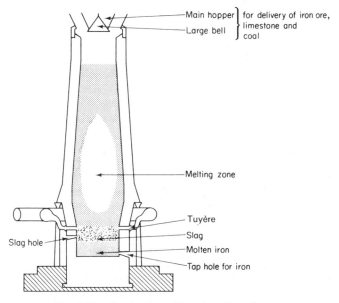

Fig. 14.7   Production of iron in a blast furnace

The ore is smelted in a **blast furnace** to produce *pig iron*. The 'charge' of iron ore, limestone and coke is fed in through the top of the furnace via a bell arrangement which prevents the escape of furnace gases (see Fig. 14.7).

Air is blown (hence the name 'blast' furnace) through pipes called *tuyères* located towards the base of the furnace. Oxygen of the air reacts with the coke to produce heat and carbon monoxide:

$$2C_{(s)} + O_{2(g)} \rightarrow 2CO_{(g)}$$

Carbon monoxide then reduces the heated iron ore to iron:

$$\underset{\text{(iron ore)}}{Fe_2^{3+}O_{3(s)}^{2-}} + 3CO_{(g)} \rightarrow 2Fe_{(l)} + 3CO_{2(g)}$$

The heated limestone decomposes to form calcium oxide, and this acts as a 'flux' which removes silica (the major impurity in the ore) as molten calcium silicate called *slag*:

$$Ca^{2+}CO_{3(s)}^{2-} \rightarrow Ca^{2+}O_{(s)}^{2-} + CO_{2(g)}$$

$$Ca^{2+}O_{(s)}^{2-} + \underset{\text{(silica)}}{SiO_{2(s)}} \rightarrow \underset{\text{(slag)}}{Ca^{2+}SiO_{3(l)}^{2-}}$$

Molten iron, containing dissolved carbon and other impurities, settles to the bottom of the furnace with the molten slag floating on the top of it.

At regular intervals the molten iron and slag are run off ('tapped') and more raw materials are fed in at the top. The whole process is continuous.

A modern blast furnace is a steel cylinder lined with firebrick. Massive quantities of pre-heated air are blown through the tuyères so that the temperature at the base of the furnace is maintained at approximately 1400 °C, whereas at the top of the furnace the temperature falls to near 200 °C.

Most of the iron produced is used to make steel by reducing its carbon content. Pig iron contains about 4% carbon, whereas steels have a carbon content in the 0·1–2% range. After suitable treatment the slag finds widespread use as a lightweight building material, in cement manufacture and as a road-building material: it is not the waste product that it used to be.

14.12 Aluminium

Aluminium, the most abundant metal in the earth's crust, occurs in clay, slate and silicate rocks. However, compounds of aluminium are so strongly bonded that it is difficult to extract the metal from its ore: a high-energy process such as electrolysis is necessary. The problem here is that aluminium ores are neither easily melted nor readily soluble, so it is difficult to find a suitable electrolyte. Fortunately, C. M. Hall, an American chemist, discovered that aluminium oxide will dissolve in molten *cryolite* ($Na_3^+AlF_6^{3-}$). The Hall cell used for the production of aluminium nowadays is shown in Fig. 14.8.

Electrolysis takes place at 900–1000 °C in a graphite-lined iron tank. The graphite lining acts as the cathode, while blocks of a specially formulated mixture

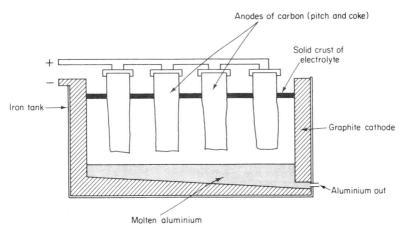

Fig. 14.8   Production of aluminium in a Hall cell

of pitch and coke (carbon) are used as anodes. Cryolite is first added to the cell and melted by the current. Purified *bauxite* ore (aluminium oxide) is dissolved in the molten cryolite and more is added at intervals while electrolysis takes place.

Cathode reaction:

$$2Al^{3+} + 6e \rightarrow 2Al$$

Anode reaction:

$$3C + 3O^{2-} \rightarrow 3CO + 6e$$
$$\text{(from the anode)}$$

Vast quantities of electricity are consumed during the electrolytic process, and for this reason aluminium plants are situated where electricity is cheap and plentiful. At present large quantities of aluminium are imported into the United Kingdom, but with the advent of cheap electricity from nuclear power the production of local aluminium will increase.

Deposits of bauxite are found in many parts of the world, including France, India, Malaysia, Africa, the United States and Canada. Natural cryolite is found only in Greenland. Much of the cryolite used in the production of aluminium is manufactured synthetically from bauxite.

14.13 Soaps and Detergents

The term 'detergent' is used to describe cleansing agents in general, and under this heading we include both **soap** and **synthetic detergents**. A detergent has two main functions: (*a*) it makes water 'wetter' and (*b*) it helps in the removal of grease and dirt.

(*a*) Detergents as Wetting Agents

Water without detergent is poor at wetting things. If you dip your hand into a bowl of water and remove it you will notice that your hand is not uniformly wet.

Now add some detergent to the water and mix it by stirring with your hand. Remove your hand and notice that now it is covered with a layer of water. It is uniformly wet. The detergent lowers the *surface tension* of the water and thus enables the water to wet the surface of the skin.

Surface tension is caused by the forces of attraction between molecules in a liquid. At the surface this produces an effect like a stretched elastic film or 'skin' because there is a tendency for molecules to be pulled into the body of the liquid. When a detergent is added to water the 'skin' is weakened and the water tends to spread out.

Fig. 14.9 shows this effect when pure water and water containing detergent are placed on a piece of fabric.

Fig. 14.9   Left: forces (represented by arrows) acting in the surface of a water droplet keep it almost spherical. Right: addition of 'head-and-tail' detergent molecules reduces the surface tension, allowing the droplet to spread out and wet the fabric

(b) Removal of Grease Particles

The removal of grease by a detergent (soap) has already been described in Section 6.16, and the nature of 'head-and-tail' detergency is illustrated in Fig. 6.15.

14.14 Manufacture of Soap

The raw materials for soap-making are animal and vegetable fats and oils. Vegetable oils include ground-nut, palm, coconut and many others, while animal fats include whale, beef and mutton tallow. Many millions of tons of these oils and fats are produced each year for a variety of uses, of which soap manufacture is only one.

Oils and fats are **esters**. An ester is a product of reaction between a carboxylic acid and an alcohol (see Section 11.10*d*):

$$\text{ACID} + \text{ALCOHOL} \rightarrow \text{ESTER} + \text{WATER}$$

The parent acid of an ester from an oil or fat has a long carbon-chain backbone, e.g. octadecanoic (stearic) acid, $C_{17}H_{35}COOH$, while the alcohol is glycerol

$$CH_2OH$$
$$|$$
$$CHOH$$
$$|$$
$$CH_2OH$$

Thus the ester formed between glycerol and octadecanoic (stearic) acid has the formula

$$C_{17}H_{35}COOCH_2$$
$$|$$
$$C_{17}H_{35}COOCH$$
$$|$$
$$C_{17}H_{35}COOCH_2$$

and is one of the main components of beef and other animal fats.

In the soap-making process a blend of oils and fats is treated with alkali (usually sodium hydroxide) in a huge steam-heated vat. The esters are converted to the parent alcohol and **soap** (the sodium salt of the acid) in a process called **saponification**.

$$
\begin{array}{l}
C_{17}H_{35}COOCH_2 \\
| \\
C_{17}H_{35}COOCH + 3Na^+OH^-_{(aq)} \rightarrow 3C_{17}H_{35}COO^-Na^+_{(s)} + \\
| \qquad\qquad\qquad\qquad\qquad\qquad \text{(sodium stearate} \\
C_{17}H_{35}COOCH_{2(s)} \qquad\qquad\qquad\qquad \text{soap)} \\
\text{(glyceryl ester of stearic acid)}
\end{array}
\qquad
\begin{array}{l}
CH_2OH \\
| \\
CHOH \\
| \\
CH_2OH_{(aq)} \\
\text{(glycerol)}
\end{array}
$$

During the saponification process a great deal of heat is generated, but to complete the reaction the mixture is finally boiled. The mixture of soap, glycerol and water is separated by adding sodium chloride (salt). This 'salting-out' process solidifies the soap since it is only slightly soluble in salty water. Brine and glycerine form a single solution while the soap floats to the top. After removal from the solution, the soap is processed and perfumes, colouring agents and fillers are added.

14.15 Synthetic Detergents

Liquid and solid synthetic detergents, sometimes called 'soapless soaps', are manufactured from petrochemicals and *not* from animal and vegetable fats and oils. They are an improvement on, rather than a substitute for, ordinary soap, producing an abundant lather. In addition they are not alkaline like many soaps and have excellent cleansing properties.

Soap in solution has an anionic *carboxylate* group attached to a hydrocarbon chain (R):

$$
R-C
\begin{array}{l}
\diagup \! \! \diagup O \\
\diagdown O^-
\end{array}
$$

One type of soapless detergent is also anionic with a long hydrocarbon chain attached to a negative *sulphonate* group. The sulphonate group is introduced by treating an alkylbenzene with a sulphonating agent such as sulphuric acid. Neutralization then produces the sodium salt. A typical alkylbenzene might be

(alkylbenzene)

The sulphonated alkylbenzene could thus be

Once again there is a long hydrocarbon chain with an anionic group at the end.

An important difference between soapless detergents and soap is their reaction with hard water containing calcium and magnesium ions. Soap produces an insoluble scum of calcium or magnesium stearate, whereas soapless detergents form no scum because the calcium and magnesium alkylbenzene sulphonates are soluble.

In the manufacture of soapless detergents, oleum (fuming sulphuric acid) reacts with the alkylbenzene to produce the sulphonated alkylbenzene and sulphuric acid. The addition of water dilutes the sulphuric acid, which separates and is removed. The remaining sulphonated alkylbenzene is neutralized with sodium hydroxide solution to form sodium alkylbenzenesulphonate—the synthetic detergent.

Before packaging, bleaches, fluorescent materials, perfumes, phosphates and other chemicals are added to improve the properties of the detergent.

Summary of Unit 14

1. **Heavy chemicals** are those manufactured on a large scale.
2. **Sodium chloride** occurs extensively and is used for the manufacture of the heavy chemicals *sodium, chlorine, sodium hydroxide* and *sodium carbonate*.
3. **Sodium** is manufactured in the **Downs cell** by the electrolysis of *molten* sodium chloride.
4. **Sodium hydroxide** is manufactured by the electrolysis of an *aqueous solution* of sodium chloride in a *mercury cell*.
5. **Chlorine** is obtained from both the Downs cell and mercury cell. Demand for chlorine exceeds that for either sodium or sodium hydroxide.
6. **Hydrochloric acid** is produced either by the direct combination of hydrogen and chlorine, or as a by-product in the chlorination of organic compounds.

7. **Sodium carbonate** is manufactured in the **Solvay process** from sodium chloride, calcium carbonate (limestone) and coal.
8. Limestone is the source of calcium oxide (quicklime), calcium hydroxide (slaked lime) and cement.
9. **Sulphuric acid** is manufactured in the **contact process** by the catalytic oxidation of sulphur dioxide. Oxygen is obtained from the air, and sulphur dioxide from sulphur.
10. **Ammonia** is synthesized from nitrogen and hydrogen in the **Haber process**.
11. **Nitric acid** is produced from ammonia by catalytic oxidation.
12. Ammonia, nitric acid and sulphuric acid are used in the manufacture of **fertilizers**.
13. The blast furnace produces **iron** from iron ore. Most of the iron produced is used to make steel.
14. The electrolysis of purified *bauxite* dissolved in molten *cryolite* produces **aluminium**.
15. A **detergent** is a cleansing agent and includes both soap and synthetic detergents.
16. **Soap** is manufactured by the *saponification* of animal and vegetable fats and oils, whereas *synthetic* detergents are manufactured from petrochemicals.

Test Yourself on Unit 14

1. Sodium is manufactured by the electrolysis of molten sodium chloride containing calcium chloride in the Downs cell.
   (*a*) Does this process use alternating or direct current?
   (*b*) Why is calcium chloride added?
   (*c*) Name the product at (i) the cathode and (ii) the anode. Write equations showing their discharge.

2. Ammonia, a heavy chemical, is manufactured by the combination of nitrogen and hydrogen in the presence of a finely divided iron catalyst at a temperature of 500 °C and 150–300 atmospheres pressure.
   (*a*) What is meant by the term 'heavy chemical'?
   (*b*) What is the name given to this method of producing ammonia?
   (*c*) Name the sources of nitrogen and hydrogen.
   (*d*) What is the purpose of the catalyst?
   (*e*) Why is the catalyst finely divided?
   (*f*) Write a balanced equation for the reaction, including state symbols.
   (*g*) In this process the conversion of nitrogen and hydrogen to ammonia is incomplete. How is the ammonia separated from the unchanged nitrogen and hydrogen?

3. Complete the following sentences on the smelting of iron ore.
   (*a*) The charge in the blast furnace is usually

    (i) iron ore and coke
    (ii) coke, iron ore and air
    (iii) limestone, coke and iron ore
    (iv) limestone, sand and iron ore
(b) Slag is
    (i) molten sand
    (ii) molten iron
    (iii) iron pyrites
    (iv) molten calcium silicate
(c) The main reducing agent in the blast furnace is
    (i) iron ore
    (ii) carbon monoxide
    (iii) hydrogen
    (iv) limestone

(d) The impure iron that comes from the blast furnace contains about
    (i) 0·2% carbon
    (ii) 4% carbon
    (iii) 10% carbon
    (iv) 20% carbon
(e) Steel contains
    (i) 0·1 2% carbon
    (ii) 5–10% carbon
    (iii) no carbon
    (iv) more than 10% carbon

4. In the contact process for the manufacture of sulphuric acid:
    (a) What are the raw materials used?
    (b) Name the product formed when these raw materials are burned together.
    (c) Write an equation for the reaction taking place in the catalyst chamber.
    (d) Name a suitable catalyst for the reaction.
    (e) How is the gas formed in the catalyst chamber converted into sulphuric acid?
    (f) State one important use of sulphuric acid.

5. Mark the following statements true or false:
    (a) The only important by-product in the Solvay process is sodium hydrogencarbonate.
    (b) Nitric acid is manufactured by the oxidation of ammonia.
    (c) Nitrogenous fertilizers include ammonium nitrate, urea, ammonium sulphate and ammonium phosphate.
    (d) Oils and fats are converted to soap in a process called esterification.
    (e) Detergents include both soap and synthetic detergents.

Mark this test out of 30 with the answers provided on page 383.

Unit Fifteen

Matter, Society and the Environment

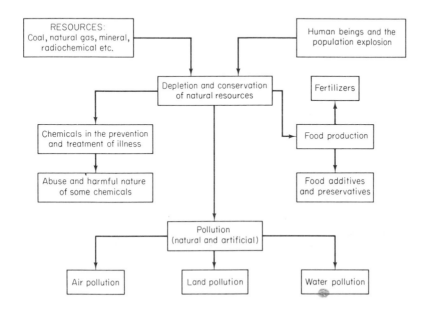

In recent years bad publicity accorded to the chemical industry by the media has generated concern in the minds of both the public and some scientists. The hazards associated with large chemical plants have been forcibly brought to people's attention through, for example, the explosion at Flixborough and by the escape of poisonous fumes at both Seveso and Bhopal, while the nuclear leak at Three Mile Island and more particularly the accident at Chernobyl have raised anxiety concerning the safety of nuclear plants.

The great importance of the chemical industry cannot be overstated, however. For example, the chlor-alkali industry manufactures chlorine, sodium hydroxide and sodium carbonate (see Unit 14) used for making soap, detergents, solvents, glass and ceramics. The organic chemical industry, based first on coal tar and then on petroleum, manufactures a vast range of products including plastics, fibres, solvents, explosives, pharmaceuticals, pesticides, dyestuffs and paints.

The economic importance of the technological revolution in manufacturing industry has to be balanced against the inevitable hazards accompanying the manufacturing processes. Because there is such potential danger in many of these processes, responsible chemical industry takes stringent precautions to avoid disaster, and bodies such as the United Kingdom's Health and Safety Executive and the regional water authorities administer appropriate

Government legislation. Such precautions cost money, however, and a balance has to be struck between cost and safety.

The chemical industry uses vast quantities of raw materials. Recently fears have arisen concerning the possiblity of the exhaustion of natural resources such as coal, petroleum and other minerals. A consequence of the increasing consumption of these resources is the steadily increasing fear of the pollution of the environment caused by industrial production, intensive farming and transport.

This unit aims to identify some of the natural resources available, and looks at their use and abuse in our increasingly technological society.

15.1 The Population Explosion

It took hundreds of thousands of years for the world population to reach 300 million, which it did about two thousand years ago. Only sixteen centuries were then needed for it to double to about 600 million. From then onwards the time taken for the world population to double has decreased rapidly. Fig. 15.1 predicts the population in the year 2000 to be approximately 7000 million.

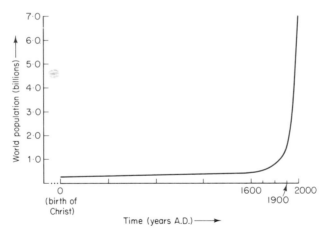

Fig. 15.1   The population explosion

More people require more food, fuel and energy, together with more of the basic natural resources, and the fear is that all these will be consumed at a faster rate than they can be replenished.

15.2 Energy Resources

(a) Fossil Fuels
Our society today depends on the availability of reliable sources of energy. World energy reserves are assessed in terms of a unit Q, where 1 Q is equi-

valent to the energy content of 25 000 million tonnes of oil—a staggeringly large quantity, which is greater than the energy content of the oil transported by 100 000 large oil tankers.

Crude oil, like coal and natural gas, is a fossil fuel, largely made up of the fossilized remains of minute creatures. The refining of crude oil (petroleum) is described in Section 11.5. The fossil fuels are by far the most important of our energy reserves at the present time. It has been estimated that the total reserves of these fuels amount to about 260 Q.

The annual world consumption of energy is less that 0·5 Q. So how long will the fossil fuels last? This is not an easy question to answer as there are so many variables to be considered. The efficiency for recovering most oil reserves is only of the order of 50%; the world energy demand may increase markedly with industrial development; population growth will increase the demand for energy. The projected annual energy demand related to population growth is illustrated in Fig. 15.2. The graph shows how long reserves

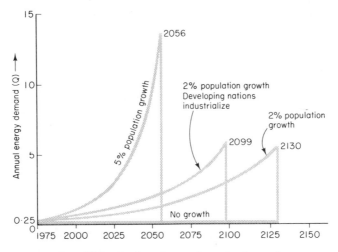

Fig. 15.2   Projected annual energy demand

would last if fossil fuels were the only means of supplying the energy demand (it assumes 100% recovery of oil reserves). Fossil fuels are thus a *finite energy source*—that is, they cannot be replaced once they have been used. Alternative energy sources are therefore being sought.

(b) Alternative Energy Sources

There are essentially only two ways of meeting our future energy demand as the availability of fossil fuels diminishes. One is the exploitation of natural resources such as solar energy, tidal energy and geothermal energy, energy from sea waves and the use of wind-powered generators, together with more traditional hydroelectric power schemes (we can deal with only some of these in this book). The other is the use of nuclear fuels.

(i) *Energy from the sun*

Section 13.6 describes two types of nuclear reaction: nuclear fission and nuclear fusion. The origin of the sun's energy is believed to be a fusion reaction in which two hydrogen nuclei fuse to form a helium nucleus with a loss of mass and corresponding liberation of an immense quantity of energy. The sun converts over 500 million tonnes of hydrogen into helium every second, resulting in the radiation of 9 billion Q of energy in a year. Of this the earth receives only some 3000 Q per year. Some is absorbed by the earth's atmosphere, and some warms the surface of the earth and is returned to the atmosphere by radiation or by evaporation of water. Only 1·2 Q is harnessed through photosynthesis for storage into chemical energy (see Sections 6.10 and 12.8).

Currently much effort is being directed towards the production of electricity from the sun's energy using photoelectric solar cells; the most promising direct use of solar energy is in the use of solar panels for heating water for domestic and industrial use. Modern buildings are now being designed to use direct solar heating more efficiently.

(ii) *Tidal energy*

Tides are caused by the gravitational attraction of the moon and to a lesser extent the sun. The rise and fall of the water between high and low tides is a source of energy, and the conversion of this tidal energy into a usable form is currently receiving considerable attention. Although a tidal power station is now operating in France, the problems and expense of building such power stations at present far outweigh the gain which such effort produces.

(iii) *Geothermal energy*

The centre of the earth consists of molten rock. This vast heat energy source beneath the earth's crust is as yet largely unused. Some of this heat energy is transferred to the surface averaging 0·05 W m^{-2} over the whole of the earth's surface, a figure which remains remarkably constant and is largely lost to the atmosphere.

Where the molten rock comes close to the surface there is the possibility of it breaking out through the crust as in volcanoes. When underground water comes close to the hot rock it can be expelled as steam and hot water in geysers and hot springs, and in certain places this steam has been harnessed to drive electric generators and provide local heating. At the time of writing, studies are also being carried out on the practical possibility of pumping water down to considerable depths to be heated by hot dry rocks and returned to the surface for use. The quantity of useful energy derived from this source is at present minute, however. Such useful energy is called *hydrothermal energy* (meaning 'energy from hot water').

Potentially this is a vast source of energy as yet unexploited which has none of the hazards associated with radioactivity. In the future, geothermal energy could become more important than energy derived from fossil fuels.

(iv) *Nuclear energy*

Almost 20% of the United Kingdom's electricity is generated by nuclear power. Nevertheless, the nuclear debate continues.

A nuclear power station differs from a coal- or oil-fired station principally in the source of heat. In a nuclear station the heat comes from the energy released in a fission reaction involving usually uranium or plutonium.

Two basic types of reactor use nuclear fission: thermal reactors and fast reactors. In a *thermal reactor*, atoms of uranium or plutonium are split by neutrons, releasing energy as heat and more neutrons to sustain a controlled chain reaction. Natural uranium consists mainly of uranium-238 with only 0·6% of uranium-235, but it is this small amount of uranium-235 that is easily split by slow neutrons. The probability of the fission of ^{235}U by slow neutrons is in fact much greater than the fission of ^{238}U by fast neutrons. Thermal reactors use a *moderator* to slow down the neutrons so they are able to react with the uranium fuel (often enriched with ^{235}U). Magnox reactors and advanced gas-cooled thermal reactors (AGR) use graphite as a moderator, and in both the heat is transferred from the reactors to the generators by gas. Pressurized water thermal reactors (PWR) use water as both moderator and coolant. An AGR and a PWR are illustrated in Figs. 15.3 and 15.4 respectively.

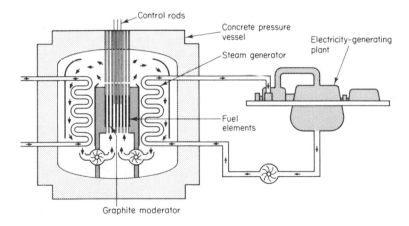

Fig. 15.3   Diagram of an advanced gas-cooled reactor (AGR). Fuel:
^{235}U-*enriched uranium dioxide. Moderator: graphite. Heat extraction: carbon dioxide gas transfers heat to water in a steam generator.*

There is a worldwide expectation that *fast reactors*, using plutonium fuel produced in the thermal reactors, will be a major source of electricity production in the future. The plutonium requires no moderator, being readily split by fast neutrons. Liquid sodium is often used as a coolant. (Fig. 15.5).

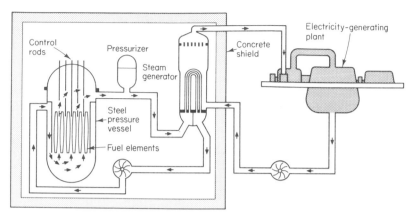

Fig. 15.4   Diagram of a pressurized water reactor (PWR). Fuel:
^{235}U*-enriched uranium dioxide. Moderator: water. Heat extraction: water in*
reactor transfers heat to water in a separate circuit.

Reserves of uranium are finite, but 'spent' fuel from a thermal nuclear
reactor contains a high proportion of unused uranium together with plutonium produced in the reactors. Spent fuel is therefore reprocessed to recover
uranium and plutonium, leaving only comparatively small amounts of nuclear
waste. This waste contains fission products such as radioisotopes $^{90}_{38}Sr$, $^{135}_{55}Cs$,
$^{137}_{55}Cs$ and $^{129}_{53}I$. These have different half-lives and decay to stable isotopes with
the emission of dangerous beta and gamma radiation. These damage living cells,

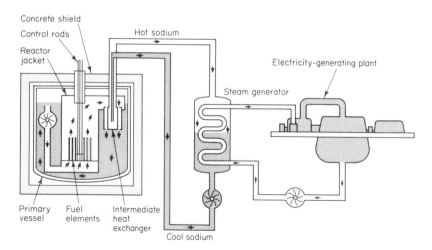

Fig. 15.5   Diagram of a sodium-cooled fast reactor. Fuel: plutonium and
uranium dioxide. Moderator: none. Heat extraction: molten sodium transfers
heat to water in a steam generator.

and in large doses this can be fatal to the organism; even in smaller doses they can cause cancer and genetic mutations. The toxicity of these radioactive materials is enhanced by the body concentrating specific isotopes; for instance, $^{90}_{38}Sr$ collects in bones because it resembles calcium in its chemistry (bone tissue is made up largely of calcium salts). Human beings and other species towards the end of a food chain are particularly vulnerable to the dangers of such radioisotope concentration. The safe disposal of nuclear waste therefore poses one of the major problems in a nuclear power station.

The electricity-generating industry is investigating new methods for the disposal of fission products produced in power stations. One method that has been studied in detail is the disposal of liquid waste in glass blocks that may be either stored or buried.

The possible environmental consequences of an operational nuclear accident became starkly evident after the 1986 Chernobyl disaster in the USSR. The RMBK water-cooled graphite-moderated reactor (see Fig. 15.6) exploded and caught

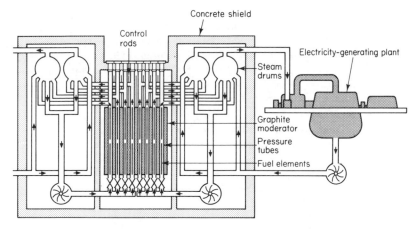

Fig. 15.6   Diagram of the Chernobyl RMBK reactor. Fuel: uranium dioxide. Moderator: graphite. Heat extraction: water.

fire, the graphite burned and part of the core melted. The radioactive plume that passed over Europe contained volatile fission products, particularly iodine and caesium, and there was widespread contamination of land, water and livestock. Incidents such as this, together with the accident at the Three Mile Island power station in the USA and the 1957 Sellafield (Windscale) leakage, have led to an increasingly active lobby in many countries against nuclear power.

15.3 Mineral Resources

With the growth of world industrial production there has been a corresponding increase in the consumption of mineral resources. For example, since 1900 steel output has risen twenty-five times and aluminium production almost two thousand times. This ever-increasing use of our mineral resources has recently given rise to concern about their possible exhaustion.

Consider, for instance, the extraction of aluminium from bauxite (see Section 14.12), iron from haematite (see Section 14.11) and chemicals from rock salt (see Unit 14). In any such process, account has to be taken of the following factors:

(a) the location and quantity of the mineral depositis,

(b) the economics of obtaining the mineral from its ore,

(c) the rate of consumption of the mineral,

(d) the possibility of recycling (particularly in the case of metals),

(e) the environmental impact caused by the holes and scars resulting from mining and quarrying,

(f) the enormous requirement of many mining and extraction processes for water, and the consequent pollution problems, and

(g) the use of large amounts of energy.

One of our major sources of minerals is the sea. Sea water is an aqueous solution of dissolved salts, gases and organic molecules, containing also suspended solids of both organic and inorganic origin. The major constituents include eleven ionic species (see Table 15.1) which together make up over 99·9% of the dissolved solids in sea water. Of these, bromine and magnesium are of particular importance for extraction.

Table 15.1 Ionic constituents of sea water

Constituent	Concentration (g/kg sea water)
Chloride, Cl^-	19·35
Sodium, Na^+	10·76
Sulphate, SO_4^{2-}	2·71
Magnesium, Mg^{2+}	1·29
Calcium, Ca^{2+}	0·41
Potassium, K^+	0·39
Hydrogencarbonate, HCO_3^-	0·14
Bromide, Br^-	0·067
Strontium, Sr^{2+}	0·008
Boron, as H_3BO_3	0·004
Fluoride, F^-	0·001

Many other elements, including iron, manganese, copper and cobalt, are present in very small concentrations, but these represent significant amounts when considering the vast quantities of water in the seas. In the future, extraction of some of these elements may perhaps become economically viable as other sources become depleted.

15.4 Pollution

A definition of *pollution* is not easy, because it depends to a large extent upon those environmental qualities which a person considers to be of value. We can say, however, that the environment has been polluted when a substance, or a form of energy (such as heat or radiation), or a disturbance such as sound is added to the environment at a rate faster than the environment can accommodate it, so that it adversely affects not just human beings but life in general.

Pollution is therefore not a new phenomenon. Nor is it necessarily caused by human beings. Even nowadays the pollution produced by a large volcanic eruption (such as that of Mount St Helens in the USA) can dwarf the human contribution. Pollution by people has only been significant since their numbers increased to such an extent that the environment became unable to assimilate and process all of the waste products that result.

There is a limit to the earth's size, energy reserves and non-renewable mineral resources. The more people there are, the greater is the demand for space, energy and mineral resources, while the greater our use of these, the greater become the environmental threats to humanity's survival. Some of the more significant environmental threats are discussed below.

(a) Air Pollution

To maintain health an individual needs each day about 2 kg of water and around 1·3 kg of food, but approaching 14 kg of air. He or she can survive without water for five days and without food for five weeks, but in the absence of air death ensues after about five minutes. And yet the air we are breathing is becoming increasingly polluted.

Most of the air pollution that arises from human activities results from the products of combustion of fuels, and five types of substance account for more than 90% of our air pollution problem. These are:

(a) carbon monoxide (CO)

(b) oxides of nitrogen (NO, NO_2, N_2O), collectively referred to as NO_x

(c) hydrocarbons (methane, ethane, ethene, etc.)

(d) sulphur oxides (SO_2, SO_3)

(e) particulates (dust, pollen, metals, etc.)

What are the major sources of these air pollutants?

(i) *Carbon monoxide*

Carbon monoxide is an odourless, colourless gas which is highly toxic (see Section 10.33): 1000 parts per million (p.p.m.) of carbon monoxide will kill a person quickly, while 100 p.p.m. causes symptoms such as headaches and upset stomach. The effects of low doses of carbon monoxide do not build up with time but, even so, repeated exposure to even these small amounts can be dangerous, particularly to someone suffering from anaemia. Exhaust emission from motor vehicles accounts for most of the carbon monoxide pollution generated. Daily concentrations of it correlate well with traffic volume because much of it is produced by incomplete combustion of the carbon in fossil fuels:

$$2\underset{\text{(in fuel)}}{C} + O_{2(g)} \rightarrow 2CO_{(g)}$$

It has been estimated that the three million vehicles in Los Angeles alone produce about 8000 tonnes of carbon monoxide in a day – about 2 kilograms of carbon monoxide per vehicle per day. High vehicle concentrations result in high carbon monoxide levels which can become dangerous in cities on still days and in tunnels and garages without efficient ventilation.

Most attempts to control carbon monoxide pollution are directed towards its elimination from vehicle exhaust emission. This is complicated by the presence of other pollutants (hydrocarbons, nitrogen oxides and particulates) that must be dealt with at the same time. Possible solutions would be:

(1) modification of the internal combustion engine to reduce the amounts of pollutants formed during combustion of fuel;

(2) development of new exhaust systems that will complete the combustion process and change potential pollutants into less harmful materials;

(3) development of new fuels and engines.

If all fossil fuels were burnt more efficiently in a complete combustion reaction then no carbon monoxide would form. All the carbon in the fuel would be converted to the less hazardous carbon dioxide:

$$\underset{\text{(in fuel)}}{C} + O_{2(g)} \rightarrow CO_{2(g)}$$

Many scientists believe, however, that even an increase in the carbon dioxide content of the air could prove a hazard. They argue that such an increase could trap more of the sun's energy and result in a significant temperature rise over the earth as a whole. This is termed 'the greenhouse effect'. It could have marked effects on climates, resulting in an increase in the desert areas and melting of the polar ice caps.

(ii) *Oxides of nitrogen*

When coal, oil and petrol are burned at high temperatures, appreciable amounts of the oxides of nitrogen are formed. A Russian study has indicated that subjection to concentrations as low as 2·8 p.p.m. of oxides of nitrogen over a three-to five-year period could result in permanent lung damage.

In Los Angeles there is an alert level for oxides of nitrogen of 3·0 p.p.m. It is estimated that the daily output of nitrogen oxides in Los Angeles is 750 tonnes.The problem has assumed major proportions because of this city's geographical position. The city is in a valley. At certain times cool air in the valley becomes trapped by a layer of warmer air above. Under these conditions the sun acting on the nitrogen dioxide and oxygen can produce nitric oxide and ozone. These can react with hydrocarbon pollutants to produce secondary pollutants such as peroxyacyl nitrates.

This *photochemical pollution* results in the yellow-brown smogs characteristic of such large cities; Tokyo and Chicago suffer similarly. This form of pollution results in plant damage, the deterioration of materials and decreased visibility, together with severe eye irritation and many other health problems.

As for carbon monoxide control, strenuous efforts are being made to develop internal combustion engines that have considerably reduced hazardous emissions. But whereas the removal of carbon monoxide requires oxidation to carbon dioxide, the removal of oxides of nitrogen requires reduction to nitrogen and oxygen. A dual catalyst system is thus required.

(iii) *Hydrocarbons*
Methane (CH_4) is emitted into the atmosphere in quantities larger than any other hydrocarbon as a result of bacterial decomposition. Human activities contribute only about 15% of the hydrocarbon emissions, most of which come from petroleum. Most of the harmful effects of hydrocarbon pollution are caused not by the hydrocarbons themselves but by products formed when the hydrocarbons undergo chemical reactions in the atmosphere (such as peroxyacyl nitrate, mentioned in (ii) above).

(iv) *Sulphur oxides*
Sulphur dioxide is the only sulphur oxide contributing significantly to air pollution. It results from the combustion of fossil fuels. Of these the burning of coal produces the most sulphur dioxide, because sulphur is found in all coals in amounts ranging between 0·2–7·0%. The sulphur was present in the proteins of the once-living material from which coal was formed. The protein sulphur survived the carbonization process and became part of the resulting coal. Petroleum and natural gas originated in similar ways and therefore also contain sulphur. Natural-gas processing and petroleum refining remove most of this sulphur, however, so these fuels are not as serious sources of sulphur dioxide pollution as is coal.

Burning sulphur containing fuels produces sulphur dioxide:

$$S + O_2 \rightarrow SO_2$$

Some of this dissolves in water to produce sulphurous acid:

$$SO_2 + H_2O \rightarrow H_2SO_3$$

Some of the sulphur dioxide is oxidized, either to sulphur trioxide:

$$2SO_2 + O_2 \rightarrow 2SO_3$$

or to sulphuric acid:

$$2SO_2 + 2H_2O + O_2 \rightarrow 2H_2SO_4$$

The acid mist so produced is responsible for much deterioration of the fabric of buildings and many other materials, particularly metals. The much-publicized *acid rain* affecting the foliage and retarding the growth of plants is believed to be due to this acid mist.

The obvious solutions to this problem are:

(1) to reduce the sulphur content of coal before burning it, and

(2) to remove the oxide gases after the fuel has burnt.

(v) *Particulates*

Natural phenomena such as windborne dust from dry land, volcanic action, forest fires and salt from sea spray account for most of the particulate matter in the atmosphere. Thus natural sources completely dominate anthropogenic ('man-made') sources, which include losses from industrial processes and emissions from the combustion of coal and other fuels.

The ash from these emissions contains carbon and silicon(IV) oxide, as well as trace elements. Of these, *lead* in the atmosphere in both particulate and gaseous forms has given cause for concern and much publicity, especially with regard to possible mental retardation of children. The first visible symptom of lead poisoning is often the onset of anaemia, but brain damage, convulsions, behavioural disorders and eventual death are the results of prolonged exposure to lead and its compounds.

Combustion of petrol accounts for by far the greater part of all atmospheric lead pollution. The sources of lead in the exhaust products of petrol combustion are the 'anti-knock' additives, tetraethyl-lead, $Pb(C_2H_5)_4$, and tetramethyl-lead, $Pb(CH_3)_4$. In addition, 1,2-dichloroethane, CH_2ClCH_2Cl, and 1,2-dibromoethane, CH_2BrCH_2Br, are added to petrol. These compounds react with the lead to form volatile lead compounds which are released into the atmosphere with other exhaust gases. Such removal of the lead is necessary for smooth running of the engine. Vigorous efforts are at present being made to develop petrol engines which do not require such lead additives.

At one time *smoke* was a serious factor in air particulate pollution, but in the United Kingdom the 'Clean Air Acts' have almost eliminated smoke as a significant pollutant. It is interesting to note that the first Alkali Act was the result of the very unpleasant pollution caused by the manufacture of soda ash (sodium carbonate) in the Cheshire and Lancashire towns in the nineteenth century.

(b) **Water and Land Pollution**

(i) *What is water pollution?*

'Pure water' as supplied to our homes by the water authorities contains dissolved carbon dioxide, oxygen and nitrogen absorbed when rain falls through the air. In addition this water running over and through the land dissolves salts of metals such as sodium, magnesium and calcium, the latter two metals being responsible for 'hardness' of water (see Unit 6.16.) Such drinking water is thus not pure in a chemical sense. After treatment at a water purification plant to remove bacteria and suspended solids it still contains these dissolved substances which give water its characteristic 'taste'.

Polluted water is that which cannot be used for the public water supply, recreational use, agricultural purposes or industry, and which will not support wildlife, fish and so on.

(ii) *Sources of water pollution*

Rivers and streams receive effluents from many sources, including:

(1) run-off from farm lands which have been treated with pesticides and fertilizers,

(2) raw and treated sewage, which add plant nutrients including nitrogen and phosphorus together with detergents etc.,

(3) industry, which adds inorganic and organic chemicals, sediments, radioactive material and *warm* water,

(4) tankers, offshore rigs and refineries, which together with accidental spills add oil and oil residues.

(iii) *Effects of water pollution*

Plant and animal life in water depend upon the presence of dissolved oxygen. Their survival depends on the ability of the water to maintain a certain minimum amount of this gas in solution. Pollution, particularly in lakes, results from oxygen depletion where the ecological balance is disturbed. This happens when bacteria digest organic waste matter (such as untreated sewage), nitrogenous chemicals (particularly nitrates from fertilizers) and phosphates from detergents in a process which uses up dissolved oxygen:

$$\text{bacteria} + \frac{\text{waste}}{\text{matter}} + \frac{\text{dissolved}}{\text{oxygen}} \rightarrow \frac{\text{carbon}}{\text{dioxide}} + \frac{\text{more}}{\text{bacteria}}.$$

The carbon dioxide stimulates the growth of algae, water weeds and plants in aquatic daytime photosynthesis:

$$\text{algae} + \text{carbon dioxide} + \text{sun's energy} \xrightarrow{\text{photosynthesis}} \text{more algae} + \text{oxygen}$$

This rapid production of plant life is known as *eutrophication*. It results in dissolved oxygen being consumed more rapidly than it can be replaced. As the oxygen content falls, fish and other forms of aquatic life die. When the oxygen disappears completely the water becomes grossly polluted and foul-smelling as anaerobic fermentation becomes the predominant process.

The population explosion linked with its dependence upon an oil-based technology has, as an inevitable consequence, led to *oil pollution* problems. Much attention has been focussed on the pollution of beaches and seas (whether intentional or accidental) with oil. The *Torrey Canyon* disaster in 1967 was one of the first major accidents involving a large super-tanker. This ship ran aground off the coast of Cornwall and discharged over 100 000 tonnes of crude oil, causing severe pollution of the sea and nearby beaches. Further spills from oil tankers and offshore oil rigs have followed, causing increasing public awareness and concern.

Apart from these accidental discharges of oil, many tankers take in sea water to provide ballast for 'empty' tanks on their return voyage to the oil-fields. Subsequently they flush these out into the sea together with oil residues before docking. This practice is illegal in most countries and stringent measures are being taken in an attempt to prevent it.

Following an oil spillage, the oil has first to be contained as much as possible by using different types of floating barriers, and it is then either skimmed off for subsequent reclamation or dispersed using detergents. Care needs to be taken to ensure that the detergents used are not also harmful to plant and animal life. All oil discharges contain volatile components, and within a few days almost one-quarter of the spilt oil is lost through evaporation. Of the remainder most forms an emulsion with sea water, and a relatively small amount finds its way on to beaches in black tarry lumps.

The long-term effects of oil spillages are currently under investigation and have yet to be fully appreciated. Short-term effects include those caused by coating and those due to the toxicity of the oil. The latter are immediately obvious when sea birds become coated with oil, reducing their ability to fly, their buoyancy and the thermal insulation of their feathers. In addition, oil spillages hinder photosynthesis and have an adverse effect on plant and animal life along shore lines.

Strict regulations apply to the discharge of *industrial effluents*—waste containing organic and inorganic material from industry. The toxic properties of the heavy metals, such as lead and mercury, is well known. Most of these metallic pollutants eventually end up in surface waters. Unlike organic pollutants, metals cannot be degraded biologically or chemically in nature. They do collect and become concentrated in food chains, however, so that the higher members of the food chain can contain concentrations of these metals very much higher than those found in water, with consequent damage to health and fertility that sometimes leads to death.

In particular the pollution due to mercury has received much publicity. It is used extensively in the production of electrical apparatus, in the chlor-alkali industry (see Section 14.3) and in fungicides. There is no known effective treatment for mercury poisoning and the damage done to the body is usually permanent. In 1953, fishermen and their families in Minamata Bay in Japan were stricken with a neurological disease. More than forty people died and many became paralysed. Eventually the cause was traced to methylmercury(II) $[(CH_3)_2Hg]$ discharged into the sea by a plastics factory; the compound became concentrated in fish and shellfish which the people ate.

Domestic effluents contain sewage and detergents. The harmful effect of untreated sewage is discussed above. The first noted pollution effect of detergents was the foam produced in streams and rivers by the surfactant component. This was solved by making the surfactant *biodegradable* (that is, capable of being broken down by the micro-organisms in the water). Today the area of most concern is the presence in the water of phosphate ions from domestic washing powders, because of the relationship of phosphates with eutrophication.

15.5 Food Supply

In Section 15.1 we pointed out that currently the world's population is increasing at an unprecedented rate. More people need more food. Increased food requirements put pressure on food producers to improve their efficiency. Efficiency has undoubtedly improved but, although food production has increased enormously, this increase is almost nullified by the increase in population. In addition the increase in food production has been much more marked in the rich developed countries than in the developing countries, and this discrepancy has aggravated the growing gap between rich and poor.

(a) Growing Food

Food, like oxygen, is necessary for life. All living creatures require food as a source of energy and for the growth and replacement of tissues. Carbohydrates (see Section 12.8) and fats (see Section 14.14) are the components of food that provide energy, while proteins (see Section 12.10) are used for growth and replacement of tissue. In addition vitamins and minerals, which regulate body processes, and water are essential components of food.

Plants are the ultimate source of our food supply, being the starting point in most food chains. Thus efficient plant growth is essential for food production. To grow efficiently, plants need the sun's energy, water and carbon dioxide for photosynthesis (see Section 6.10). In addition efficient growth requires certain major nutrients such as nitrogen for the synthesis of amino-acids and subsequently plant protein, together with minor nutrients. If such nutrients are not present in sufficient quantity in the soil, the crops produced will be of low yield and quality. Nutrients must therefore be added to the soil, either in the form of organic material such as manure or, more commonly, as chemical fertilizer (see Sections 6.7 and 14.10). The required amount of each nutrient (particularly nitrogen, phosphorus and potassium) can be calculated so as to give maximum growth. The addition of too much fertilizer, especially during periods of heavy rainfall, can lead to leaching out of these nutrients into water supplies, with the consequent pollution problems outlined in Section 15.4(b)(iii).

(b) Crop Protection

Natural pests, including insects, rodents, weeds and fungi, compete with us for our means of survival, including our food supply. Chemicals used for destroying them are called *pesticides*. DDT, for example, which belongs to a group of compounds called chlorinated hydrocarbons, was used extensively and effectively in the early 1950s for controlling insect pests. It became apparent, however, that it was causing serious damage to birds and fish and also that some insects were building up an immunity to it. Its use therefore declined, and in many countries is now prohibited altogether. More recently, pesticides such as the pyrethrins have been developed from natural sources. These are non-toxic to bees and are degraded harmlessly in warm-blooded animals, though they are toxic to fish. Hence they are used particularly as insecticidal sprays in households, food storage warehouses and factories and on domestic and farm animals.

The problems associated with the use of chemical pesticides have created research interests in alternative pest control methods. These include:

(i) the use of *biological controls* where natural predators such as parasitic wasps are used to destroy insect pests;

(ii) *genetic control*, such as the release of sexually sterile male insects to mate with female flies, which then lay infertile eggs;

(iii) the use of *attractants* to lure insects into traps where they can be killed—a degree of control of the tsetse fly has been achieved in this way.

(c) Food Preservation

Producing food can benefit no one if the food spoils before it can be eaten. There are two major causes of food deterioration:

(i) the effects of enzyme-catalysed chemical reactions, such as the browning of peeled potatoes and apples left exposed to the air, and

(ii) the action of micro-organisms such as *Salmonella* and *Clostridium botulinum*.

Enzymes can be decomposed by boiling the food. Such 'blanched' food can then be preserved for long periods by freezing. *Micro-organisms*, which may infect the eater with disease or which may produce poisonous substances (*toxins*), feed on foodstuffs and flourish particularly in damp, warm (body temperature), non-acidic conditions. Methods of food preservation must take these conditions into account. Some are listed below.

(i) *Heating*, as used in bottling and canning, not only decomposes enzymes but also destroys micro-organisms responsible for food deterioration.

(ii) *Refrigeration* (usually below $-15°C$) slows down enzyme reaction and inhibits micro-organism growth.

(iii) *Dehydration* (drying) of, for example, fruit removes the moisture necessary for micro-organism growth. Dried fruit is in addition often treated with sulphur dioxide gas to form an acidic environment which further inhibits bacterial action.

(iv) *Preservatives* such as salt for meat and fish, sulphur dioxide for fruit juices, and ethanoic acid (acetic acid) for preserving (pickling) vegetables such as onions, all prevent micro-organism growth.

(v) *Irradiation* of food to kill micro-organisms is likely to play a more important role in food preservation in the future.

15.6 Chemicals and the Human Body

The use of chemicals to heal the body and the misuse of chemicals to intoxicate are both far older than any written record of civilization. The effects of some chemicals can be either beneficial or disastrous, depending on the circumstances in which they are administered: heroin, for instance, has eased the pain of many terminally ill patients, and through its addictive nature has also tragically damaged many others who turned to it for excitement or support even though they were in good health.

(a) Chemicals in the Prevention and Treatment of Illness

Few of the medicines used today were available before World War II, and indeed many were not introduced until after 1960. Even as late as the 1930s, reliance was often placed on natural products such as morphine and quinine. The war of 1939–45 stimulated research into the production of pharmaceutical chemicals because of the need to treat wounded soldiers. Additionally much research into tropical medicines took place to help combat infectious diseases which were prevalent in those parts of Africa, the Pacific islands and Asia where the fighting forces needed protection.

The 1960s was a period of rapid growth in the production of medicines to combat heart disease, gastric ulcers, mental disorders, arthritis, cancer and allergies. Considerable success has been achieved in the chemical treatment of many of these disorders. This has led to a higher proportion of people surviving into old age and a change in the whole pattern of disease and mortality; diseases associated with ageing now dominate.

The pharmaceutical industry is compelled to be dynamic because new infections such as Legionnaires' disease and AIDS have appeared in recent years and a vigorous research programme needs to be maintained to combat these. Because of its involvement with the health of us all, the industry attracts much more debate and controversy (consider, for instance, the thalidomide disaster) than other high-technology industries.

(b) Chemicals – their Abuse in Society

All medicines are drugs, but not all drugs are medicines. Coffee, tea, tobacco and alcohol are drugs but would not be classified as medicines. A *drug* is defined as any substance which alters the structure or function of a living organism. A *medicine* is a drug which possesses useful properties in the prevention or relief of physical or mental symptoms.

Alcohol (ethanol) is one of the most widely used of all drugs. Its social acceptance should not mislead us into minimizing its potential dangers. It produces drunkenness, disorientation, blurred vision and often vomiting. Intoxication with alcohol is often followed by a 'hangover' which makes the sufferer feel sick, weak and dizzy. Even small amounts of alcohol impair driving ability.

Inhalation of volatile substances such as glues, petroleum products and ether has been known for many years and causes stimulation at first followed by slurred speech, headache, nausea and vomiting. Finally the user becomes unconscious. The misuse of other drugs such as LSD (D-lysergic acid diethylamide or 'acid'), cocaine, morphine, opiates, cannabis and tobacco is currently a cause for considerable concern in our society. Neither the reasons for drug misuse, nor its remedy, are yet fully understood.

15.7 Conclusion

Today the population explosion is producing problems associated with our energy needs, food resources and consequent pollution hazards—problems that were not considered to be serious thirty years ago. Chemists of the immediate future will undoubtedly pay more attention to the personal and social aspects of technology, so that society may develop in a cleaner and healthier environment.

Summary of Unit 15

1. The **chemical industry** is essential to modern society.
2. The **world population** is increasing dramatically, so more people require more *food, fuel and energy*, together with other natural resources.
3. Of our present energy resources, the **fossil fuels** are far the most important. *Oil, coal and natural gas* are all fossil fuels.
4. Fossil fuels are a *finite* energy source, so **alternative energy sources** are required for the future.
5. *Nuclear energy* is at present the most important alternative energy source. Others include *solar energy*, *tidal energy* and *geothermal energy*.
6. The **sea** is one of our major sources of minerals, particularly *bromine* and *magnesium*.
7. A consequence of a modern technological society is a tendency to pollute the environment.
8. **Air pollution** is caused by *carbon monoxide, oxides of nitrogen, hydrocarbons, sulphur oxides,* and *particulates.*
9. **Water pollution** results from *pesticides, fertilizers, raw and treated sewage, radioactive material* and *oil spillages.*
10. **Eutrophication** results in oxygen depletion in lakes.
11. Increased food requirements put pressure on food producers to improve their efficiency.
12. **Fertilizers** can be added (as sources of nitrogen, phosphorus and potassium) to improve plant growth.
13. **Pesticides** such as the pyrethrins (and previously DDT) are used to control pests.
14. **Food preservation** techniques are used to prevent enzyme and micro-organism deterioration of food.
15. The chemical industry is constantly developing new **medicines** for the prevention and treatment of disease.

Test Yourself on Unit 15

1. Mark the following statements true or false:
    (a) The world's population has doubled every hundred years since 1600 A.D.
    (b) All fossil fuels contain carbon.
    (c) Natural uranium consists mainly of the isotope uranium-235.
    (d) Sea water is a valuable source of bromine and magnesium.
    (e) Uranium is an example of a finite energy source.

2. Petrol is a mixture of liquid hydrocarbons. One of these is octane C_8H_{18}.
(a) Write an equation for the complete combustion of C_8H_{18} in a car engine.
(b) Name one gas (formed during incomplete combustion of C_8H_{18}) which is very poisonous in car exhaust fumes.
(c) Why does a black solid collect in the exhaust pipe of a car using octane as a fuel?
(d) In what form is lead added to petrol?
(e) Why is the addition of lead to petrol a dangerous source of air pollution?

3. River water often contains nitrates as a result of pollution.
(a) Name a source of nitrate pollution.
(b) What effect does this have on the growth of water weeds and aquatic plants?
(c) What is the eventual effect on the fish population?

4. The following is a list of gases:

ammonia	methane
carbon dioxide	nitrogen
carbon monoxide	nitrogen dioxide
hydrogen	oxygen
sulphur dioxide	

Answer the following questions, using only the gases in the list. Any gas may be used once, more than once or not at all.
Give the name of the gas which:
(a) is used by green plants in photosynthesis;
(b) is an important raw material used in the manufacture of fertilizers;
(c) forms when a yellow solid is burned in air and is often a pollutant in acid rain;
(d) is formed when carbon burns in a limited supply of air;
(e) is emitted into the atmosphere as a result of bacterial decomposition and is also used as a household fuel;
(f) is responsible for producing photochemical smogs in cities such as Los Angeles;
(g) may be sprayed on to dried fruit to help preserve it;
(h) is depleted in lakes and rivers during the process of eutrophication.

Mark this test out of 24 with the answers provided on page 384.

Appendix: The Elements

Atomic number	Name	Symbol	Relative atomic mass		Electron configuration
			Approx.	Accurate	
1	Hydrogen	H	1·0	1·008	1
2	Helium	He	4·0	4·003	2
3	Lithium	Li	7·0	6·939	2.1
4	Beryllium	Be	9·0	9·012	2.2
5	Boron	B	11	10·81	2.3
6	Carbon	C	12	12·01	2.4
7	Nitrogen	N	14	14·01	2.5
8	Oxygen	O	16	16·00	2.6
9	Fluorine	F	19	19·00	2.7
10	Neon	Ne	20	20·18	2.8
11	Sodium	Na	23	22·99	2.8.1
12	Magnesium	Mg	24	24.31	2.8.2
13	Aluminium	Al	27	26·98	2.8.3
14	Silicon	Si	28	28·09	2.8.4
15	Phosphorus	P	31	30·99	2.8.5
16	Sulphur	S	32	32·06	2.8.6
17	Chlorine	Cl	35·5	35·45	2.8.7
18	Argon	Ar	40	39·95	2.8.8
19	Potassium	K	39	39·10	2.8.8.1
20	Calcium	Ca	40	40·08	2.8.8.2
21	Scandium	Sc	45	44·96	2.8.9.2
22	Titanium	Ti	48	47·90	2.8.10.2
23	Vanadium	V	51	50·94	2.8.11.2
24	Chromium	Cr	52	52·00	2.8.12.2
25	Manganese	Mn	55	54·94	2.8.13.2
26	Iron	Fe	56	55·85	2.8.14.2
27	Cobalt	Co	59	58·93	2.8.15.2
28	Nickel	Ni	59	58·71	2.8.16.2
29	Copper	Cu	63·5	63·54	2.8.17.2
30	Zinc	Zn	65	65·37	2.8.18.2
31	Gallium	Ga	70	69·72	2.8.18.3
32	Germanium	Ge	72·5	72·59	2.8.18.4
33	Arsenic	As	75	74·92	2.8.18.5

Atomic number	Name	Symbol	Relative atomic mass Approx.	Relative atomic mass Accurate	Electron configuration
34	Selenium	Se	79	78·96	2.8.18.6
35	Bromine	Br	80	79·91	2.8.18.7
36	Krypton	Kr	84	83·80	2.8.18.8
37	Rubidium	Rb	85·5	85·47	2.8.18.8.1
38	Strontium	Sr	87·5	87·62	2.8.18.8.2
39	Yttrium	Y	89	88·91	2.8.18.9.2
40	Zirconium	Zr	91	91·22	2.8.18.10.2
41	Niobium	Nb	93	92·91	2.8.18.12.1
42	Molybdenum	Mo	96	95·94	2.8.18.13.1
43	Technetium	Tc	(99)	(99)	2.8.18.14.1
44	Ruthenium	Ru	101	101·1	2.8.18.15.1
45	Rhodium	Rh	103	102·9	2.8.18.16.1
46	Palladium	Pd	106·5	106·4	2.8.18.18
47	Silver	Ag	108	107·9	2.8.18.18.1
48	Cadmium	Cd	112·5	112·4	2.8.18.18.2
49	Indium	In	115	114·8	2.8.18.18.3
50	Tin	Sn	119	118·7	2.8.18.18.4
51	Antimony	Sb	122	121·8	2.8.18.18.5
52	Tellurium	Te	127·5	127·6	2.8.18.18.6
53	Iodine	I	127	126·9	2.8.18.18.7
54	Xenon	Xe	131	131·3	2.8.18.18.8
55	Caesium	Cs	133	132·9	2.8.18.18.8.1
56	Barium	Ba	137·5	137·3	2.8.18.18.8.2
57	Lanthanum	La	139	138·9	2.8.18.18.9.2
58	Cerium	Ce	140	140·1	2.8.18.20.8.2
59	Praseodymium	Pr	141	140·9	2.8.18.21.8.2
60	Neodymium	Nd	144	144·2	2.8.18.22.8.2
61	Promethium	Pm	(145)	(145)	2.8.18.23.8.2
62	Samarium	Sm	150·5	150·4	2.8.18.24.8.2
63	Europium	Eu	152	152·0	2.8.18.25.8.2
64	Gadolinium	Gd	157	157·3	2.8.18.25.9.2
65	Terbium	Tb	159	158·9	2.8.18.27.8.2
66	Dysprosium	Dy	162·5	162·5	2.8.18.28.8.2
67	Holmium	Ho	165	164·9	2.8.18.29.8.2
68	Erbium	Er	167	167·3	2.8.18.30.8.2
69	Thulium	Tm	169	168·9	2.8.18.31.8.2
70	Ytterbium	Yb	173	173·0	2.8.18.32.8.2
71	Lutetium	Lu	175	175·0	2.8.18.32.9.2
72	Hafnium	Hf	178·5	178·5	2.8.18.32.10.2
73	Tantalum	Ta	181	180·9	2.8.18.32.11.2

Atomic number	Name	Symbol	Relative atomic mass		Electron configuration
			Approx.	Accurate	
74	Tungsten	W	184	183·9	2.8.18.32.12.2
75	Rhenium	Re	186	186·2	2.8.18.32.13.2
76	Osmium	Os	190	190·2	2.8.18.32.14.2
77	Iridium	Ir	192	192·2	2.8.18.32.17
78	Platinum	Pt	195	195·1	2.8.18.32.17.1
79	Gold	Au	197	197·0	2.8.18.32.18.1
80	Mercury	Hg	200·5	200·6	2.8.18.32.18.2
81	Thallium	Tl	204·5	204·4	2.8.18.32.18.3
82	Lead	Pb	207	207·2	2.8.18.32.18.4
83	Bismuth	Bi	209	209·0	2.8.18.32.18.5
84	Polonium	Po	(209)	(209)	2.8.18.32.18.6
85	Astatine	At	(210)	(210)	2.8.18.32.18.7
86	Radon	Rn	(222)	(222)	2.8.18.32.18.8
87	Francium	Fr	(223)	(223)	2.8.18.32.18.8.1
88	Radium	Ra	(226)	(226)	2.8.18.32.18.8.2
89	Actinium	Ac	(227)	(227)	2.8.18.32.18.9.2
90	Thorium	Th	232	232·0	2.8.18.32.18.10.2
91	Protactinium	Pa	(231)	(231)	2.8.18.32.20.9.2
92	Uranium	U	238	238·0	2.8.18.32.21.9.2
93	Neptunium	Np	(237)	(237)	2.8.18.32.22.9.2
94	Plutonium	Pu	(244)	(244)	2.8.18.32.24.8.2
95	Americium	Am	(243)	(243)	2.8.18.32.25.8.2
96	Curium	Cm	(247)	(247)	2.8.18.32.25.9.2
97	Berkelium	Bk	(247)	(247)	2.8.18.32.26.9.2
98	Californium	Cf	(251)	(251)	2.8.18.32.28.8.2
99	Einsteinium	Es	(254)	(254)	2.8.18.32.29.8.2
100	Fermium	Fm	(253)	(253)	2.8.18.32.30.8.2
101	Mendelevium	Md	(256)	(256)	2.8.18.32.31.8.2
102	Nobelium	No	(253)	(253)	2.8.18.32.32.8.2
103	Lawrencium	Lr	(257)	(257)	2.8.18.32.32.9.2

Note
Some relative atomic masses are cited in parentheses, to indicate that they refer to the most stable or best-known isotope of the element concerned.

Examination Questions

Questions have been included from many examination boards. The authors wish to thank the undermentioned authorities for granting permission for the reproduction of examination questions:

Joint Matriculation Board   [JMB]
University of London Schools Examinations Department   [L]
Northern Ireland Schools Examination Council   [NI]
Southern Examining Group (GCSE specimen questions)   [S]
Northern Examining Association (Associated Lancashire Schools Examining Board, Joint Matriculation Board, North Regional Examinations Board, North West Regional Examinations Board, Yorkshire and Humberside Regional Examinations Board) specimen questions for 1988 GCSE syllabus 'A'   [N]
London and East Anglian Examining Board   [L & EA]

Objective Questions

(a) Matching Pairs

In the following 'matching pairs' questions, each numbered question is accompanied by a set of alternative answers. Within each question each letter may be used once, more than once or not at all.

1. From the list A to D below
    A   condensation
    B   distillation
    C   evaporation
    D   filtration
select the process which
   (a) can be used to separate sand from water most quickly and cheaply,
   (b) causes water to collect on classroom windows in cold weather,
   (c) involves the change of a liquid to a gas and of the gas back to a liquid.
[N]

2. The alternatives A to E describe some of various states in which matter may be found.
    A   solid only
    B   solid and liquid
    C   liquid only
    D   liquid and gas
    E   gas only

Select, from the list A to E, the alternative which best describes the state or states of matter
(a) consisting only of particles which move randomly and are held together by forces of attraction,
(b) present when a solution is being distilled,
(c) leaving the catalyst chamber in the contact process.   [JMB]

3. Questions (a) to (d) concern the following practical methods:
    A   chromatography
    B   crystallization
    C   distillation
    D   electrolysis
    E   filtration
Choose, from A to E, the method which would be used to
    (a) isolate nitrogen from liquid air,
    (b) separate coloured substances in a sample of coloured soft drink,
    (c) separate petrol from crude oil,
    (d) separate a drug which has been precipitated from a solution.   [L & EA]

4. Using the letters A to L, match the descriptions given with the substances shown below them.

Description
    A   yellow-green gas   G   red-brown gas
    B   brown powder   H   black powder
    C   blue crystals   I   silver liquid
    D   red-brown liquid   J   yellow powder
    E   white powder   K   pale green crystals
    F   red crystals   L   blue liquid

Substance
    copper sulphate
    zinc oxide
    iron(II) sulphate
    carbon (graphite)
    sulphur
    bromine
    chlorine
    mercury   [NI]

5.
    A $^{14}_{6}$ J
    B $^{23}_{11}$ L
    C $^{32}_{16}$ Q
    D $^{35}_{17}$ T
    E $^{38}_{18}$ Z

Which one of the atoms listed A to E
   (*a*) can form an ion with a single negative charge?
   (*b*) does not react with either oxygen or hydrogen?
   (*c*) could form a covalent compound of the type XH_4?
   (*d*) has the same number of occupied principal energy levels as an atom of
   oxygen?   [JMB]

6. The following substances are well-known laboratory reagents or
catalysts:
   A   hydrogen peroxide
   B   iron
   C   nickel
   D   manganese(IV) oxide
   E   vanadium(V) oxide
For each of the processes in questions (*a*) to (*d*), select from the list A to E
the substance which is used for the purpose stated:
   (*a*) the reagent in the common laboratory preparation of oxygen,
   (*b*) the catalyst in the manufacture of sulphuric acid,
   (*c*) the catalyst in the manufacture of ammonia from nitrogen and hydrogen by the Haber process,
   (*d*) an oxidizing agent in the laboratory preparation of chlorine.   [JMB]

7. Questions (*a*) to (*d*) concern the following graphs:

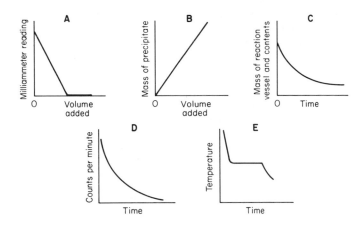

Select from A to E the graph which best represents a change taking place in
each reaction represented below.
   (*a*) $^{212}_{83}Bi \rightarrow {}^{212}_{84}Po + \beta$
   (*b*) $H_2O_{(l)} \rightarrow H_2O_{(s)}$
   (*c*) $2H_2O_{2(aq)} \rightarrow 2H_2O_{(1)} + O_{2(g)}$
   (*d*) $BaCl_{2(aq)} + Na_2SO_{4(aq)} \rightarrow BaSO_{4(s)} + 2NaCl_{(aq)}$ [L]

8.
   A   addition
   B   combustion
   C   cracking
   D   fermentation
   E   dehydration

Which one of the above processes, A to E, is involved in the conversion of
   (a) butane to carbon dioxide and water vapour?
   (b) $C_{20}H_{42}$ to C_2H_4?
   (c) $C_6H_{12}O_6$ to C_2H_5OH?
   (d) ethene to ethanol?   [JMB]

9.   Questions (a) to (d) concern the diagram below, which shows parts of
the periodic table divided into five sections labelled A, B, C, D and E.

Select, from A to E, the section in which you would find a
   (a) gas which forms no compounds,
   (b) metal which produces hydrogen when added to water,
   (c) metal which floats on water,
   (d) metal which is used for making the water pipes in a domestic hot water
system.   [L & EA]

(b) **Multiple Choice Questions**
In each of the following questions, a statement or question is followed by five
written alternative responses, A, B, C, D and E. You are required to decide
which of these alternatives is the best answer.

10.   The diagram represents the results obtained when paper chromatography was used to identify the two substances present in a mixture.

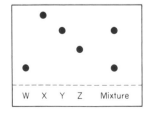

Which two substances are most likely to be present in the mixture?

A   X and Z
B   W and X
C   W and Y
D   Y and Z
E   W and Z   [L & EA]

11. If fine pollen grains on the surface of water are examined under a
microscope, it will be seen that the pollen grains are in random motion, fre-
quently changing direction. The movement is most likely to be due to
   A   air draughts blowing on the water,
   B   chemical reaction between the pollen and the water,
   C   attraction and repulsion between charged particles,
   D   collisions between water molecules and pollen grains,
   E   diffusion of pollen grains.   [L & EA]

12. What is the maximum mass of iron, in grams, which can be extracted
from 80 g of iron(III) oxide, Fe_2O_3? (Relative atomic masses: Fe = 56, O =
16)
   A   28
   B   32
   C   40
   D   48
   E   56   [L]

13. Which of the following contains the same number of atoms as 6 g of
magnesium? (Relative atomic masses: C = 12, Mg = 24, S = 32, Ca = 40,
Fe = 56, Zn = 65.)
   A   4 g of carbon
   B   8 g of sulphur
   C   5 g of calcium
   D   28 g of iron
   E   13 g of zinc

14. 18 g of water is liberated when 0·2 mole of a crystalline salt is heated
to constant mass. Which one of the following represents the number of moles
of water of crystallization combined with 1 mole of the salt? (Relative atomic
masses: H = 1, O = 16.)
   A   1
   B   2
   C   5
   D   7
   E   8

15. The element antimony has the symbol Sb. If the formula for antimony
oxide is Sb_2O_3, what is the correct formula for antimony sulphate?

   A $Sb_2(SO_4)_3$
   B $SbSO_4$
   C Sb_2SO_4
   D $Sb(SO_4)_2$
   E $Sb_3(SO_4)_2$

16.   The relative molecular mass of an oxide MO is 223. What is the relative molecular mass of the oxide M_3O_4?
   A   685
   B   700
   C   707
   D   717
   E   885
(Relative atomic mass: O = 16)   [NI]

17.   The equation for the reaction occurring when butane burns is

$$C_4H_{10} + 6\tfrac{1}{2}O_2 \rightarrow 4CO_2 + 5H_2O; \quad \Delta H = -2877 \text{ kJ mol}^{-1}$$

How much butane will liberate 2877 kJ on burning completely in oxygen?
   A 1 dm^3
   B   1 mole
   C   6·5 mole
   D   7·5 mole
   E   1 kg

18.   How many grams of sodium hydroxide, NaOH, must be dissolved to make 250 cm^3 of 1·0 M solution?
   A   10 g
   B   20 g
   C   40 g
   D   80 g
   E   250 g
(Relative atomic masses: H = 1, O = 16, Na = 23)   [NI]

19.   When chlorine gas reacts with iodine monochloride, the following equilibrium is set up:

$$ICl_{(l)} + Cl_{2(g)} \rightleftharpoons ICl_{3(s)}$$

Which one of the following statements about this equilibrium is NOT correct?
   A   Removing the chlorine gas causes the amount of solid to decrease.
   B   When all the solid is removed, the forward reaction stops.
   C   The forward and backward reactions are going on at the same rate.
   D   The proportions of the three substances remain constant.
   E   Both forward and backward reactions continue to take place.
                                                                                 [L]

20. 0·01 mole of sodium and 0·01 mole of potassium were reacted separately with excess water at the same temperature and pressure. Which of the following is true?

	Initial reaction rate	Volume of hydrogen
A	faster with potassium	same in each case
B	same in each case	greater with potassium
C	faster with sodium	same in each case
D	faster with potassium	greater with potassium
E	faster with potassium	greater with sodium

21. The reaction taking place at the negative electrode during the electroplating of copper can be represented by the equation

A $Cu \rightarrow Cu^{2+} + 2e$

B $Cu^{2+} + e \rightarrow Cu$

C $Cu^{2+} + 2e \rightarrow Cu$

D $Cu^{2+} + 2e \rightarrow 2Cu$ [N]

22. Which of the following substances will conduct electricity by movement of ions?

A mercury

B lead(II) bromide crystals

C graphite

D sodium

E molten sodium hydroxide

23. A substance must be an acid if

A it liberates ammonia from an ammonium salt

B it is the oxide of a metal

C it forms hydrogen ions when added to water

D it reacts with a base to form a salt and hydrogen

E an aqueous solution of it produces oxygen at the anode during electrolysis

24. Which of the following $1·0$ mol dm^{-3} solutions would have the lowest pH value?

A potassium hydroxide

B sodium nitrate

C aqueous ammonia (ammonium hydroxide solution)

D ethanoic acid (acetic acid)

E nitric acid

25. Which of the following oxides is amphoteric?

A aluminium oxide

B sodium oxide

C   carbon dioxide
D   magnesium oxide
E   phosphorus(v) oxide (phosphorus pentoxide)

26.   The element W reduces the oxide of element X, but has no reaction with the oxide of element Z. Element Y, however, will reduce the oxide of both elements X and Z.
The order of reactivity of these elements, with the most reactive first, is
A $X\ Y\ W\ Z$
B $Y\ X\ W\ Z$
C $W\ Z\ X\ Y$
D $W\ X\ Y\ Z$
E $Y\ Z\ W\ X$ [L]

27.   Which of the following statements about temporarily hard water is INCORRECT?
A   it contains metallic cations, usually calcium and magnesium
B   it contains hydrogencarbonates which can be removed by boiling
C   it can be softened by adding sodium carbonate
D   it does not cause a scum with soap
E   it can be softened by using Permutit ion-exchange resins

28.   Which one of the following gases is often used in the laboratory as a reducing agent?
A   oxygen
B   carbon dioxide
C   hydrogen
D   nitrogen
E   chlorine   [NI]

29.   Aluminium is obtained industrially by
A   reaction of aluminium oxide with carbon in a blast furnace
B   reaction of aluminium oxide with carbon monoxide in a blast furnace
C   electrolysis using aluminium chloride
D   electrolysis using aluminium sulphate
E   electrolysis using aluminium oxide   [L & EA]

30.   Which one of the following is a true description of the industrial manufacture of ammonia by the Haber process?
A   combination of nitrogen and hydrogen; no catalyst; high pressure
B   combination of nitrogen and hydrogen; iron catalyst; low pressure
C   combination of nitrogen and hydrogen; iron catalyst; high pressure
D   combination of nitrogen, hydrogen and oxygen; iron catalyst; low pressure
E   combination of nitrogen, hydrogen and oxygen; iron catalyst; high pressure

31. Which of the following is an incorrect statement about the halogens (Group 7) as the group is descended (F → I)?
A   the oxidizing power of the elements decreases
B   the reactivity of the elements decreases
C   the size of the atom increases
D   the boiling points of the elements increase
E   the ability to gain an electron increases

32. Which of the following would help to decide whether a liquid was a saturated or unsaturated hydrocarbon?
A   Test to see whether it will decolorize bromine water.
B   Test to see if it will burn.
C   Test to see whether it mixes with water.
D   Heat a drop strongly with excess copper(II) oxide to see whether it gives carbon dioxide.
E   See if it can be cracked to give a gas   [L]

33. What type of process occurs when sugar is converted into alcohol (ethanol)?
A   hydrogenation
B   fermentation
C   cracking
D   polymerization
E   hydrolysis

34. Each of the following compounds contains nitrogen in a form suitable for application as a fertilizer. Which compound would supply the greatest mass of the element nitrogen for each kilogram of the compound applied to the soil?

	Compound	Formula	Relative molecular mass
A	ammonium bromide	NH_4Br	98
B	ammonium sulphate	$(NH_4)_2SO_4$	132
C	ammonium nitrate	NH_4NO_3	80
D	sodium nitrate	$NaNO_3$	85
E	potassium nitrate	KNO_3	101 [L & EA]

35. In the manufacture of pure copper, copper(II) sulphate solution is electrolysed using a pure copper cathode and a pure copper anode. What mass of copper would be deposited on the cathode by the passing of 2 faradays of electricity? (Relative atomic mass of Cu = 64.)
A   2 g
B   32 g
C   64 g
D   128 g
E   256 g

36.   An atom of an element X contains 19 protons and 20 neutrons. Which of the following represents an atom of X?

A $^{19}_{20}X$
B $^{39}_{20}X$
C $^{39}_{19}X$
D $^{20}_{19}X$
E $^{38}_{19}X$

37.   A mixture of water and crushed ice was heated gradually, with a constant flame and efficient stirring, until the mixture boiled for a short time.

If the temperature of the mixture was plotted against time (t), which graph would be obtained?   [L]

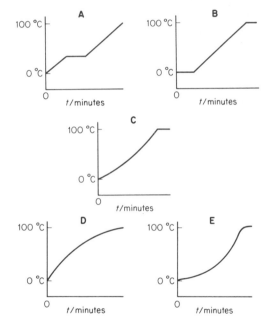

38.   Which of the following pollutants is thought to be the most likely to cause mental retardation in children?

A   carbon monoxide (from the incomplete combustion of fossil fuels)
B   lead compounds (found in car engine exhaust gases)
C   carbon dioxide (from the combustion of fossil fuels)
D   oxides of nitrogen (from the burning of coal, oil and petrol)
E   sulphur oxides (from the combustion of fossil fuels)

Structured Questions

39.   The table shows the melting points and boiling points of substances A, B, C and D.

Substance	Melting point in °C	Boiling point in °C
A	645	1300
B	−7	59
C	−165	−92
D	27	98

(a) Which substance is a liquid at room temperature?
(b) Which substance is a gas at room temperature?
(c) Which substance *cannot* be a metal?
(d) Which substance is likely to be an ionic compound?   [s]

40. This question is about elements, their properties, and their positions in the periodic table.

Group	I	II						III	IV	V	VI	VII	VIII (0)
	X	Be						B	C	N	O	F	
		Mg							Y				
		Ca											
		Sr											
		Ba											

Block *S*   Block *D*   Block *P*

Use the skeleton periodic table above and the elements listed below, and no others, to answer the questions which follow.

Elements to be used   Argon Ar   Iron Fe
                      Bromine Br   Lithium Li
                      Copper Cu   Neon Ne
                      Helium He   Potassium K
                      Iodine I   Silicon Si

(a) Name TWO elements from the list which should be placed in Group I.
(b) Name TWO elements from the list which should be placed in Block *D*.
(c) Name TWO elements which should be placed in Group 0.
(d) Name TWO elements which should be placed in Group VII.

(e) Name the elements X and Y.

(f) Which one of the elements in Group II will be the most reactive? State a reason for your choice.

Some data about an element and its chloride are:

	Melting point/°C	Boiling point/°C
Element	120	444
Chloride	−80	136

(g) Is this chloride likely to conduct electricity? Give a reason for your answer.

(h) Draw and label a diagram of the apparatus you would use to attempt the preparation and collection of a sample of this chloride using the element and chlorine only. (You may assume that a supply of pure dry chlorine is available.)

(i) State ONE important safety precaution which must be taken when performing the above experiment.   [L]

41. Fizzy drinks can be made by dissolving carbon dioxide gas, under pressure, into water. The drink is kept in a tightly stoppered bottle until needed. When the stopper is removed bubbles of carbon dioxide form making the drink fizzy.

The carbon dioxide first dissolves in the water and some of the dissolved gas may react with water to form carbonic acid.

$$CO_{2(g)} + aq \rightleftharpoons CO_{2(aq)} \qquad \textbf{Equation 1}$$

$$CO_{2(aq)} + H_2O_{(l)} \rightleftharpoons H_2CO_{3(aq)} \qquad \textbf{Equation 2}$$

(a) Use Equation 1 to explain the appearance of bubbles of carbon dioxide when the stopper of a fizzy drink bottle is loosened.

(b) Unless the bottles are tightly stoppered, these drinks tend to lose their fizziness quickly on warm days. Use Equation 1 to explain why this happens.

(c) What advantage may be gained by keeping bottles of water in the fridge before making fizzy drinks?

(d) Write an equation to show how H_2CO_3 can behave as an acid in water.

When carbon dioxide is bubbled into lime-water a white precipitate of calcium carbonate forms. An equation for the reaction may be written as follows.

$$H_2CO_{3(aq)} + Ca(OH)_{2(aq)} \rightleftharpoons CaCO_{3(s)} + 2H_2O_{(l)} \qquad \textbf{Equation 3}$$

If more carbon dioxide is bubbled into the mixture, the cloudiness disappears and a solution of calcium hydrogencarbonate is formed as shown below.

$$H_2CO_{3(aq)} + CaCO_{3(s)} \rightleftharpoons Ca(HCO_3)_{2(aq)} \qquad \textbf{Equation 4}$$

(*e*) Write TWO separate ionic equations to summarize the reactions indicated by Equation 3.

(*f*) Use Equation 4 to explain how domestic water supplies in many places contain calcium ions (Ca^{2+}).

(*g*) How do the equations given above explain the formation of white deposits (such as kettle fur) in vessels used to boil tap water in hard water districts?

(*h*) Sodium stearate ($C_{17}H_{35}COONa$) is a typical simple soap. Why is its use as a soapy detergent limited by the calcium ions which are often present in water? [S]

42. When copper(II) sulphate solution is electrolysed between platinum electrodes, the equation for the overall changes taking place can be written as

$$2CuSO_{4(aq)} + 2H_2O_{(1)} \rightarrow 2Cu_{(s)} + O_{2(g)} + 2H_2SO_{4(aq)}$$

(*a*) (i) Describe what you would SEE at the cathode.

(ii) Write an ionic half-equation for the reaction occurring at the cathode.

(*b*) (i) Describe what you SEE at the anode.

(ii) Describe a test for the anode product.

(iii) Write an ionic half-equation for the reaction occurring at the anode.

(*c*) One mole of gas molecules occupy 24 000 cm^3 at room temperature and atmospheric pressure. The relative atomic mass of copper is 63·5 and the charge carried by one mole of electrons (faraday) is 96 500 coulombs.

If 1·27 g of copper are produced, calculate the

(i) volume of oxygen, measured at room temperature and atmospheric pressure, that would be produced in the same time,

(ii) length of time for which a current of 0·5 A would have to be maintained to form these amounts of products.

(*d*) Copper can be purified using electrolysis. State what you would use as the

(i) cathode,

(ii) anode,

(iii) electrolyte,

(iv) During this process a slime or sludge forms under one electrode. Give the name of the electrode under which it forms.

(v) Suggest a reason why a part of the slime may be valuable. [L & EA]

43. (*a*) When ammonia dissolves in water, a reversible reaction occurs which can be represented by the equation

$$NH_{3(g)} + H_2O_{(1)} \rightleftharpoons NH_{4(aq)}^+ + OH_{(aq)}^-$$

(i) State which two particles present are acting as bases.

(ii) Give an explanation of your answer in (i).

(*b*) When a solution of hydrochloric acid is mixed with a solution of ammonia, another reaction occurs.

364   Success in Chemistry

(i) Name the type of reaction.
(ii) Write the simplest ionic equation to represent the reaction.
(iii) Name the solid compound which could be obtained fron the resulting solution.

(c) Hydrochloric acid is a strong acid and ethanoic acid is a weak acid. Explain the terms (i) strong acid, (ii) weak acid. Describe a simple experiment which you could use to demonstrate the difference in the strengths of the two acids.   [JMB]

44. (a) Nitric acid is produced from ammonia by the sequence of changes shown in the diagram.

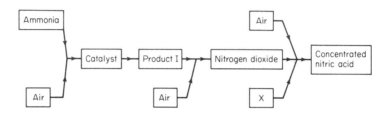

Name:
    (i) the catalyst,
    (ii) the intermediate product, I, formed in the catalyst chamber,
    (iii) substance X, added at the final stage.
(b) Cadmium (Cd) is a metal. What would be the products of the reaction between dilute nitric acid and cadmium oxide, CdO?
(c) A green powder is added to dilute nitric acid. It fizzes, giving off carbon dioxide and forming a blue solution. Identify the green powder.
(d) The diagram of the nitrogen cycle in Fig. 6.8 (page 122) may help you to answer this part. Nitrates are used as fertilizers.
    (i) What class of compounds is formed from nitrates in plants?
    (ii) Why cannot atmospheric nitrogen be used as a fertilizer by most green plants?
    (iii) Plants are not able to absorb ammonium compounds, so how are these compounds able to act as fertilizers?
    (iv) How can the use of too much nitrate fertilizer cause water pollution?
(e) Calculate the percentage by mass of nitrogen in ammonium sulphate, $(NH_4)_2SO_4$. (Relative atomic masses: N = 14, H = 1, S = 32, O = 16)   [S]

45. Titanium is the eighth most abundant element in the earth's crust. It is more abundant than copper. It occurs in titanium(IV) oxide which is present to the extent of 0.01 per cent in much sea-shore sand. The metal is extracted by first converting the titanium(IV) oxide to titanium(IV) chloride. The titanium(IV) chloride is then heated with sodium.
Titanium has a lower density than iron, is stronger than steel and is also

resistant to corrosion. Information about the abundance (in parts per million) and cost of titanium and copper are given in the table below.

	Titanium	Copper
Abundance (p.p.m.)	4400	55
Cost (£/tonne)	3700	800

(a) Write the formula of the naturally occurring oxide of titanium.

(b) (i) What reagent might be used to convert titanium(IV) oxide to titanium(IV) chloride?

(ii) Complete the symbol equation for the reaction between titanium(IV) chloride and sodium.

$$TiCl_4 + 4Na \rightarrow$$

(c) Suggest *two* reasons why titanium metal is more expensive than copper.

(d) Describe how simple cells could be set up and used to place titanium in the reactivity series of metals.

(e) Give TWO reasons why titanium is particularly suitable for replacement hip joints.   [N]

46. Below is a diagram of a blast furnace in which iron is extracted from iron ore.

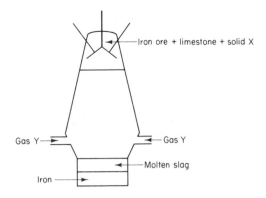

(a) Give the name of (i) solid X, (ii) gas Y.

(b) Why does the molten slag float on top of the molten iron?

(c) The inside of the furnace is hot. What chemical reaction produces this heat?

(d) Most iron is made into steel. What is the difference between iron and steel?

(e) Name ONE metal other than iron used in the alloy stainless steel.

(f) (i) What is the advantage of using stainless steel instead of iron?

  (ii) Name ONE object made of stainless steel.

(g) Oxygen is used to convert iron into steel. What is the oxygen used for?

(h) (i) From what raw material is the oxygen obtained?

  (ii) What method is used to obtain the oxygen in (h)(i)?

(i) Name ONE alloy containing *no* iron.   [L & EA]

47. Rock salt is an important worldwide resource. It may be crushed and used without further processing, purified to give cooking salt or used as a source of raw materials for the chemical industry.

(a) State ONE advantage and ONE disadvantage of using salt in cooking.

(b) Draw a diagram to show how you could make and collect a small quantity of chlorine gas in the laboratory by the electrolysis of brine.

(c) Write an ionic equation for the formation of chlorine. Explain why sodium ions and hydroxide ions collect at the other electrode.

(d) Chlorine is used in the manufacture of many useful products. Name TWO such products or groups of products and give an example of the use of each substance (or group of substances) you have named.   [N]

48. The substance ethanediol ($C_2H_6O_2$) is used as an antifreeze. It is made from ethene according to this equation:

$$C_2H_4 \xrightarrow{\text{water, air, catalyst}} C_2H_6O_2$$

ethene   ethanediol

The relative atomic masses are C = 12; H = 1; O = 16.

(a) State ONE use of antifreeze.

(b) What is a catalyst?

(c) Calculate the relative molecular mass of ethene.

(d) Calculate the relative molecular mass of ethanediol.

(e) Calculate the mass of ethanediol that could be made from 5.6 g of ethene.   [s]

49. The general formula for members of the alkane series of hydrocarbons is C_nH_{2n+2}. Methane is the first member.

(a) Give ONE source of methane.

(b) (i) Name the products of complete burning of any alkane.

  (ii) Write an equation for the complete burning of methane.

(c) (i) Under what conditions could methane burn to form carbon monoxide?

  (ii) Why is carbon monoxide dangerous?

(d) The fifth member of the alkane series is called pentane. Work out its molecular formula.   [s]

50. Read the passage below and then answer the questions which follow.

Hydrogen as a fuel

With the world's fossil fuel resources rapidly dwindling, considerable research is being directed towards finding a replacement for them. One such fuel is hydrogen, despite the difficulties associated with its manufacture from water, potentially its major source. The main problem of its economical extraction is associated with the strength of the hydrogen/oxygen bonds, which require about 460 kJ mol^{-1} to break them.

Apart from its manufacture, there are also many problems attached to the use of hydrogen. How can it be transported safely? Can it be used in modern engines without causing too many re-design problems? Are the hazards associated with its use acceptable? On the credit side, it has much to commend it. Its combustion product, when it is burnt with pure oxygen, is virtually pollution free; although if it were mixed with air and used in internal combustion engines in the same way as conventional fuel, this mixture would still produce oxides of nitrogen. Furthermore, because of its much faster burning rate, a different carburettor system would be needed. One way of overcoming this problem would be to use the gas in fuel cells which, unlike petrol, are able to generate electricity very efficiently. However, with present-day technology, fuel cells are unable to cope with the relatively large amount of power needed to drive a vehicle.

As an aircraft fuel, mass for mass, hydrogen is able to deliver about three times as much energy as kerosene. This would reduce the problem of having to dump fuel if an aircraft were to turn back after take-off. On the other hand, the volume needed to store the fuel is greater than the corresponding volume needed for kerosene, mainly on account of its different state. Liquefaction may be the answer, but then the maintenance of a very low temperature is a serious problem. However, in spite of all the difficulties associated with its use, the future prospects of hydrogen are encouraging, and, if the main obstacle to its extraction can be overcome, it has great potential as a replacement for fossil fuels.

(a) What factor makes the extraction of hydrogen from water a particularly difficult problem?

(b) How much energy is needed to separate all the atoms in one mole of water?

(c) (i) Why is the combusion product 'virtually pollution free'?

    (ii) What pollutants could still be produced if hydrogen were used as a fuel in an internal combustion engine?

    (iii) Explain briefly how these pollutants would be formed.

    (iv) What other property of hydrogen presents a problem when hydrogen is used in internal combustion engines?

    (v) What advantage would a hydrogen-driven fuel cell have over petrol as an energy source for motor vehicles?

    (vi) What prevents a fuel cell being used in present day vehicles?

(d) (i) What particular property of hydrogen gives it an advantage over conventional aircraft fuels?

(ii) Why would the use of hydrogen as a fuel reduce the wastage in the event of a fault developing soon after take-off?   [JMB]

Essay-type Questions

51. (a) Describe briefly how oxygen is obtained for large-scale use from the air.

(b) Explain the involvement of atmospheric oxygen in the carbon cycle by describing

(i) one process by which oxygen is removed from the air, and

(ii) the replacement of oxygen into the air.

(c) What changes, if any, would be seen if the following were placed in separate jars or tubes of oxygen? Write equations for any reactions which occur.

(i) burning sulphur

(ii) pH indicator

(iii) burning magnesium ribbon   [JMB]

52. (a) Representing each covalent bond by a single line between two atoms, use diagrams to show the bonding in one molecule of (i) methane, (ii) ammonia, (iii) water. Beneath each structure, give the physical state of each compound at room temperature and pressure.

(b) Account for the ionic bonding between sodium and chlorine in sodium chloride and for its physical state at room temperature and pressure.

(c) Give three general properties in which simple ionic and covalent compounds differ, and state how they differ in each case.

(d) Magnesium oxide is an ionic compound. Explain why the melting point of magnesium oxide is much greater than that of sodium chloride.

(e) Explain why the melting point of diamond is greater than 3000 °C.   [JMB]

53. Identify the substances A to J in the following statements, giving equations for any two reactions in each section.

(a) A colourless solution A was alkaline to indicator paper and produced a brown precipitate B on addition to iron(III) chloride solution. Passage of carbon dioxide through A followed by crystallization gave a white solid C which on heating decomposed to leave no solid residue.

(b) When small pieces of a metal D were added in excess to copper(II) sulphate solution, a brown precipitate E and a colourless solution were formed. After filtration, the colourless solution was treated with sodium hydroxide solution which was added dropwise. A white precipitate F was formed which dissolved when an excess of the alkali was added.

(c) 10 cm³ of 1·0 M sulphuric acid were added to 10 cm³ of 1·0 M sodium hydroxide solution. The resulting solution was crystallized and a white solid G separated from the solution. Separate portions of a solution of G gave (i) a white precipitate, H, on treatment with barium chloride solution and (ii) a gas, J, on addition to sodium carbonate.   [JMB]

Answers to Test Questions

Unit 1

1. A = sublimation   (1 mark)
   B = melting   (1 ,,)
   C = freezing   (1 ,,)
   D = vaporization/evaporation   (1 ,,)
   E = condensation   (1 ,,) (5)

2. (a) evaporation   (1 ,,)
   (b) fractional distillation   (1 ,,)
   (c) sublimation   (1 ,,)
   (d) filtration   (1 ,,)
   (e) chromatography   (1 ,,) (5)

3. (a) true   (1 ,,)
   (b) false (it gives only an order of magnitude) (1 ,,)
   (c) true   (1 ,,)
   (d) true   (1 ,,) (4)

4. (a) (iv)   (1 ,,)
   (b) (i)   (1 ,,)
   (c) (v)   (1 ,,)
   (d) (vi)   (1 ,,)
   (e) (iii)   (1 ,,)
   (f) (ii)   (1 ,,) (6)

5.

Particle	Mass	Charge
	1	+1
Neutron	1	
Electron		

(5)

6. The nucleus of the potassium atom contains 19 protons
   (1 mark) and 20 neutrons (1 mark)   (2)

7. Nitrogen isotopes have the same atomic number of 7
   (1 mark) and different mass numbers (1 mark). One isotope
   contains seven neutrons, the other eight neutrons (1 mark) (3)

TOTAL: 30

Unit 2

1. $2n^2$ (1 mark) (1)

2. (a) true (1 ,,)
   (b) false (1 ,,)
   (c) true (1 ,,)
   (d) true (1 ,,) (4)

3. (b) atomic number (2 marks) (2)

4. (i) set (c) (1 mark)
   (ii) set (b) (1 ,,)
   (iii) set (a) (1 ,,)
   (iv) set (d) (1 ,,)
   (v) set (e) (1 ,,) (5)

5. (a) Magnesium   2.8.2 (1 ,,)
      Chlorine   2.8.7 (1 ,,)
   (b) To attain the noble-gas structure of argon (1 mark);
      by gaining one electron it completes its outer stable
      octet of electrons (1 mark).
   (c) Mg^{2+} (1 ,,)
   (d) Mg^{2+} Cl^-_2 (1 ,,)
   (e) High melting point (1 ,,)
      Forms a giant strongly bonded ionic lattice. (1 ,,) (8)

6. Because sodium chloride exists as ions within a giant
   lattice structure. No separate molecules exist in such
   a structure. (1 ,,)
   Ammonia exists as separate small covalently bonded
   molecules. (1 ,,) (2)

7. (a) true (1 ,,)
   (b) false (1 ,,)
   (c) true (1 ,,)
   (d) true (once formed, the dative covalent bond is indis-
      tinguishable from an ordinary covalent bond) (1 ,,) (4)

8. Methane exists as separate covalently bonded CH_4
   molecules (1 mark). The forces *between* these molecules
   are weak and hence methane is a gas (1 mark).
   Diamond exists as a giant covalently bonded molecule
   (1 mark). Each carbon atom is tetrahedrally bonded to four
   other carbon atoms by strong covalent bonds. Much
   energy is necessary to break these bonds and melt the solid
   (1 mark). (4)

TOTAL: 30

Unit 3

1. (a) Mass of oxide − mass of lead = mass of oxygen

$$14\cdot34\text{g} - 12\cdot42\text{g} = 1\cdot92\text{g}$$ (1 mark)

   (b) Number of moles of oxygen atoms

$$= \frac{1\cdot92}{16}$$

$$= 0\cdot12$$ (1 ,,)

   (c) Number of moles of lead $\quad = \dfrac{12\cdot42}{207}$

$$= 0\cdot06$$ (1 ,,)

   (d) 0·06 mole of lead combines with 0·12 mole of oxygen atoms

$$\therefore \text{1 mole of lead combines with } \frac{0\cdot12}{0\cdot06} = 2 \text{ moles of}$$

   oxygen atoms   (1 ,,)

   (e) ∴ formula of oxide $= \text{PbO}_2$ (1 ,,) (5)

2. (a) $\text{Na}_2^+ \text{CO}_3^{2-}$ (1 ,,)
   (b) $\text{Fe}^{2+} \text{SO}_4^{2-}$ (1 ,,)
   (c) $\text{Zn}^{2+} (\text{NO}_3^-)_2$ (1 ,,)
   (d) $\text{Al}^{3+} (\text{OH}^-)_3$ (1 ,,)
   (e) $\text{Ag}^+ \text{Cl}^-$ (1 ,,)
   (f) $\text{K}_2^+ \text{SO}_3^{2-}$ (1 ,,) (6)

3. (a) sodium hydrogensulphate   (1 ,,)
   (b) iron(III) nitrate   (1 ,,)
   (c) magnesium oxide   (1 ,,)
   (d) ammonium sulphide   (1 ,,)
   (e) mercury(II) sulphate   (1 ,,) (5)

4. (a) The hydrocarbon contains (100 − 14·3)% by mass of
   carbon $\qquad\qquad = 85\cdot7\%$ (1 ,,)

   (b) ∴ 85·7g of carbon will combine with 14·3g of hydrogen
   ∴ number of moles of carbon atoms

$$= \frac{85\cdot7}{12}$$

$$= 7\cdot14$$ (1 ,,)

   Number of moles of hydrogen atoms

$$= \frac{14\cdot3}{1}$$

$$= 14\cdot3$$ (1 ,,)

   ∴ 7·14 moles of carbon atoms combine with 14·3 moles
   of hydrogen atoms
   ∴ 1 mole of carbon atoms combines with 2 moles of
   hydrogen atoms
   ∴ empirical formula $= \text{CH}_2$ (1 ,,)

The sum of the relative atomic masses of this formula CH_2 is 14. But the relative atomic mass of the hydrocarbon is 28, i.e. 2 times 14. Thus the molecular formula is 2 times CH_2, which is C_2H_4 .... (1 mark) (5)

5. (i) 6 .... (1 ,,)
   (ii) 2 .... (1 ,,)
   (iii) $2Ag^+NO_{3(aq)}^- + K_2^+CrO_{4(aq)}^{2-} \rightarrow$
   $$Ag_2^+CrO_{4(s)}^{2-} + 2K^+NO_{3(aq)}^-$$ (2 marks)
   (iv) Because all the chromate(VI) ion is precipitated as silver chromate(VI) after the addition of 6 cm³ of silver nitrate .... (1 mark) (5)
6. (a) iv (1 mark), (b) iii (1 mark) .... (2)
7. 1·27g (2 marks) .... (2)

TOTAL: 30

Unit 4

1. (c) 200 cm³ .... (1 mark) (1)

2. (d) 28·3 cm³ .... (1 ,,) (1)

3. (a) 1 mole of $Na^+HCO_3^-$ has a mass of 84g

   $\therefore$ 8·4g $Na^+HCO_3^-$ is $\dfrac{8\cdot4}{84}$ moles $= 0\cdot1$ mole .... (1 ,,)

   (b) 1 mole of $Na_2^+CO_3^{2-}$ has a mass of 106g

   $\therefore$ 5·30g of $Na_2^+CO_3^{2-}$ is $\dfrac{5\cdot30}{106}$ mole $= 0\cdot05$ mole .... (1 ,,)

   (c) 0·1 mole of $Na^+HCO_3^-$ give 0·05 mole $Na_2^+CO_3^{2-}$
   $\therefore$ 2 moles of $Na^+HCO_3^-$ give 1 mole $Na_2^+CO_3^{2-}$
   $\therefore$ Answer $=$ 2 moles .... (1 ,,)
   (d) $2Na^+HCO_{3(s)}^- \rightarrow Na_2^+CO_{3(s)}^{2-} + CO_{2(g)} + H_2O_{(g)}$ .... (1 ,,)
   (e) 1 mole of CO_2 occupies 22 400 cm³ at STP

   $\therefore$ 112 cm³ of CO_2 at STP is $\dfrac{112}{22\,400}$ mole $= 0\cdot005$ mole .... (1 ,,)

   (f) Yes, 1 mole of $Na^+HCO_3^-$ gives 1 mole of CO_2
   $\therefore$ 0·005 mole $Na^+HCO_3^-$ gives 0·005 mole CO_2 .... (1 ,,) (6)

4. (a) 4480 cm³ .... (1 ,,) (1)

5. 1 mole of methane has a mass of 16g
   $\therefore$ 32g is the mass of 2 moles of methane
   (a) 4 moles .... (1 ,,)
   (b) 2 moles .... (1 ,,)
   (c) $2 \times 22\,400 = 44\,800$ cm³ .... (1 ,,)
   (d) $4 \times 22\,400 = 89\,600$ cm³ .... (1 ,,) (4)

6. (a) $C_4H_{10(g)} + 6\frac{1}{2}O_{2(g)} \rightarrow 4CO_{2(g)} + 5H_2O_{(g)}$ .... (1 ,,)

(b) 1 volume of C_4H_{10} requires $6\frac{1}{2}$ volumes O_2

∴ 100 cm³ of C_4H_{10} requires $6\frac{1}{2} \times 100$ cm³ $O_2 =$ 650 cm³   (1 mark)

(c) 1 volume of C_4H_{10} produces 4 volumes $CO_2 + 5$ volumes H_2O

∴ 100 cm³ of C_4H_{10} produces 400 cm³ $CO^2 + 500$ cm³ H_2O (1 ,,) (3)

7. (a) Mass of nitrogen $= 84{\cdot}920 - 84{\cdot}640 = 0{\cdot}280$ g

∴ number of moles of nitrogen $(N_2) = \dfrac{0{\cdot}280}{28} = 0{\cdot}01$ mole   (1 ,,)

(b) 0·01 mole of nitrogen occupies 240 cm³ at room temperature and pressure

∴ 1 mole of nitrogen occupies 24 000 cm³   (1 ,,)

(c) Mass of hydrocarbon $= 85{\cdot}420 - 84{\cdot}640 = 0{\cdot}780$ g

240 cm³ of hydrocarbon has a mass of 0·780 g

∴ 24 000 cm³ of hydrocarbon has a mass of

$$\frac{0{\cdot}780 \times 24\,000}{240}$$

∴ relative molecular mass of hydrocarbon $= 78$ (2 marks) (4)

TOTAL: 20

Unit 5

1. Solid lead(II) bromide will not conduct an electric current. The *ions* are held in a rigid crystal *lattice* and are not free to move to the *electrodes*. When the solid is *heated*, it *melts* and allows the passage of an electric current. *Lead* is liberated at the cathode and *bromine* at the anode. The decomposition of an electrolyte by an electric current is called *electrolysis*.

(Allow 1 mark for each word in italics)   (8)

2. (a) false   (1 mark)

(b) false   (1 ,,)

(c) false   (1 ,,)

(d) true   (1 ,,)

(e) true   (1 ,,) (5)

3. (a) 1·6 amp × (10 × 60 seconds) = 960 coulombs   (1 ,,)

(b) Number of faradays $= \dfrac{960}{96\,000} = 0{\cdot}01$ (1 ,,)

(c) The silver ion has a charge of $+1$

∴ 1 faraday (1 mole of electrons) will liberate 1 mole of silver   (1 ,,)

0·01 faraday liberates 1·08g of silver

$\therefore$ 1 faraday liberates $\dfrac{1·08}{0·01}$g of silver

$= 108$g

$\therefore$ relative atomic mass of silver $= 108$ (1 mark)

(d) Number of moles of X $= \dfrac{0·09}{27} = \dfrac{0·01}{3}$ or 0·0033   (1 ,,)

(e) $\dfrac{0·01}{3}$ moles of X is liberated by 0·01 faraday

$\therefore$ 1 mole of X is liberated by $\dfrac{0·01}{0·01} \times 3$

$= 3$ faradays   (1 ,,)

3 faradays (3 moles of electrons) are required to liberate 1 mole of X

$\therefore$ Charge on an ion of X $= +3$, i.e. the ion is X^{3+} (1 ,,)

(f) $\left.\begin{array}{l} Ag^{+} + e \rightarrow Ag \\ X^{3+} + 3e \rightarrow X \end{array}\right\}$ (1 ,,)

(g) The metal ion is *reduced* to the metal by electron gain   (1 ,,) (9)

4. (a) Metal A must be higher than B and D but lower than C
Metal B must be lower than C and D
Metal C must be higher than A and B
Metal D must be higher than B
Therefore the most reactive metal is C; next is A, then
D, and least reactive is B.   (2 marks)

(b) W = no reaction   (1 mark)
    X = no reaction   (1 ,,)
    Y = no reaction   (1 ,,)
    Z = reaction   (1 ,,) (6)

5. (a) (iv)   (1 ,,)
(b) (iii) $+0·93$ volts   (1 ,,) (2)

TOTAL: 30

Unit 6

1. (b) oxygen (c) carbon dioxide (d) nitrogen (e) argon
    (4 marks: deduct 1 mark for each incorrect answer)   (4)

2. (a) oxygen and carbon dioxide   (2 marks)
(b) nitrogen   (1 mark)
(c) one of the noble-gas family, e.g. argon   (1 ,,)
(d) 158 cm³ (allow 160 cm³)   (1 ,,) (5)

3. (a) (i) $2H_2O_{2(aq)} \rightarrow 2H_2O_{(l)} + O_{2(g)}$ (1 ,,)
    (ii) $2Pb_3O_{4(s)} \rightarrow 6Pb^{2+}O^{2-}_{(s)} + O_{2(g)}$ (1 ,,)
    (iii) $2K^{+}ClO^{-}_{3(s)} \rightarrow 2K^{+}Cl^{-}_{(s)} + 3O_{2(g)}$ (1 ,,)
    (iv) $2K^{+}NO^{-}_{3(s)} \rightarrow 2K^{+}NO^{-}_{2(s)} + O_{2(g)}$ (1 ,,)

(b) manganese(IV) oxide ($Mn^{4+}O_2^{2-}$) (1 mark)
(c) reaction (i) (1 ,,) (6)

4. (a) sulphur gives sulphur dioxide; (1 ,,)
magnesium gives magnesium oxide (1 ,,)
(b) sulphur burns with a *blue flame* and *gradually disappears*; (2 ,,)
magnesium burns with a *brilliant white light* and a *white solid* remains (2 ,,) (6)

5. (a) any two of the following: (i) lightning, (ii) root nodules of certain plants, (iii) nitrogen-fixing bacteria (2 marks)
(b) any two of the following: (i) ammonium sulphate, (ii) ammonium nitrate, (iii) ammonia, (iv) ammonium phosphate, (v) urea (2 ,,)
(c) ammonia (1 mark)
(d) photosynthesis (1 ,,) (6)

6. (a) true (1 ,,)
(b) true (1 ,,)
(c) false (1 ,,)
(d) true (1 ,,) (4)

7. (a) both (1 ,,)
(b) boiling removes temporary hardness; (1 ,,)
calcium hydrogencarbonate (1 ,,)
(accept magnesium hydrogencarbonate)
(c) the water still contains permanent hardness which is not removed by boiling (1 ,,)
(d) $Ca^{2+}CO_{3(s)}^{2-}+H_2O_{(l)}+CO_{2(g)} \rightarrow Ca^{2+}(HCO_3^-)_{2(aq)}$ (1 ,,)
(e) an ion-exchange resin removes both temporary and permanent hardness (1 ,,)
(f) the ions causing hardness in water are $Ca_{(aq)}^{2+}$ and/or $Mg_{(aq)}^{2+}$ (2 marks)
(g) $Ca_{(aq)}^{2+}+2Na^+St_{(aq)}^- \rightarrow Ca^{2+}St_{2(s)}^-+2Na_{(aq)}^+$ (1 mark)
(or $Mg_{(aq)}^{2+}+2Na^+St_{(aq)}^- \rightarrow Mg^{2+}St_{2(s)}^-+2Na_{(aq)}^+$) (9)

TOTAL: 40

Unit 7

1. (c) endothermic (1 mark) (1)

2. A large quantity of heat is liberated during the hydration of the ions. This is much greater than the energy required to break down the crystal lattice. (1 ,,) (1)

3. Any three of the following (1 mark each): (a) temperature, (b) concentration of reactants, (c) catalyst, (d) physical state of the reactants. (3)

4. (a) $Zn_{(s)} + 2\,H^+Cl^-_{(aq)} \rightarrow Zn^{2+}Cl^-_{2(aq)} + H_{2(g)}$ (1 mark)

(b) X   (1 ,,)

(c) at Z reaction is complete   (1 ,,)

(d) $\dfrac{0\cdot13}{65} = 0\cdot002$ mole   (1 ,,)

(e) 2 moles of hydrochloric acid react with 1 mole zinc
∴ $2 \times 0\cdot002$ moles of hydrochloric acid react with 0·002 mole zinc
0·004 mole of hydrochloric acid react with 0·002 mole zinc
∴ volume required $= 4\cdot0\,cm^3$ of 1M hydrochloric acid   (1 ,,)

(f) true   (1 ,,)

(g) false   (1 ,,)

(h) true   (1 ,,)

(i) false   (1 ,,)

(j) false   (1 ,,)

(k) false   (1 ,,)

(l) false   (1 ,,) (12)

5. (a) (iv) 2, 2, 1, 2, 1   (1 ,,)

(b) The solution would go yellow (1 mark) due to the formation of chromate(VI) ion. The addition of sodium hydroxide effectively neutralizes the acid and the equilibrium is pushed to the left (1 mark).   (2 marks)(3)

TOTAL: 20

Unit 8

1. (a) base   (1 mark)

(b) salt   (1 ,,)

(c) acid   (1 ,,)

(d) base   (1 ,,)

(e) salt   (1 ,,) (5)

2. (d) less than 7   (1 ,,) (1)

3. (b) 7   (1 ,,) (1)

4. (a) false   (1 ,,)

(b) true   (1 ,,)

(c) false   (1 ,,)

(d) false   (1 ,,)

(e) false   (1 ,,) (5)

5. (i) (*b*) efflorescence (1 mark)
   (ii) (*e*) deliquescence (1 mark) (2)

6. (*a*) 1 dm^3 (1000 cm^3) (1 ,,)
   (*b*) $Na^+OH^-_{(aq)} + H^+Cl^-_{(aq)} \rightarrow Na^+Cl^-_{(aq)} + H_2O_{(l)}$

   $$\frac{\text{Volume }(H^+Cl^-) \times \text{Molarity }(H^+Cl^-)}{\text{Volume }(Na^+OH^-) \times \text{Molarity }(Na^+OH^-)} =$$

   $$\frac{\text{Moles of } H^+Cl^- \text{ in equation}}{\text{Moles of } Na^+OH^- \text{ in equation}}$$

   $$\frac{25 \times 0.05}{V \times 0.10} = \frac{1}{1}$$

   $$\therefore V = \frac{25 \times 0.05}{0.10} = 12.5 \text{ cm}^3 \qquad \text{(2 marks)}$$

   (*c*) $2Na^+OH^-_{(aq)} + H^+_2SO^{2-}_{4(aq)} \rightarrow Na^+_2SO^{2-}_{4(aq)} + 2H_2O_{(l)}$ (1 mark)
   By similar formula to that in answer (*b*):

   $$\frac{25 \times 0.05}{V \times 0.10} = \frac{1}{2}$$

   $$\therefore V = \frac{25 \times 0.05 \times 2}{0.10} = 25.0 \text{ cm}^3 \qquad \text{(1 ,,)}$$

   (*d*) 0.5 mole of hydrogen ions from an acid reacts exactly
   with 0.5 mole of sodium hydroxide solution.
   1000 cm^3 of 0.1M sodium hydroxide solution contains 0.1 mole sodium hydroxide
   $\therefore$ 5000 cm^3 of 0.1M sodium hydroxide solution contains 0.5 mole sodium hydroxide
   $\therefore$ volume required $= 5000 \text{ cm}^3$ (1 ,,) (6)

7. Approximately 50cm^3 of *dilute* (1 mark) *sulphuric
   acid* (1 mark) is warmed with an *excess* (1 mark) copper(II)
   oxide. When no more of the oxide reacts the mixture is
   filtered (1 mark) into an evaporating basin and the
   filtrate (or solution) (1 mark) evaporated to *crystallization
   point* (1 mark). After cooling, the crystals are filtered,
   washed with a *little water* (1 mark) and dried *between filter
   papers* (1 mark). (8)

8. (*a*) sodium chloride (1 mark)
   (*b*) potassium nitrate (1 ,,)
   (*c*) 70°C (1 ,,)
   (*d*) sodium nitrate (1 ,,)
   (*e*) $150\text{g} - 80\text{g} = 70\text{g}$ (2 marks) (6)

9.

Classification	Oxide	
Acidic	phosphorus(v) oxide	(1 mark)
Basic	copper(II) oxide	(1 ,,)
Neutral	carbon monoxide	(1 ,,)
Amphoteric	aluminium oxide	(1 ,,)
Peroxide	sodium peroxide	(1 ,,)
Compound oxide	dilead(II) lead(IV) oxide	(1 ,,) (6)

TOTAL: 40

Unit 9

1. (a) copper (1 mark)
   (b) iron (1 ,,)
   (c) sodium (1 ,,)
   (d) calcium (1 ,,)
   (e) magnesium (1 ,,)
   (f) aluminium (1 ,,)
   (g) iron (1 ,,)
   (h) sodium (1 ,,)
   (i) copper (1 ,,)
   (j) copper (1 ,,) (10)

2. A is potassium (1 ,,)
   B is lead (1 ,,)
   C is copper (1 ,,)
   D is zinc (1 ,,)
   E is lithium (1 ,,) (5)

3. (a) is X (1 ,,)
   (b) is W and Z (1 ,,)
   (c) is X (1 ,,)
   (d) is W and Z (1 ,,)
   (e) is Y (1 ,,) (5)
   (deduct one mark for each incorrect answer)

4. A is $Fe^{2+}SO_4^{2-} . 7H_2O$ (1 ,,)
      or hydrated iron(II) sulphate
   B is $Ba^{2+}SO_4^{2-}$ (1 ,,)
      or barium sulphate

C and D are SO_2 and SO_3 (2 marks)
  or sulphur dioxide and sulphur trioxide

E is $Fe_2^{3+}O_3^{2-}$ (1 mark)
  or iron(III) oxide

F is $Fe^{3+}Cl_3^-$ (1 ,,)
  or iron(III) chloride solution

G is sulphur (S)   (1 ,,)

H is $Fe^{2+}Cl_2^-$
  or iron(II) chloride solution   (1 ,,)

(i) With aqueous ammonia A gives *green* iron(II)
    hydroxide   (2 marks)

(ii) With nitric acid A is oxidized to an iron(III) salt and
    this with aqueous ammonia gives *red-brown* iron(III)
    hydroxide   (2 ,,) (12)

5. (*a*) true   (1 mark)
  (*b*) true   (1 ,,)
  (*c*) false   (1 ,,)
  (*d*) false   (1 ,,)
  (*e*) false   (1 ,,)
  (*f*) false (sodium carbonate does not decompose on
    heating)   (1 ,,)
  (*g*) true   (1 ,,)
  (*h*) true   (1 ,,) (8)

TOTAL: 40

Unit 10

1. (*a*) Sulphuric acid, ammonium chloride   (1 mark)
    $H_2SO_{4(l)}+NH_4^+Cl_{(s)}^- \rightarrow NH_4^+HSO_{4(s)}^-+HCl_{(g)}$ (1 ,,)

  (*b*) Manganese(IV) oxide, sulphuric acid, ammonium
    chloride   (1 ,,)
    $Mn^{4+}O_{2(s)}^{2-}+3H_2SO_{4(l)}+2NH_4^+Cl_{(s)}^- \rightarrow$
    $Mn^{2+}SO_{4(aq)}^{2-}+2NH_4^+HSO_{4(aq)}^-+2H_2O_{(l)}+Cl_{2(g)}$ (1 ,,)

  (*c*) Ammonium chloride, sodium hydroxide   (1 ,,)
    $NH_4^+Cl_{(s)}^-+Na^+OH_{(aq)}^- \rightarrow Na^+Cl_{(aq)}^-+NH_{3(g)}+H_2O_{(l)}$ (1 ,,)

  (*d*) Copper, sulphuric acid   (1 ,,)
    $Cu_{(s)}+2H_2SO_{4(l)} \rightarrow Cu^{2+}SO_{4(aq)}^{2-}+SO_{2(g)}+2H_2O_{(l)}$ (1 ,,) (8)

2. (*a*) 2.8.18.7   (1 ,,)
  (*b*) fluorine   (1 ,,)
  (*c*) fluorine   (1 ,,)
  (*d*) iron(III) chloride   (1 ,,)
  (*e*) $F+e \rightarrow F^-$ (1 ,,)
    this is a *reduction* process (electron gain)   (1 ,,)
  (*f*) bromine   (1 ,,)
  (*g*) iodine   (1 ,,) (8)

3. (a) as a dehydrating agent   (1 mark)
   (b) as an oxidizing agent   (1 ,,)
   (c) as a dehydrating agent   (1 ,,)
   (d) as an acid   (1 ,,)
   (e) as an oxidizing agent   (1 ,,) (5)

4. (a) ammonia   (1 ,,)
   (b) sulphur trioxide   (1 ,,)
   (c) phosphorus(v) oxide (phosphorus pentoxide)   (1 ,,)
   (d) carbon monoxide   (1 ,,) (4)

5. A = sodium iodide   (1 ,,)
   B = sodium sulphite   (1 ,,)
   C = sodium sulphide   (1 ,,)
   D = silver iodide   (1 ,,)
   E = sulphur dioxide   (1 ,,)
   F = hydrogen sulphide   (1 ,,) (6)

6. (a) true   (1 ,,)
   (b) true   (1 ,,)
   (c) false   (1 ,,)
   (d) false   (1 ,,)
   (e) true   (1 ,,)
   (f) true   (1 ,,)
   (g) false   (1 ,,)
   (h) true   (1 ,,)
   (i) true   (1 ,,) (9)

                                                                 TOTAL: 40

Unit 11

1. (a) pentane   (1 mark)
   (b) propene (prop-1-ene)   (1 ,,)
   (c) 1,2-dibromopropane   (1 ,,)
   (d) 1,1-dichloropropane   (1 ,,)
   (e) ethylamine (1-aminoethane)   (1 ,,) (5)

2. (d) C_7H_{14} is an *alkene*   (1 ,,) (1)

3. (b) two:

                                                                 (1 ,,) (1)

and

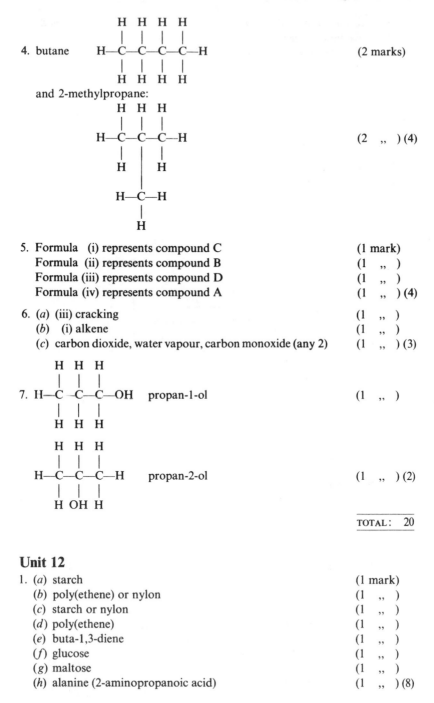

4. butane (2 marks)

and 2-methylpropane:

(2 ,,) (4)

5. Formula (i) represents compound C   (1 mark)
   Formula (ii) represents compound B   (1 ,,)
   Formula (iii) represents compound D   (1 ,,)
   Formula (iv) represents compound A   (1 ,,) (4)

6. (a) (iii) cracking   (1 ,,)
   (b) (i) alkene   (1 ,,)
   (c) carbon dioxide, water vapour, carbon monoxide (any 2)   (1 ,,) (3)

7. H—C—C—C—OH   propan-1-ol   (1 ,,)

H—C—C—C—H   propan-2-ol   (1 ,,) (2)

TOTAL: 20

Unit 12

1. (a) starch   (1 mark)
   (b) poly(ethene) or nylon   (1 ,,)
   (c) starch or nylon   (1 ,,)
   (d) poly(ethene)   (1 ,,)
   (e) buta-1,3-diene   (1 ,,)
   (f) glucose   (1 ,,)
   (g) maltose   (1 ,,)
   (h) alanine (2-aminopropanoic acid)   (1 ,,) (8)

2. (*a*) addition　　　　　　　　　　　　　　　　　　(1 mark)
　　(*b*) polytetrafluoroethene)　　　　　　　　　　　(1　,,　)
　　(*c*)

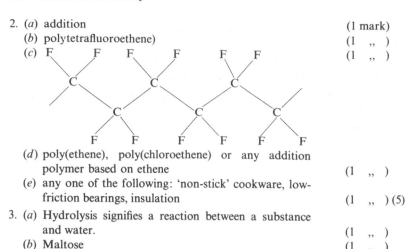

　　(*d*) poly(ethene), poly(chloroethene) or any addition
　　　　polymer based on ethene　　　　　　　　　　(1　,,　)
　　(*e*) any one of the following: 'non-stick' cookware, low-
　　　　friction bearings, insulation　　　　　　　　(1　,,　) (5)
3. (*a*) Hydrolysis signifies a reaction between a substance
　　　　and water.　　　　　　　　　　　　　　　　(1　,,　)
　　(*b*) Maltose　　　　　　　　　　　　　　　　　(1　,,　)
　　(*c*) Accept either *amylase* or *enzyme*　　　　　(1　,,　) (3)

4. (*a*)

Polymer	Column A	Column B
Terylene	polyester	synthetic
Cellulose	carbohydrate	natural
Poly(phenylethene)	hydrocarbon	synthetic
Insulin	protein	natural

(8 marks)

　　(*b*) Poly(phenylethene) is *addition*, the other three *con-
　　　　densation*　　　　　　　　　　　　　　　　(4 marks)
　　(*c*) Insulin　　　　　　　　　　　　　　　　　(1 mark)
　　(*d*) Insulin　　　　　　　　　　　　　　　　　(1　,,　) (14)

TOTAL:　30

Unit 13

1. (*a*) An alpha particle is a helium nucleus consisting of 2
　　　　protons and 2 neutrons　　　　　　　　　　(1 mark)
　　(*b*) A beta particle is an electron　　　　　　　(1　,,　)
　　(*c*) A gamma ray is electromagnetic radiation　　(1　,,　) (3)
2. (*a*) 228 is the Mass Number = sum of protons+neutrons
　　　　88 is the Atomic Number = number of protons　(1　,,　)
　　　　Number of neutrons = 228−88 = 140　　　　　(1　,,　)
　　(*b*)　(i) β　　　　　　　　　　　　　　　　(1　,,　)
　　　　　(ii) β　　　　　　　　　　　　　　　(1　,,　)
　　　　　(iii) α　　　　　　　　　　　　　　(1　,,　)

(c) They are *isotopes* (same atomic number)   (1 mark) (6)

3. X has atomic number 88 and mass number 228   (2 marks)
   Y has atomic number 89 and mass number 228   (2 ,,)
   Z has atomic number 90 and mass number 228   (2 ,,) (6)

4. (a) false ($\frac{1}{8}$ remains after three half-lives, i.e. 30 minutes)   (1 mark)
   (b) false   (1 ,,)
   (c) true   (1 ,,)
   (d) true   (1 ,,)
   (e) false   (1 ,,) (5)

   TOTAL: 20

Unit 14

1. (a) Direct current   (1 mark)
   (b) To lower the melting point of the sodium(I) chloride   (1 ,,)
   (c) (i) sodium: $Na^+ + e \rightarrow Na$ (2 marks)
       (ii) chlorine: $2Cl^- - 2e \rightarrow Cl_2$ (2 ,,) (6)

2. (a) A 'heavy' chemical is one produced in large quantities   (1 mark)
   (b) The Haber process   (1 ,,)
   (c) Nitrogen from the air   (1 ,,)
       Hydrogen from methane (or from water gas)   (1 ,,)
   (d) It increases the rate of the reaction   (1 ,,)
   (e) A finely divided catalyst provides a large surface area   (1 ,,)
   (f) $N_{2(g)} + 3H_{2(g)} \rightleftharpoons 2NH_{3(g)}$ (1 ,,)
   (g) It is either liquefied or dissolved in water   (1 ,,) (8)

3. (a) (iii) limestone, iron ore and coke   (1 ,,)
   (b) (iv) molten calcium silicate   (1 ,,)
   (c) (ii) carbon monoxide   (1 ,,)
   (d) (ii) 4% carbon   (1 ,,)
   (e) (i) $0 \cdot 1 - 2\%$ carbon   (1 ,,) (5)

4. (a) Sulphur and air   (1 ,,)
   (b) Sulphur dioxide   (1 ,,)
   (c) $2SO_{2(g)} + O_{2(g)} \rightleftharpoons 2SO_{3(g)}$ (1 ,,)
   (d) Vanadium(V) oxide (vanadium pentoxide)   (1 ,,)
   (e) By dissolving sulphur trioxide in concentrated sulphuric acid and diluting the product   (1 ,,)
   (f) There are numerous uses, e.g. manufacture of fertilizers   (1 ,,) (6)

5. (a) false (the by-product is calcium chloride)   (1 ,,)
   (b) true   (1 ,,)
   (c) true   (1 ,,)
   (d) false (the process is saponification)   (1 ,,)
   (e) true   (1 ,,) (5)

   TOTAL: 30

Unit 15

1. (a) false   (1 mark)
   (b) true   (1 ,,)
   (c) false   (1 ,,)
   (d) true   (1 ,,)
   (e) true   (1 ,,)

2. (a) $C_8H_{18} + 12\frac{1}{2}O_2 \rightarrow 8\ CO_2 + 9H_2O$

   or

   $2C_8H_{18} + 25O_2 \rightarrow 16CO_2 + 18H_2O$ (2 marks)
   (b) Carbon monoxide (CO)   (1 mark)
   (c) Black solid is carbon   (1 ,,)
       Formed by incomplete combustion of petrol   (1 ,,)
   (d) Either tetraethyl-lead or tetramethyl-lead   (1 ,,)
   (c) Because volatile lead compounds are released
       into the atmosphere with the exhaust gases   (1 ,,)
       These lead compounds are poisonous   (1 ,,) (8)

3. (a) Fertilizers (or sewage)   (1 mark)
   (b) Plant growth is speeded up   (1 ,,)
   (c) The fish eventually die due to oxygen depletion   (1 ,,) (3)

4. (a) Carbon dioxide   (1 mark)
   b) Ammonia   (1 ,,)
   (c) Sulphur dioxide   (1 ,,)
   (d) Carbon monoxide   (1 ,,)
   (e) Methane   (1 ,,)
   (f) Nitrogen dioxide   (1 ,,)
   (g) Sulphur dioxide   (1 ,,)
   (h) Oxygen   (1 ,,) (8)

TOTAL: 24

Index